U.S. Marine Corps Physical Readiness Training for Combat

Plus

Desert Operations

Department of Defense

Pentagon
Publishing

UNITED STATES MARINE CORPS
Marine Corps Combat Development Command
Quantico, Virginia 22134-5001

29 January 1988

FLEET MARINE FORCE REFERENCE PUBLICATION............0-1B

MARINE PHYSICAL READINESS TRAINING FOR COMBAT

1. PURPOSE

Fleet Marine Force Reference Publication (FMFRP) 0-1Marine Physical Readiness Training for Combat is intended for use by all Marines. It provides the information and references necessary to establish and conduct physical conditioning programs to prepare Marines for the physical demands of combat.

2. APPLICABILITY

Although not directive, this handbook is intended for use as a reference by all Marines in developing unit physical readiness training for combat. This FMFRP does not establish policy for the conduct of physical fitness training. Refer to Marine Corps Order 6300.3_ for applicable Marine Corps policy on physical fitness. It may also be used as a basis for instruction conducted by unit leaders or as a reference for instruction or training.

3. RECOMMENDATIONS

To improve this manual, the MAGTF Warfighting Center seeks comments and recommendations from the user. Officers and noncommissioned officers with constructive comments are encouraged to respond. Lack of response will be considered as concurrence. The User Suggestion Form which follows can be detached and forwarded to - -

Commanding General
Marine Corps Combat Development Command (WF 06)
Quantico, Virginia 22134-5001

4. DISTRIBUTION

Initial distribution of this FMFRP 0-1B is shown below. Addressees on this list should ensure that appropriate subordinate organizations are provided copies.

JOHN I. HOPKINS
Brigadier General, U.S. Marine Corps
Assistant Chief of Staff, Warfighting
Marine Corps Combat Development Command
Quantico, Virginia

DISTRIBUTION: TDJ

From:

To: Commanding General, Marine Corps Combat Development Command (WF 06), Quantico, Virginia 22134-5001

Subj: FMFRP 0-1B, MARINE PHYSICAL READINESS TRAINING FOR COMBAT, RECOMMENDATIONS CONCERNING

1. In accordance with the Promulgation Page to FMFRP 0-1B, which invites individuals to submit suggestions concerning this FMFRP directly to the above addressee, the following unclassified recommendation is forwarded:

Page________Article/Para No.________Line No.________Figure/Table No.________

Nature of Change:_____Add_____Delete_____Change_____Correct

2. Proposed New Verbatim Text: (Verbatim, double spaced; continue on additional pages as necessary.)

3. Justification/Source: (Need not be double spaced.)

NOTE: Only one recommendation per page.

PREFACE

MARINE PHYSICAL READINESS TRAINING FOR COMBAT

Chapter 1. Physical Readiness Leadership
Section I. Role of Physical Fitness In Combat

Paragraph

1101 Purpose of Physical Fitness in Training
1102 Scope of the Manual
1103 Commanders Role
1104 History of Military Physical Readiness
1105 Physical Demands of Combat
1106 Physical Effects of Combat Stress
1107 Limits of Physical Readiness

Section II Fundamentals of Physical Fitness
1201 Fitness for Marines
1202 Components of Physical Fitness
1203 Types of Exercise
1204 Principles of Physical Conditioning
1205 Three Stages of Physical Conditioning
1206 Effects of Climatic Conditions
1207 Warming Up and Cooling Off
1208 Physical Activity as Age Increases

Section III Goals of Physical Readiness Training
1301 Necessity for Physical Readiness Training
1302 Objective of Physical Readiness Training
1303 Benefits of Exercise

Section IV Leadership Roles
1401 Psychological Leadership
1402 Command and Supervisory Functions
1403 Small-Unit Leaders and Instructors

Chapter 2. Physical Readiness Training Programs
Section I. Development of a Program
2101 Guidance for Planners
2102 Marine Corps Mission
2103 Evaluation of Fitness
2104 Integration of Training
2105 Activities at the Unit Level
2106 Steps in Planning
2107 Selection of Activities and Systems

2108 Sample Weekly Programs

Section II. Remedial Physical Training
2201 Need for Attention
2202 Identification of Individuals
2203 Administration of Remedial Action

Chapter 3. Physical Conditioning Activities
Section I. Marching Under Load
3101 General
3102 Training Goals
3103 Movement Planning
3104 Individual Load
3105 March Techniques and Procedures
3106 Tactical Marches
3107 Administrative Marches
3108 Forces Marches
3109 Night Marches

Section II. Conditioning Drills One, Two, and Three
3201 General
3202 Conditioning Drill One
3203 Conditioning Drill Two
3204 Conditioning Drill Three

Section III. Rifle and Log Drills
3301 Rifle Drill
3302 Exercises Performed with Rifles
3303 Log Exercises
3304 Exercise with Logs

Section IV. Grass Drills
3401 General
3402 Grass Drill One and Two

Section V. Guerrilla Exercises
3501 General
3502 Guerrilla Tables

Section VI. Running
3601 Running
3602 Types of Running
3603 Orienteering

Section VII. Circuit Training

3701 General
3702 Fixed Strength Circuit
3703 Moveable Strength Circuit
3704 Circuit-Interval Table

Section VIII. Basic Physical Skills and Obstacle Courses
3801 General
3802 Basic Physical Skills
3803 Obstacle Courses
3804 Conditioning Obstacle Courses
3805 Confidence Obstacle Courses

Section IX. Individual Exercise Programs
3901 General
3902 The Bench Conditioner
3903 The 6-12 Plan
3904 Weight Training
3905 Weight Lifting
3906 Isometric Exercising

Chapter 4. Combat Water Survival
Section I. Marine Corps Water Survival Program
4101 Military Swimming
4102 Program Development
4103 Qualification Standards and Training Guidelines
4104 Water Survival/Swimming Qualification Standards and Test Procedures

Section II. Water Sports and Competitive Activities
4201 General
4202 Water Relays
4203 Water Polo

Section III. The Battle Swimming Test
4301 General
4302 The Battle Swimming Test

Chapter 5. Competitive Conditioning Activities
Section I. Organization of Competitive Activities
5101 Leadership of Competitive Activities
5102 Military Field Meets

Section II. Combatives
5201 Description and Objectives
5202 Combative Tables

Section III. Relays

5301 Description and Objective

5302 Relay Tables

Section IV. Team Contests and Athletics

5401 Description and Objectives

5402 Team Contests

5403 Cross-Country and Distance Running

5404 Team Athletics

Chapter 6. Evaluation of Performance During Training

6001 Objectives of Combat Readiness

6002 The Physical Fitness Test

6003 Other Methods of Evaluation

6004 Responsibility

6005 Commander's Physical Fitness Program

6006 Other Formal Tests

6007 Physical Readiness Test

6008 Airborne Trainee Physical Fitness Test

6009 Battle Fitness Test

Chapter 7. The Human Body

7001 General

7002 Systems of the Body

7003 The Skeleton

7004 The Muscles

7005 Functioning of the Skeleton and Muscles

7006 The Circulatory and Respiratory Systems

7007 Cardiovascular Functioning

Appendix

A Instructor Hints and Instructor Training

Chapter 1
PHYSICAL READINESS LEADERSHIP
Section I. Role of Physical Fitness in Combat

1101. PURPOSE OF PHYSICAL FITNESS TRAINING

Physical fitness training in the Marine Corps has one purpose: to prepare Marines to physically withstand the rigors of combat. All other goals of physical fitness training are subordinate to and must support attainment of this goal. The idea that only infantry or reconnaissance units and their attachments normally face physically demanding combat is wrong. This error must not influence the priority commanders of combat support, combat service support, aviation, and headquarters units give to physical fitness for combat. Physical fitness for combat has a high priority for all Marines. A sound, effective unit program, requiring limited time and material, offers a greater payoff in combat than many more expensive and time-consuming training programs. This handbook describes the concept and provides the ingredients of a program for commanders to use in carrying out one of their most serious responsibilities: ensuring that their Marines are physically ready for combat.

1102. SCOPE OF THE MANUAL

This handbook provides guidance for all leaders, trainers, and planners of physical training programs. It describes unit physical fitness for combat training in the following chapters:

a. Physical Readiness Leadership. Chapter 1 provides guidance to leaders in the conduct of physical combat readiness training.

b. Physical Readiness Training Programs. Chapter 2 discusses how to structure programs to meet physical readiness goals in a variety of combat training situations.

c. Physical Conditioning Activities. Chapter 3 discusses the primary physical conditioning activities which commanders may use to attain readiness for combat goals. Sections in this chapter address foot marches under load, strength-building activities, and activities which build confidence and the aggressive spirit needed in combat.

d. Combat Water Survival. As an amphibious force, Marines cannot ignore the importance of combat water survival training. Chapter 4 describes a program for commanders to conduct this training.

e. Competitive Conditioning Activities. Chapter 5 describes competitive activities which stress the relationship to combat actions, featuring team-building types of competition.

f. Evaluation of Performance During Training. Chapter 6 describes tests which measure individual and unit physical fitness for combat.

g. The Human Body. Chapter 7 covers the structure and functioning of the body.

1103. COMMANDER'S ROLE

Major General Commandant John A. Lejeune in the 1921 edition of Marine Corps Manual expressed our philosophy of leadership. Among other things, General Lejeune stated that "it will be necessary for officers ... to endeavor to enlist the interest of their men in building up and maintaining their bodies in the finest physical condition." It is significant that the Commandant prefaced these writings in 1921 with the statement that his thoughts were greatly influenced by the experience of World War I. We have never varied as a Corps from this belief that one of a commander's most serious responsibilities is to have Marines physically prepared for combat. Senior commanders and their staffs carry out this role primarily by providing command emphasis, including planning, support, and supervision. Commanders at the battalion/squadron and company/battery level execute a program of physical fitness for combat by close supervision and, most importantly, leadership by example.

1104. HISTORY OF MILITARY PHYSICAL READINESS

Every war has revealed our military physical deficiencies during the initial periods of mobilization. This realization followed the Civil War and has recurred regularly with each national emergency.

a. Training programs in each war were geared to the physical need of the era. Success was dependent upon the amount of time available during training to physically prepare Marines for battle conditions. Frequently, casualties in initial engagements were attributed to the inability of our Marines to physically withstand the rigors of combat over rugged terrain and under unfavorable climatic conditions. With adequate preparation, Marines have always handled the test of battle.

b. During World War II, the first physical conditioning doctrine that could be scientifically justified by testing was introduced. As the war progressed, this program was effective in the physical conditioning of millions of Marines for combat.

c. Postwar periods have traditionally been a time of consolidation. Unfortunately, some leaders considered the conditioning phase of training to be a wartime tool. With such a philosophy prevalent between wars, physical readiness was relegated to secondary importance resulting in a serious lowering of combat effectiveness. The initial commitment of Marines in Korea dramatically displayed this failure to recognize the extreme physical demands of warfare. Call-up of large numbers of reserve Marines and pressing them into the war within weeks after call-up mainly caused the loss in effectiveness.

d. Costly lessons learned from our military experiences over a period of years and the course of several wars led to an increasing interest in the physical conditioning of the individual Marine. No longer can we afford to emphasize physical fitness during wartime and de-emphasize it during peacetime. It is evident that, in spite of increased mechanization and modern weapons, physical readiness retains a vital place in the life of each individual Marine and in every unit within the Marine Corps regardless of mission.

e. Commanders are well aware of the need for rugged and well conditioned Marines. The daily demands of housekeeping, maintenance, support, training, operations, and other time-consuming tasks make it necessary for commanders to set aside time for frequent, regular, and vigorous exercise periods.

1105. PHYSICAL DEMANDS OF COMBAT

There are three primary elements of effective physical fitness for combat: lower body strength and stamina; upper body strength and stamina; and a competitive, combative spirit. Cardiovascular functioning is not treated as a separate element since it is a necessary foundation for the other three. Exercises and activities designed to improve Marines' condition in the three primary elements will foster good cardiovascular and respiratory condition.

a. Lower Body Strength and Stamina. Some common demands that combat places on lower body strength and endurance are marching long distances under load and functioning effectively at the destination; moving quickly and evasively under fire; and carrying wounded Marines to safety.

b. Upper Body Strength and Stamina. Some common demands that combat places on upper body strength and stamina are rapidly emplacing crew-served weapons; handling large-caliber ammunition for extended periods; climbing walls, cliffs, and other high obstacles; and performing field maintenance on air craft or heavy machinery.

c. Competitive, Combative Spirit. While part of this critical aspect of combat readiness is mental and emotional, robust physical condition and a training program which promotes physical aggressiveness greatly enhance a competitive and combative spirit. Activities which force Marines to overcome natural physical fear, which directly leads to fatigue, are particularly valuable.

1106. PHYSICAL EFFECTS OF COMBAT STRESS

It has been thoroughly documented that the added uncertainty and stress of combat have a major physical effect on Marines as well as the acknowledged psychological effect. In other words, in combat, fear equals fatigue. In training, we cannot easily reproduce this combat stress which reduces the effectiveness of individuals and units. However, we can produce fatigue and psychological doubt in training situations by developing a challenging physical training program which stresses the participants. This compels Marines to exercise their ability to continue to function under difficult and trying conditions. Such activities as long foot marches under load and difficult confidence courses train Marines to overcome their own fear and fatigue. Rugged competitive activities such as martial arts and pugil stick fighting prepare Marines psychologically to overcome an opponent as well as their own fear and fatigue. This sort of physically demanding training builds self-confidence and unit morale. The adage that "the unit that sweats more in peacetime will bleed less in battle" certainly applies to its physical training program. Athletic coaches have long followed this approach by such practices as having teams shoot foul shots after wind sprints in order to strengthen their mental conditioning and toughness. Although pushing Marines to their limit is necessary, commanders often set lower standards initially than ultimately desired. The program should then call for steadily increasing the difficulty. A good guide for a starting point is a standard that will be physically and psychologically demanding for no less than three-fourths of the unit. Expressed another way, no more than one-fourth of the unit should be working at less than their limit. The program should progress quickly, augmented, if needed, by remedial training for those unable to meet the unit standards, until the training is challenging for all but the most exceptional Marines in the command. Commanders must never forget that Marines expect to be challenged and thrive on being pushed to their limit.

1107. LIMITS OF PHYSICAL READINESS

A well-conditioned Marine unit has a significant advantage in combat, but this does not mean that physical conditioning can substitute for technical knowledge or good planning. For instance, units must, as discussed in chapter 3, pursue a continuing and challenging program of marches under loads. This does not disprove the axiom that fighting men, if required to carry over one-third of body weight, will generally become ineffective in battle. Leaders and planners must decide what is essential to be carried and must use an effective logistic distribution system, such as unit trains, to make remaining items available when they are needed. In a similar vein, the valid requirement to foster a competitive and combative spirit in Marines does not excuse leaders from the responsibility to practice tactics which will not needlessly risk lives. Physical readiness for combat is a vital goal, but it is only one part of a unit's overall readiness for the lest of combat.

Section II. Fundamentals of Physical Fitness

1201. FITNESS FOR MARINES

Total combat readiness includes technical proficiency and mental, emotional, and physical fitness. If any one of these attributes is lacking, combat effectiveness suffers proportionately. Without technical fitness, Marines lack the knowledge and skill to fight; without mental and emotional fitness, they lack the incentive and will to fight; and without physical fitness, they lack the physical ability and confidence to fight. Physical fitness in a Marine means a whole and healthy body, the capacity for skillful and sustained performance, the ability to recover from exertion rapidly, the desire to complete a designated task, and the confidence to face any eventuality.

1202. COMPONENTS OF PHYSICAL FITNESS

A sound body, free of disease and defect, does not in itself constitute physical fitness. Before an untrained Marine can be considered physically fit for combat, the following traits must be developed:

a. Strength. Every Marine needs enough strength to perform the heaviest task encountered in routine and emergency activities. The basic areas where strength is required are in the arm and shoulder girdle, abdomen, back, and legs. Muscles increase in size, strength, and firmness with regular and strenuous exercises. Without work, they grow flabby and weak.

b. Endurance. Each Marine needs sustaining power to maintain the maximum ability without undue fatigue. There are two types of endurance:

(1) Muscular Endurance. The Marine needs muscular endurance to fight the enemy under the most tiring combat conditions. Muscular endurance is the ability to perform continuous work over long periods of time. Endurance depends on the bloodstream's ability to deliver large amounts of oxygen and nutrition to the muscle masses and then carry away the waste products quickly.

(2) Cardiovascular-Respiratory Endurance. The development of cardiovascular-respiratory endurance ("wind") is necessary to maintain muscular endurance. Cardiovascular-respiratory endurance depends on the efficiency of the blood vessels, lungs, and heart. The maximum effort a Marine can exert over a period of time is limited by the capacity to absorb oxygen and expel carbon dioxide. The average Marine's cardiovascular-respiratory capacity can be greatly increased by exercise.

c. Agility. A Marine must be able to change direction quickly and as faultlessly as possible. The ability to react instantly and to maintain orientation during rapid changes of body position is important to survival. This important ability may be developed by conditioning exercises that require varied and rapid changes of body position on the ground and in the air, such as obstacle courses.

d. Coordination. Coordination is the ability to move all parts of the body in a smooth, efficient, and concerted effort (commonly called timing). A well-coordinated individual does not make useless movements. An individual who moves with precision and accuracy saves energy. Coordination is best developed by practicing diversified muscular activities and skills affecting all body parts.

1203. TYPES OF EXERCISE

Basically, there are two forms of exercise: isotonic and isometric. Both forms are contained in the activities found in this manual.

a. Isotonic. Isotonic exercises are those in which the expenditure of energy is regulated and released during concentrated efforts. The regulated expenditure of energy is controlled by both the mode of exercise and the individual's effort. This type of effort is common to most exercise and sports. To develop endurance, coordination, and flexibility, isotonic exercise must be used. Strength can also be developed or increased through this type of exercise.

b. Isometric. Isometric exercises are those in which maximum effort is applied and held until the engaged muscle energy is depleted during a single contraction effort. The individual exerts full force against an immovable object for a relatively short period of time (6 to 10 seconds) and then repeats the contraction several times with a short period of rest between each contraction. This type of exercise develops only strength; therefore, it has limited application.

1204. PRINCIPLES OF PHYSICAL CONDITIONING

Men and women vary in their physical makeup. Physiological (body) function and reaction also vary in proportion to heavy demands placed upon the body. To attain the maximum benefit without sacrificing Marines' welfare, fitness must grow with a careful program of conditioning. To allow for adjustments in body functioning as the conditioning program progresses and to ensure attainment of objectives, the following principles of physical conditioning must be applied:

a. Overload. As strength and endurance increase, the physical load must be increased until the desired level of fitness is reached.

b. Progression. In beginning stages, the load must be moderate. Gradual progression from this low state of fitness to a higher state is possible through application of a progressive program.

c. Balance. An effective program utilizes various types of activities and provides for the concurrent development of strength, endurance, and coordination as well as basic physical skills.

d. Variety. Some programs fail because the routine becomes boring. The most successful programs always include conditioning activities, competitive events, and military physical skill development.

e. Regularity. There is no easy or occasional way to develop physical fitness. Regularity of exercise is a must, with daily exercise preferred.

1205. THREE STAGES OF PHYSICAL CONDITIONING

Unconditioned or poorly conditioned Marines pass through the following stages in reaching the desired state of physical condition.

a. Toughening Stage. This stage is approximately 2 weeks long and is usually characterized by muscular stiffness and soreness followed by recovery.

b. Slow Improvement Stage. This stage is approximately 6 to 10 weeks long and is characterized by slow and steady improvement until the desired level, or a high level, of fitness is attained.

c. Sustaining Stage. This stage goes on indefinitely in order to maintain the level of conditioning achieved by passage through the previous stages.

1206. EFFECTS OF CLIMATIC CONDITIONS

Temperature, both atmospheric and body, affects the physical performance of Marines. Proper maintenance of body temperature through warm up exercise, proper dress in cold weather, and removal or adjustment of clothing in hot weather is necessary for effective performance and health. Climatic factors to be considered are the following:

a. Exercise In High Temperatures and High Humidity. Marines can endure strenuous physical activity in extremely hot temperatures if they are given an opportunity to become acclimated and if they take enough salt and water. It is essential to continue physical training programs in hot climates. Marines can better withstand high temperatures when they are well-conditioned. High humidity combined with high temperatures presents a serious danger. These conditions prevent the natural cooling of the body by the evaporation of perspiration. Training schedules should conform to the provisions of the base commander in respect to wet bulb conditions. Those who conduct training under these conditions should monitor weight loss and be careful to make fluid replacement adjustments.

b. Exercise at High Altitudes. Certain problems are encountered in conditioning Marines stationed in high altitudes because the heart undergoes greater exertion during exercise. It is particularly important that only light exercise be given initially at such altitudes. A Marine's body gradually adjusts to high altitudes within a few weeks. After this adjustment, progressively greater amounts of exercise are possible.

c. Exercise in Arctic Regions. Military duty in the arctic is so arduous that a high level of physical conditioning is essential. Because of the difficulties of carrying on physical conditioning exercises in extreme cold, Marines should be conditioned to the highest level possible before they arrive. A sustaining program will then maintain that level. When exercising in cold weather, Marines should be required to remove excess clothing to prevent them from becoming damp with perspiration.

1207. WARMING UP AND COOLING OFF

It is a fundamental physiological principle that Marines should warm up gradually before taking strenuous exercise. Such action speeds up the circulation to prepare the body to take an overload and helps to prevent injury to muscles and joints after exercising. Marines should be kept mildly active, walking, stretching, or performing some other mild muscular activity until breathing and temperature have returned to normal. Marines should never be allowed to cool off too rapidly; in cool or cold weather, they should put on additional clothing during the cooling-off period.

1208. PHYSICAL ACTIVITY AS AGE INCREASES

Combat makes severe physical demands on Marines. All Marines, regardless of age, must be physically ready to meet these demands. There is no physiological reason to cease exercise or exertion with age. Increased age usually brings increased responsibility which, in many instances, leads to a routine that can become almost devoid of physical activity. The key to fitness with increased age is to continue exercising at a reasonable level and to include exercise of a vigorous type in the daily routine. Older persons who have not regularly maintained a reasonable state of physical fitness require more time than younger persons to become fit. Such individuals usually require a longer period of time to recover from physical effort than younger Marines. If general health is good, evidence shows that older Marines can develop and maintain a degree of fitness which permits vigorous activity and proper performance of duties. It is both the individual's and the commander's responsibility that all Marines continue a daily sustaining exercise program.

Section III. Goals of Physical Readiness Training

1301. NECESSITY FOR PHYSICAL READINESS TRAINING

A very important objective of training is attainment and maintenance of operational readiness. Marines must be physically ready for operations at any time and under all conditions of climate and environment. A combination of training to develop proficiency in physical skills and conditioning to improve strength and endurance results in physical readiness for combat. The degree of physical fitness required of Marines can be acquired through physical exercises in a normal well-coordinated and closely monitored training routine. The performance of purely military duties, such as drills and marching, is not enough to build all the desired areas of fitness. Few recruits are physically fit for the arduous duties ahead of them. The softening influences of our mechanized civilization and the difficulties of conditioning Marines make physical fitness more important than ever before. If Marines are to be fully developed to and maintained at the desired standard of physical fitness, a well-conceived plan of physical readiness training must be a basic part of every training program. Marines cannot be adequately prepared in any other way for the hard work and arduous demands required on the battlefield.

1302. OBJECTIVE OF PHYSICAL READINESS TRAINING

The overall objective of the physical readiness training program is to develop individuals and units who are physically able and ready to perform their duty assignments or missions during training and in combat. To attain the objective of physical readiness, exercise activities must develop the following areas:

a. Strength and Endurance. Developing adequate strength to perform required duties and adequate endurance to sustain activity over a long period of time.

b. Muscle Tone. Developing muscle tone adequate to maintain proper posture and reasonable weight control.

c. Skills. Developing military physical skills which are essential to personal safety and effective combat performance. As skill is developed, agility and coordination are attained. The essential skills are-

(1) Marching Under Load. Marching with individual and unit weapons and equipment.

(2) Running. Distance and sprint running on roads and cross-country.

(3) Jumping. Broad jumping and vertical jumping downward from a height.

(4) Dodging. Changing body direction rapidly while running.

(5) Climbing and Traversing. Vertical climbing of ropes, poles, walls, and cargo nets. Traversing horizontal objects such as ropes, pipes, and ladders.

(6) Crawling. High crawling and low crawling for speed and stealth.

(7) Throwing. Propelling objects, such as grenades, for a distance with accuracy.

(8) Vaulting. Surmounting low objects, such as fences and barriers, by use of hand assists.

(9) Carrying. Carrying objects and employing man-carries.

(10) Balancing. Maintaining proper body balance on narrow walkways and at heights above normal.

(11) Falling. Contacting the ground from standing, running, and jumping postures.

(12) Surviving in Water (Or Other Specialized Situations) Using water survival techniques.

d. Character Traits. Jnstilling character traits which help accomplish military missions to include-

(1) Confidence. Confidence develops through achieving progressively more difficult tasks as physical ability grows.

(2) Aggressiveness. Participation in combative activities and contests develops desire and willingness to overcome an opponent.

(3) Reaction Under Pressure. Competitive contests and game situations train Marines to think and to act quickly while under pressure.

(4) Teamwork. Teamwork develops through competitive events in which a number of Marines must coordinate their efforts to accomplish a physical task.

1303. BENEFITS OF EXERCISE

The benefits of exercise are not always understood. Some of the more important results of exercise are listed below:

a. Improved Muscle Tone. Muscular tone improves and, at the same time, muscular strength and endurance are built up.

b. Cardiovascular-respiratory Endurance. Cardiovascular-respiratory endurance, or wind, improves through a process of opening up dormant lung capacity to absorb greater amounts of oxygen.

c. Circulation. Circulation of the blood speeds up and extends to a greater portion of the body as exercise forces the blood to service all parts of the body. The efficiency and effectiveness of the heart, lungs, and blood vessels improve.

d. Flexibility. A wider range of muscular movement is possible and rapidity in physical skills grows.

e. Elimination of Body Waste. Bending and twisting the body and the general speedup of body processes caused by exercise regulate and help eliminate body wastes.

f. Tension. Working off excess nervous energy and relief from daily worries and cares relieve tension. Participation in exercises leaves little time for worry.

g. Sleep. Sleep improves because muscles are healthfully tired after about of exercise. A by-product of sound sleep is relief of tension.

h. Obesity Control.. Control of obesity (fat) is made possible by using up excessive amounts of fat-producing food elements.

i. Injury Susceptibility. Susceptibility to injury is reduced through exercise. Muscles, tendons, and joints are strengthened. Injuries such as hernia, back strain, and joint sprains are less likely to occur if muscles are maintained in proper tone.

Section IV. Leadership Roles

1401. PSYCHOLOGICAL LEADERSHIP

The full development of a Marine's resources is not all physical. To be effective in developing physical readiness, leaders must realize that mind and attitude are also important to success. The more important psychological considerations are to—

a. Promote Understanding of the Value of Physical Readiness. A desire to be physically ready should be created in all Marines. Motivation is increased and Marines take greater interest in their individual physical fitness if they understand the value and benefits of vigorous exercise. When Marines realize their efforts are an investment in their own personal welfare, it should not be difficult to obtain their cooperation. Marines should understand the objectives, the benefits, and the value of each type of exercise activity in their program. They should also understand the relation of physical readiness to survival in combat.

b. Maintain a Positive Approach. Physical readiness training for combat is strenuous and demanding. It is a responsibility of leadership to create an atmosphere where all desire to participate fully. This attitude should be fostered. A negative approach must not be identified with physical readiness training, even with those having difficulty. Only in unusual cases should fear of punishment be the motivating factor behind good performance. For those few who cannot keep up or attempt to malinger, an effective remedial program is essential.

c. Seek Cooperation and Develop Morale. In a program placing maximum physical stress upon individuals, it is necessary to gain their cooperation. Favorable reaction is enhanced by proper planning and organization, challenging requirements, use of competition, and application of a progressive program resulting in physical fitness. As physical fitness grows, morale also grows.

1402. COMMAND AND SUPERVISORY FUNCTIONS

a. Command Functions. Commanders should take the following actions to support physical readiness training:

(1) Lead by personal example.

(2) Instill command interest and indicate to subordinate personnel the importance of this training to the welfare of the organization.

(3) Allot sufficient time for the achievement of objectives and monitor the use of such allotted time. The substitution of other training or routine duties for scheduled physical readiness training is unsound and unwise.

(4) Assign and properly utilize qualified personnel to supervise and conduct physical readiness training. If leaders are not competent, take action to ensure they become competent quickly. Reassign those who do not meet this standard to other duties.

(5) Make necessary facilities and funds available to support a program to develop physical readiness within all personnel.

(6) Measure the physical fitness of individuals and units in order to evaluate progress and to determine if the program is successful.

b. Supervisory Functions. Leaders responsible for planning, conducting, and supervising physical readiness training should take the following actions:

(1) Prepare physical readiness training schedules which apply the principles of physical conditioning and which aim for a particular type of program plan.

(2) Provide for wide participation of as many Marines as possible. All Marines, regardless of position or age, will benefit from regular exercise. In some instances, special efforts are necessary to overcome obstacles to regular and frequent training. Special effort is also necessary to ensure remedial conditioning. Such conditioning should occur for those who are physically substandard and after extended absence due to leave, sickness, injury, and travel.

(3) Prevent waste or unwise use of time allotted for physical readiness training. Time-wasters include unprepared instructors; assignment of one instructor to a group larger than a platoon; progression which does not keep pace with the physical development of the Marines; extreme formality; inadequate equipment or facilities which require waiting turns to exercise; and lengthy rest periods between exercises which interfere with the application of overload.

(4) Ensure that the program contains vigorous physical activity. Such activity places progressively greater demands upon the body during each exercise session and also over the duration of the training program. To be of benefit, exercise must tire the muscles and cause the heart to increase its rate of beat.

(5) Set an overall objective for each physical fitness program.Observe the training as necessary to ensure that the established objectives are being achieved.

(6) Observe physical readiness training to insure the use of a positive approach. To implement a positive attitude, small-unit leaders and instructors should personally set the example; have an understanding, fair, and sympathetic attitude; recognize individual differences; and motivate Marines toward their best effort.

(7) Guide and inform small-unit leaders and instructors concerning approved techniques, directives, and literature. As necessary, arrange for local training of instructors to include clinics, conferences, schools, and demonstrations.

(8) Determine the effectiveness of physical readiness training by personal participation in and observation of training, analysis of field inspection reports, and analysis of individual physical fitness test scores. Scores may be combined to reflect the fitness of the unit.

1403. SMALL-UNIT LEADERS AND INSTRUCTORS

a. Responsibility. The instruction and conduct of physical readiness activities are the function of company/battery commanders, platoonleaders and persons assigned as instructors. Experience has proved the effectiveness of physical fitness development when conducted in company-and platoon-size units under direct control of the leader with overall supervision by the parent-unit commander. For example, all the platoons of a company may exercise at the same time under the general supervision of the company commander, with each platoon conducting the assignment separately and under its own leadership.

b. Leader's Assignment. A small-unit leader or an instructor in a school or training activity is assigned to a combat unit or to a support unit. In this assignment, the leader is responsible for all training to include physical readiness training. In a different situation, the noncommissioned officer (NCO) or officer may beassigned as a full-time physical readiness instructor. This contrasts to a unit leader assignment where only part of the time is devoted to such training. In either case, the leader will hold an important and vital position for the physical fitness of Marines.

c. Leader's Training. Leaders may come to the assignment either fully or partially trained. It may be their first responsibility for the development of physical fitness. If they have had previous training through experience, make certain that their information is supplemented with study of this manual. If they have had professional training in physical education during civilian life, but no military experience, they should also use this manual to learn the methods used by the Marine Corps. A new leader should take advantage of various ways to learn including attendance at leader training courses, self-study, practice, and discussion with more experienced leaders.

d.Leader's Objective. As a physical readiness training instructor, the leader has two general objectives. The first is to motivate Marines to want to be physically fit. The second is to conduct a program that will develop a high degree of physical fitnss. Motivated Marines will react enthusiastically to such a program. It aids greatly in achieving local program objectives.

e. Leader's Personal Fitnes. A unit leader who must instruct and demonstrate physical activities must be in physical condition to do the job without undue physical stress.The leader should be able to do those things that must be demonstrated. The leader's strength, endurance, posture, and skill should set the example. This does not mean that the leader must excel, as other Marines do not expect championship performance. However, they do expect, and deserve, a creditable showing of fitness for the job.

f. Leader's Knowledge. The leader must have three types of knowledge to properly administer physical readiness training. Theyare- -

(1) Knowledge of Marines. The leader must understand Marines, know how to lead and motivate them, understand how they learn, and apply this knowledge wisely in the day-to-day training situation.

(2) Understanding of Body Functioning. A more intelligent exercise program results from understanding and applying the principles which govern physical conditioning of the body. The leader with such knowledge can better prescribe, adjust, and regulate exercise types, amounts, and progression to attain fitness.

(3) Understanding Exercise Activities. The leader needs to understand the contribution each type of physical activity makes to physical fitness, and how to use each activity to develop fitness. Skill to demonstrate and lead the various activities is a necessary part of technique and is invaluable to the instructor or small-unit leader.

Chapter 2
PHYSICAL READINESS TRAINING PROGRAMS
Section I. Development of a Program

2101.GUIDANCE FOR PLANNERS

This chapter instructs planners on physical readiness training procedures. It contains program planning guidance including factors to consider when developing programs; steps in assembling a program; definitions of activity packages and systems of exercise; selection of activity packages; and selection of systems used in implementing those packages.

2102. MARINE CORPS MISSION

The mission of the Marine Corps is to seize and defend advanced naval bases and to perform other missions as the President of the United States may direct. Traditionally, to perform this mission, Marines have been projected into the area of operations by Navy vessels and have attacked hostile beaches over the shore in landing craft, assault amphibian vehicles, and helicopters. Recently, performance of this task has been somewhat complicated by the introduction of the concepts of maritime prepositioning and the air-landed Marine expeditionary-brigade. Under these concepts, Marines, landed at airfields in foreign countries, will join their equipment at commercial ports. The distance between these airfields and ports is often 10 or more miles. In time of international crisis, it can be expected that transportation assets will not be adequate. Marines from all elements of a Marine Air-Ground Task Force may have to march from the airfield to the site of their equipment. All Marines must be physically conditioned to rigors of conducting foot marches with individual weapons and equipment.

2103. EVALUATION OF FITNESS

The Marine Corps mission is such that all Marines must constantly and consistently achieve a high level of fitness that prepares them for the demands of combat. While Marines have many different MOS's, all Marines must be prepared for the demands of marching under load and performing basic infantry tasks, such as rear security and patrolling. While the physical fitness test remains our universal measure of individual fitness, it is the commander's responsibility to observe and evaluate the unit's ability to perform effectively in combat. Often the commander's most effective evaluation tool is weekly physical training sessions. The commander should participate in these sessions and see that they are fully integrated into the unit physical conditioning program.

2104. INTEGRATION OF TRAINING

Training time must be used efficiently and wisely. Every opportunity to integrate physical training into other training activities should be seized. For example, foot marches under load can be integrated into the normal daily activities as a way to move from one training or work site to another. This integration reflects realism in training and should be used to maintain overall proficiency. To accomplish this requires imagination and enthusiasm on the part of the commander. The area of operations in which the training is conducted determines what can be integrated and how. If possible, training should also consider the terrain and climate of the area in which the unit will subsequently conduct training or operations. Training should include familiarization with special equipment and the application of specialized techniques to tactical principles. Activities such as gun drills and command post emplacement drills also have physical conditioning value, particularly if supervised and done competitively or measured against time standards.

2105. ACTIVITIES AT THE UNIT LEVEL

Physical readiness training is a command responsibility and is generally conducted in Marine units at the company and platoon levels. Battalion/squadron commanders must constantly supervise, making sure that all Marines are physically prepared for combat.

a. Physical readiness training should usually be conducted weekly by each battalion/squadron-level unit, using events designed for all personnel (e.g., marches under load, battalion/squadron runs, military field days, etc.).These events enable the commanders to demonstrate personal leadership, and observe and evaluate the physical condition and combat physical readiness of their Marines. This exercise period should be designed to elevate the unit's morale and emphasize unit identity through the wearing of uniforms and the carrying of battalion and company guidons. The battalion/squadron commander should make every effort to be a visible participant in the unit's physical readiness program on a weekly basis. These activities could be conditioning marches under load or a weekly calisthenics/unit run.The suggested times for these activities are early on Monday morning or late on Friday afternoon after the training week has been completed. The uniform may be physical training gear orutilities and boots.

b. Commanders should be cautious of long runs in boots and utilities. This is not to suggest that units should not run in utilities, but to remind commanders that boots are designed for marching, not running. If commanders choose boots and utilities, they should select off-road routes which will provide a cushioning effect.

c. Units should be made well aware of the commander's intent before the unit falls out for exercise. If the exercise period is conducted early in the morning, it need not necessitate an extensive shower/clean-up period afterward, so long as all Marines from the commander on down remain dirty. Commanders should inform their Marines not to fall out for these sessions in clean, pressed uniforms.

d. The physical readiness training of headquarters units is often difficult to manage. Often physical training is left for individuals or sections to develop and manage on their own. This approach can be effective, but if not supervised, it can result in a "paper" program. Headquarters unit commanders should endeavor to conduct weekly integrated training which will enable them to evaluate the physical readiness of their personnel. Often this training has to be scheduled before or after normal working hours to maximize participation of the entire unit.

2106. STEPS IN PLANNING

To implement workable and effective programs (as directed by CMC ALMAR 261/87) the planner must-

a. Determine the Type of Program Needed. Marine units are inherently different in organization and mission. The physical readiness program must be tailored to the mission and to the current physical condition of most unit personnel. Programs to meet this need are of the following types:

(1) **Developmental Programs.** Marines in a beginning or poor state of physical readiness need a program which will develop strength, endurance, physical skills, and character traits which are beneficial to successful accomplishment of military missions. Such programs should be applied progressively to rise gradually to a peak of fitness and skill.

(2) Maintenance Programs. Once Marines reach the sustaining stage of conditioning, their goal is then to maintain this level by participation in a maintenance program.

(3) Remedial Programs. The term "remedial" is usually applied to those individuals or groups who possess substandard physical fitness. For example, a remedial physical conditioning program could be applied to persons who are overweight, who fail to reach physical fitness test standards, or who have missed extended periods of conditioning due to illness, injury, extended hospitalization, or other absence. Saturday training sessions are designed as remedial sessions for all individuals who either miss the daily training or have failed to meet adequate standards. Note that personnel in light duty or no-duty status should be expectedto attend all training sessions, observing, supervising, or assisting as needed.

b. Determine the Time Required. The amount of time for training operations varies considerably. However, every unit can find time to conduct physical readiness training. Frequent (i.e.,daily) physical training of short duration-30 to 60 minutes-is preferred over occasional longer periods. Other demands for training time are so urgent that every minute of time allotted for physical readinesstraining should be used. Determine the time required per week anddivide it into daily blocks.

c. Organize for Various Group Size. It is essential to stress exercise rather than formality. Marines must complete the program where they are- -on the training field, in the motor pool, on the range, next to the classroom, in the office area, in the shop, aboard ship, or elsewhere. Although desirable, it is not always possible to assemble company-size units for physical training. Platoon-size groups are appropriate for the proper conduct of physical conditioning activities. Certain situations may require exercise programs for section- or squad-size units. This manual outlines programs for all situations and types of organizations.

d. Allow for Weather and Exercise Area. In programing and scheduling, the climate and terrain often govern the selection of activities.

(1) Weather changes cause differences in temperature, rainfall, wind chill, and snow. These changes should be anticipated as they dictate the type of program. Alternate plans should be part of the schedule. Nevertheless, normal weather changes and rain should not drive training inside.
(2) Local terrain and available exercise areas may also influence the selection of activities and the type of program which it is possible to support. Some activities can be completed in nearly any area.
e. Plan for Seasonal Change. As most physical readiness training is conducted outdoors, it is necessary to recognize seasonal change. A program should be divided into fall, winter, spring, and summer parts. Seasonal change also causes change in light. For example, an early morning program started in the summer will have ideal lightconditions, yet in the fall or winter, darkness will occur at that same hour and interfere with the conduct of the program. Develop programs in seasonal blocks, and make provisions for anticipated changes.

f. Consider Needed Facilities. An excellent program can be conducted with practically no facilities since there are exercises which require no equipment. However, a better program can be developed when supported by certain facilities and items of equipment. Proper command support, plus ingenuity, will solve this problem. (Items of equipment, when necessary to support the recommended exercises, are included in the chapters on exercise activities.)

g. Specify Appropriate Uniform. The uniform worn for exercising depends upon the season of the year, the state of the weather, and local regulations. Whenever possible, Marines should be dressed alike. Undershirts are preferred as the upper garment when the weather permits. A uniform that restricts the free movement of the body should not be worn when exercising.

h. Consider Availability of Instructors. Leaders who can lead and direct the scheduled activity must be available. Organizational units should train junior officers and noncommissioned officers down to squad or section leaders to instruct and lead the various activities.

i. Select Activity and System. With the type of program needed and the objectives in mind, select an activity package or an integrated training package for each day's scheduled physical training. At this same time, the system to be used in employment of the selected activity or activities must be determined. This selection affects equipment, areas, instructors, transportation, and other support requirements.

j. Secure Command Participation and Support. Prepare and brief the commander to assure full understanding of the objectives and administration of the program. The full participation and support of the commander will greatly improve the level of success.

k. Supervise Execution. Determine needs, publish the program, and supervise its execution as a necessary part of developing the unit's training schedule.

2107. SELECTION OF ACTIVITIES AND SYSTEMS

a. Exercise Activities. The planner must determine exercise activities which will be appropriate to include in the program. An exercise activity is a single means of exercise usually identified by the name applied; for example, running, log exercises, and obstacle course. Many exercise activi ties can become part of activity packages.

b. Activity Packages. Many of the physical activities described in this manual are arranged in prescribed sequences and are known as activity packages. An activity package is a number of exercises of the same type, assembled as a group or a set, and arranged in a specific sequence. Exercise packages are organized in such manner that not more than 15 minutes will be required to complete the execution of any package. Each type of activity is explained in laterchapters. The number of available drills, tables, or circuits; the manner of organization; and the contribution each makes to the total program are covered. Full understanding of this information will greatly assist in developing effective programs. Various designations are used to identify exercise packages; for example, conditioning exercises when arranged in a set order are known as drills, and other packaged activities are designated as tables or circuits. The following are activity packages:

(1) Activity packages for groups:

o Conditioning Drill Two
o Conditioning Drill Three
o Rifle Drill
o Log Drill
o Grass Drill
o Running Tables
o Guerrilla Tables
o Circuit-Interval Table
o Combatives Tables
o Relay Tables

(2) Strength Circuits:

o Fixed Circuit
o Movable Circuit
o Simplified Circuit (Circuit Interval Table)

(3) Activity packages for individuals:

o The 6-12 Plan
o Weight Training
o Isometric Exercise

c. Advantages of Using Packages. These of exercise packages simplifies scheduling and conducting of exercise and results in the following benefits:

(1) Schedule development is simplified as the planner assembles packages which will satisfy the training objective. There is no need to deal with selection of individual activities or to be concerned about the amount of time to be expended on each.

(2) Any 15-minute period, and in some cases less time, can be scheduled or used to perform an activity.

(3) If longer periods of time are available or if the objective demands, several packages can be assembled to provide a more complete period of activity.

(4) Marines are assured a balanced set of exercises or activities as each package is carefully arranged to reach all muscle groups.

(5) The instructor can concentrate on the conduct of a vigorous workout as the type and duration of the activity have already been determined.

d. Nonpackaged Activities. Several types of activity are not packaged. Activities in this category are conditioning marches, unit runs, obstacle courses, combat water survival swimming, teamcontests, and team sports. These activities can be scheduled in combination with packaged activities, or they may be scheduled separately. Often they require a longer period of time. Most nonpackaged activities require a 50-minute period to satisfactorily complete their objective. For conditioning marches, a few hours are required. The benefits of these activities should not be overlooked as some desirable objectives cannot be attained without them.

e. Systems of Exercise. Several methods or systems of organizing exercise and activity packages can be used. Each system is based upon a specific organization as follows:

(1) Single Activity System. The unit leader immediately assumes command of the unit at the beginning of the exercise period. The leader moves the unit to a predesignated exercise site at double time, forms the unit in a circle around the leader, groundsclothing and equipment as appropriate, and quickly moves into the exercise routine. There is usually no time to teach; therefore,the Marines must know the activity to be used. At the conclusion of a 5- to 15-minute period, the unit leader returns the unit to the instructional area at double time and releases the unit for the next scheduled activity.

(2) Progressive Activity System. All Marines (company or platoon) complete activities in the same order during the period. For example, Drill One is followed by dual combatives, and finally a 1-mile run. This system is usually progressive from a warm up activity, such as Drill One, to an activity which contributes in a major way to one of the objectives, such as aggressiveness development through combatives or cardiovascular development, such as running.

(3) Rotating Activity System. Set up the same number of activities or stations as there are platoons in the company. Each platoon rotates through each station in turn. With four platoons in a 50-minute period, about 10 minutes can be devoted to each station. With three platoons, approximately 15 minutes may be spent at each station. Activities must be of a type that can be covered in the time allotted. For example, with three platoons, Station 1 could be Conditioning Drill One; Station 2, running; and Station 3, a team contest.

(4) Circuit System. Set up a number of stations to provide various types of exercise equipment and/or items of apparatus. The idea is to keep all Marines busy and exercising vigorously for a short period of time at each station. The fixed strength circuits and the movable strength circuits are examples of this system. Station changes must be rapid, and the exercise must be started quickly after each change. Since the objective is to exercise at top speed, the motivation comes from frequent change of activity by moving to another station. Rotation by station groups continues until all Marines have covered all stations.

(5) Interval System. This system stresses the development of strength and endurance. It involves heavy work for a given distance within a specified time, alternated with lighter work and recovery, but never stopping during the workout. This procedure is repeated, and the intensity is increased gradually as physical condition improves, but always with adequate recovery. The important concept is stress, recover, stress, recover, and so on. This system is often applied through running, but other activities of a continuous nature may also be used. An example is the Circuit-Interval Table.

f. Steps in Selecting Activities and Systems. The planner should follow these steps.

(1) Evaluate the needs of one unit with emphasis upon unit mission, objective, and time available.

(2) For each day, select an activity package or a combination of activities which will contribute to the objective.

(3) Then determine the system to be used in implementing the selected activities. There are several possible choices and much opportunity for flexibility in program development. For example, the weekly program may contain various exercise packages, systems, and time periods. Additional guidance is contained in following chapters.

2108. SAMPLE WEEKLY PROGRAMS

The variety of weekly programs given shows available alternatives. Programs and schedules other than those illustrated may be assembled to provide schedules to fit any situation. Many other scheduling combinations are possible through use of the packaged activities, varied time periods, and different types of assigned duty.

a. Sample A. Figure 2-1 illustrates a 60-minute time allotment in which the single activity system is used with activity packages that vary from day to day. In this program, 5 hours of training are scheduled for the week. This program needs some type of integrated unit training to supplement it each week. A bimonthly conditioning march of 10 miles in 3 hours is suggested to round out this program. The Saturday session is designed for remedial physical training.

NOTE: A physical training activity must be conducted at a minimum of once every 72 hours. After 72 hours, muscles begin to atrophy.

Figure 2-1 is available in the illustration section at the end of the book

b. Sample B. A more comprehensive weekly program (fig. 2-2) illustrates the scheduling of various-length periods, a variety of activities, and the use of three systems during the week. In this schedule, 3.5 hours are included for the week. The following should be noted concerning this schedule:

Figure 2-2 is available in the illustration section at the end of the book

(1) On Monday, a 60-minute period is available for physical readiness training. Three 20-minute packages are scheduled. Each platoon, under its own leadership, will progress through each scheduled activity in turn.

(2) On Tuesday, Wednesday, and Thursday, only 30-minute periods are available. Here each platoon leader supervises each platoon in a single activity system.

(3) On Friday, the company commander directs training to exhibit personal leadership example and supervision. In this example, only one strength circuit and one cross-country course are available, and each will accommodate only one platoon. For this reason, Conditioning Drill One station is included. A platoon rotates to each station within the 60-minute period.

(4) The Saturday session once again is designed for remedial physical training.

(5) Bimonthly, a 10-mile/3-hour march under load will be necessary to round out this program.

NOTE: Cardiovascular/respiratory exercise to improve endurance must be of long enough duration (20 minutes or more) to induce and maintain oxygen debt.

c. Sample C. A third sample schedule (fig. 2-3) illustrates the use of the single activity system for 5.25 hours of training. This schedule includes sustaining-type activities for a unit that has passed through both the toughening and slow improvement stages of conditioning. The use of competitive activities is featured to holdinterest and provide self-motivation. A bimonthly 10-mile/3-hour march under load will be necessary to round out this program. The Saturday session once again is designed forremedial physical training.

Figure 2-3 is available in the illustration section at the end of the book

NOTE: Thc body will respond to hard work or physical training without soreness and stiffness provided the work or physical training is conducted regularly.

NOTE: Excrciscs of short duration increase strength and bulk but do little to effect endurance.

NOTE: Thc times shown in the charts arc participation times.

Section II. Remedial Physical Training

2201. NEED FOR ATTENTION

Remedial physical training is a process by which physically substandard individuals are conditioned to meet standard levels of performance. To achieve a full degree of operational readiness, it is necessary to bring all Marines up to the prescribed standard. Experience has demonstrated that some Marines have difficulty due to a poor state of fitness, obesity, or lack of motivation. Attention to these individual deficiencies will help improve unit combat readiness.

2202. IDENTIFICATION OF INDIVIDUALS

The company or battery commander identifies persons who cannot achieve the prescribed level of proficiency through the physical fitness test or as a result of observation during training. The commander notes particular weakness of body parts as indicated by failure of certain test events. These Marines are then placed in a special remedial program either at platoon, company/battery, or battalion/squadron level. Extra time is devoted to overcoming the weakness. These extra periods of conditioning may be during or after duty hours, as determined by the commanders. Other persons need reconditioning as result of hospitalization. Restoring physical fitness to damaged areas of the body is accomplished through progressively graded physical activities under professional supervision, not by the physical fitness trainers.

a. Types of Deficiencies. Physical deficiencies which can be corrected by exercise fall into several categories.

(1) Lack of Strength in One or More Body Parts. The major muscle areas concerned are the arms and shoulder girdle, back, abdomen, and legs.

(2) Lack of Overall Endurance. Usually, there is deficient muscular and cardiovascular endurance.

(3) Deficiency in Coordination and Agility. In these cases, physical skill is not developed to a satisfactory degree in activities such as crawling, running, jumping, climbing, traversing, vaulting, pushing, pulling, lifting, and carrying.

(4) Overweight or Underweight. Either condition may interfere with physical fitness and mission accomplishment. Lack of exercise is not always the cause. The cause may be malfunction of normal physiological functions or it may be poor health habits such as over- or under-eating, lack of adequate rest, or overconsumption of alcohol.

(5) Lack of Motivation. Not all persons are motivated to attain or maintain **a** desirable state of fitness. Some Marines do not understand the importance of physical fitness, some find proper exercise too difficult, and others find it inconvenient.

b. Causes of Deficiencies. There are several causes for Marines being deficient in physical fitness:

(1) Absence of exercise.

(2) Exercise which fails to develop all muscle groups and components of fitness.

(3) Exercise which is not vigorous enough or which lacks progression.
(4) Injury or illness which depletes fitness.
(5) Inadequate amounts of sleep or rest.

c. Methods of Detection. The commander has several means by which to detect physical deficiencies:

(1)

from those who pass. Make an analysis to determine the cause of failure as revealed by test scores.

(2) Observe Marines as they perform physical tasks-

o Marines who have difficulty during training or in physical types of work,

o Marines who have become obese and therefore experience difficulty.

(3) Be alert to those Marines who are often on sick call, returnees from hospitalization, or newly assigned.

2203. ADMINISTRATION OF REMEDIAL ACTION

a. Group. Attitude Marines who are singled out as being physically deficient are self-conscious and are not always convinced they need extra help. Within a deficient group, motivation may be low. These Marines must be convinced that a remedial program which is tailored to their needs will help them and will eliminate their deficiencies. Encouragement is often needed and desirable.

b. Leadership. The leader of this group must study individuals and know their deficiencies. The unit leader should counsel Marines individually, observe them closely as they progress through the remedial program, maintain records, and adjust the program as required.

c. Measurement.. Whenever improvement in performance is noted, Marines should be measured by the physical fitness test either individually or by group. When testing reveals an individual to be satisfactory, that individual should be released from the remedial group. There may be exceptions to this policy in the case of Marines who are overweight or in the execution of an exercise program prescribed for some special purpose.

d. Organization of a Remedial Group

(1) A remedial group is usually a small group of Marines within a company or battalion. In some situations, it may be a larger group numbering 50 to 75 Marines. In the case of a larger group, divide the Marines initially into subgroups according to ability. Prescribe exercise loads commensurate with their ability. General conditioning is sufficient in the early part of the program to qualify Marines who are on the borderline. Qualification will reduce the size of the group, permitting more individual and specialized attention to those who remain.

(2) As the program progresses, regroup individuals who have like deficiencies in order that they can concentrate on their weakness. For example—

(a) One group may be weak in the arms and shoulders as revealed by failure in the chin up event. This group, with an assistant instruction charge, can work on pull-ups, rope climb, pushups, rifle or logexercises, horizontal ladder, and similar types of exercise.

(b) Another group may be weak in general endurance as revealed by the3-mile run. This group could profit from participation in conditioning drills, running, grass drill, and strength circuit.

(3) It may be necessary to form some groups to overcome weaknesses in skills, such as an inability to throw, to quickly change direction while running, to crawl rapidly, or to carry a load. Lack of coordination or lack of practice may cause these deficiencies. In this instance, instructors must provide an opportunity to practice and correct poor form and other errors as they are noted.

Chapter 3
PHYSICAL CONDITIONING ACTIVITIES
Section I. Marching Under Load

3101. GENERAL

Few physical fitness activities are as directly related to readiness for combat as foot marches under load. In addition to the obvious physical conditioning and unit cohesiveness benefits, marches under load prepare most Marines for numerous foreseeable tasks in combat. The idea that only infantry and reconnaissance units actually require training marches to prepare them for combat is clearly a mistake. In addition to the very real possibility of Marines from all types of units being used to fill combat-depleted infantry units, it is likely in maritime prepositioning force operations that aviation, combat service support, and command element personnel will have to march with weapons and equipment, from the arrival airfield to the equipment marshalling site. In American coastal cities, the harbor is nearly always 10 or more miles from the airport. It is difficult to imagine any different condition in a lesser developed country .A further combat-related benefit is that foot marching under load is the most effective way to develop leg strength in the context of a unit training program. In his classic study of troop performance in World War II, Men Against Fire, S.L.A. Marshall noted that the greatest single weakness of replacements of all occupational fields was leg strength. Truck drivers must push trucks which are stuck in mud, and headquarters communicators must climb hills to emplace antennas. The demands on the lower body in combat apply to all Marines, and foot marching is one activity which can help build these muscles while being conveniently integrated with the rest of the training program.

3102.TRAINING GOALS

The standard for success of a foot march is very simple to measure: did the unit arrive at the destination at the prescribed time with Marines in condition and required equipment present to accomplish the mission?A progressive program can increase a unit's readiness. It can instill pride by increasing distances and rates of march and by selecting routes over increasingly challenging terrain as the Marines become better conditioned. However, the ability to execute the mission at the conclusion of the march must remain the standard of success.

3103.MOVEMENT PLANNING

The basic considerations in planning a foot march are the mission, tactical situation, terrain and weather, and the units to participate. The success of the march will depend largely upon the thoroughness with which it is planned. A successful march is characterized by adherence to prescribed routes and time schedules, the efficient employment of the means available, and the ability of the unit to accomplish its assigned mission upon arrival at the destination.

a. Movement Orders. Movement planning culminates in the preparation and issuance of an operation order prepared in the standard, five-paragraph format. Necessary annexes are attached to furnish detailed information required for the movement. Written movement orders are rarely prepared at company level. A discussion of the annexes is contained in this chapter.

b.March Planning. March planning, as discussed here, is the planning conducted at battalion level. March planning may be organized into the following steps:

(1) **Preparation and Issuance of the Warning Order.** In order to afford subordinate units the maximum possible time to prepare for a pending move, a warning order containing all available information about the march is issued. The amount of planning time available will determine the time of issuance and the content of the warning order.

(2) Estimate of the Situation. In his estimate, the commander considers the mission, terrain, weather, time, and space factors, available routes, available transportation for the movement of equipment and/or shuttling of Marines, enemy capabilities, disposition of own forces, physical condition and training of Marines, and courses of action available to the command.

(3) Organization and Dispatch of a Reconnaissance. Party very march plan is based on as thorough and complete a ground reconnaissance as time and the situation will permit. Map and aerial reconnaissance are valuable in formulating a plan, but are not a substitute for ground reconnaissance. Route reconnaissance is accomplished by a reconnaissance party which usually consists of a reconnaissance element, an engineer element from the attached or supporting engineer unit, and a traffic control element. Unit standing operating procedures generally establish the basic composition of the reconnaissance party. It is modified as necessary to meet the requirements of a particular march. A recommended method of reporting information obtained by the route reconnaissance party is shown in figure 3-1.The minimum information required from the reconnaissance party is—

(a) Available routes and conditions.(Routes may be specified by higher headquarters.)

(b) Recommended rate of march.

(c) Selection of start point and release point, or confirmation of the suitability of start point and release point previously selected by map reconnaissance.

(d) Confirmation of location of the assembly or bivouac area.

(e) Location of critical points on the route.

(f) Distance between critical points on the route and total distance from start point to release point.

Figure 3-1 is available in the illustration section at the end of the book

(g) Location of obstacles and estimation of necessary Marines and equip- ment needed to repair and maintain routes.

(h) Number of guides required and their location on the route.

(4) Development of Detailed Movement Plan. Some of the elements included in a detailed movement plan are-

(a) Organizing the Column. To facilitate control and scheduling, units will be organized into serials and march units and given an order of march. In determining the order of march, the march planner must consider the enemy situation and the desirable order of arrival of the units at the destination. Where dispersion is required, a unit may be organized into two or more columns, each assigned a different route.

(b) Using Reconnaissance Information. Results of the route reconnaissance will be used to select the route(s); determine the start point, critical points along the route(s), and the release point; and select the rate of march.

(c) Determining March Computations. March unit pass time is based on the strength, formation, and rate of march. The pass time of the marching columns, plus necessary time distance computations, will be used to determine the completion time of the march.

(d) Drafting of Road Movement Table. Using the completed march computations, a draft road movement tabic is compiled.

(e) Checking the Plan. Using the draft road movement table and a road movement graph, the movement plan is checked to ensure that it conforms to the directive of the higher headquarters and the battalion commander's instructions.

(5) Preparation and Issuance of the Road Movement Order. After the plan has been checked and approved by the commander, an operation order is prepared and issued. The operation order may be in written form or issued orally, and is accompanied by a road movement table, overlay and/or strip map, and appropriate administrative details.

(6) Road Movement Table. A road movement table, prepared as an annex to the operation order, provides serial commanders with arrival and clearance times at critical points along the route of march. It also provides the column commander with information as to the proposed location of elements of the column at various times.

(7) Overlay/Strip Map. An overlay serves the normal purpose and should show, as a minimum, the present location of units, route of march, critical points, and the new location of units at the destination. A strip map is a schematic diagram of the route of march and shows landmarks and critical points with the distances between them. A strip map may be issued as an annex to the road movement order, in addition to or in lieu of an overlay.

c. March Computations. Prior to issuance of the operation order, the S-3 must verify time and space computations as they provide him with the necessary data for the preparation of a road movement table.

(1) Time-Distance. Time distance (TD) is determined by dividing the distance to be traveled (D) by the rate of inarch (R):

TD = D/RWhere TD = hours D = distance in kilometers R = kilometers per hour

(2) Length of ColumnThe length of column (LC) is used to determine the pass time of a column. The sum total of the following two parts determines the length of column. (See pars, (a) and (b)).The two parts are the space occupied by the Marines alone (including the distance between Marines) and the sum of the distances between the units of the column (column gap).

(a) The length of column of Marines alone is determined by multiplying the number of Marines by the appropriate factor selected from the table below. The length of column does not include distances between units. LC Marines = No. of Marines x factor.

(b) The total distance in meters between units is determined as follows:

Step One. Determine the number of gaps between serials (total serials minus one).

Step Two. Multiply the number of serial gaps from Step One by the length (in meters) between respective units.

Step Three. Determine the number of gaps between march units (total march units minus one, minus the number of serial distances).

Step Four. Multiply the number of march unit gaps from Step Three by the length (in meters) between respective units.

Step Five. Add the totals from Step Two and Step Four to get the total meters for the column gap.

EXAMPLE:A battalion foot column is organized into 12 platoon-sized march units and 3 company-sized serials. REQUIRED: total column gap distances when there are 100 meters between serials and 50 meters between march units.

(3) Pass-Time.Pass-time (PT) is the time a unit takes to pass a specified point. For foot columns, the pass-time is determined by applying the following formula:PT (minutes) = LC x FACTOR (for appropriate rate of march).

PASS-TIME FACTORS - FOOT TROOPS

.0150 for 4.0 km/h
.0187 for 3.2 km/h
.0250 for 2.4 km/h
.0375 for 1.6 km/h

EXAMPLE: Determine the pass-time of a unit whose length of column is 1,500meters and is marching at a rate of 4 km/h.PT (min) = 1,500 x .0150 (the factor for 4.0 km/h) = 22.5 mm.

(4) Completion Time. Completion time is the time of day that a march will be completed. Completion time is determined by using the following formula: Completion Time = SP time + TD + PT + Scheduled Halts (other than normal hourly halts).

EXAMPLE: A column's starting point (SP) time is 0700.The time-distance is6 hours and 40 minutes. Pass-time of the column is 30 minutes. A 35-minutelunch halt has been scheduled. What is the completion time of the march?Employing the 24-hour clock system, the formula can be applied as follows for simplified addition of the times:

HrMin SP time................0700
TD......................0640
PT......................0030
Lunch halt..............0035

Completion time.........13 105or 1445

The march will be completed at1445.

(5) Experience Tables. Based on previous movements made by a unit, data is accumulated to facilitate march planning. Such data includes approximate pass-times lor various elements of the battalion. The S-3 can utilize these data rather than computing them each time a march is scheduled. Such experience tables serve to reduce the time required to complete the computation phase of march planning. Matter appropriate to the unit standing operating procedure should be integrated therein.

d. Road Movement Graph

(1) A road movement graph is a time-distance diagram used in planning, preparing, or checking road movement tables, and for controlling marches. The graph provides a visual representation of a march plan so that conflicts and discrepancies may be prevented in the planning stage, before congestion occurs on the route. It is not usually issued as a part of the order. Road movement graphs may be applied to small units, to movements of a single column, or to a large organization scheduling separate elements, moving by various means, with different rates of march, over one or more routes.

(2) To construct a road movement graph, use the following steps: (See fig. 3-2.)

Figure 3-2 is available in the illustration section at the end of the book

Step One. Determine the number of hours available for the march. Designate the lower left corner of a sheet of graph paper as the SP time or the earliest even hour before the march is to begin. Select a scale and plot the hours available in sequence from the left on the horizontal axis.

Step Two. Determine the distance to be moved in kilometers. Indicate the SP at the lower left corner of the graph sheet and, using an appropriate scale, plot the number of kilometers on the vertical scale from the SP to the release point. Indicate the location of critical points on the vertical scale.

Step Three. At the proper distance from the SP, draw a horizontal line indicating the location of the RP. Indicate by a vertical line the hour when the movement must be completed. Plot lines representing route restriction, if any, at the proper distances and times on the graph.

Step Four. Determine the pass-times of foot and motor elements of the column.

Step Five. Starting at the SP at the specified hour, plot the movement of the head of the leading element. If the rate of march is 3 kilometers per hour, the lead element will have moved 1.5 kilometers in 30 minutes, 3 kilometers in 1 hour, etc. Plot the trace of the lead man to the RP. The last Marine will cross the SP in pass-times after the lead man. Measure this time on the graph and plot the trace of the last Marine of the column. The lines describing the head and tail of a march serial are parallel. Indicate the time subsequent serials reach the SP and plot the trace of the head and tail. Check to see that the plan complies with all restrictions and orders. If there are any violations or conflicts, the plan may be made to conform by changing the order of march, the starting time of the serials, the planned rate of march, or the organization of the column.

e. Conversion to the Metric System. The adoption of the metric system for expressing linear distances requires a simple means of conversion.

(1) To accomplish this, use the following tables:

o Multiply inches x 2.54 to obtain centimeters. Multiply centimeters x .39 to obtain inches,

o Multiply yards x .91 to obtain meters. Multiply meters x 1.1 to obtain yards,

o Multiply miles x 1.6 (or 8/5) to obtain kilometers. Multiply kilometers x .62 (or 5/8) to obtain miles.

(2) Linear distances can be accurately measured on maps which have been converted to the metric system by using a straightedge graduated in centimeters (cm).The centimeter scale can be used in lieu of the graphic scale when the scale of the map is shown, as follows:

Map Scale:	One cm Equals:
1:500,000	5,000 meters
1:250,000	2,500 meters
1:100,000	1,000 meters
1:50,000	500 meters
1:25,000	250 meters

3104. INDIVIDUAL LOAD

Since the backbone of Marine ground combat forces is the infantry, tactical mobility is largely dependent on the marching ability of the individual Marine. The load that each Marine must carry is the greatest impediment to mobility. In general, the commander must make every effort to reduce the individual load of his Marines to the absolute minimum. While the load has the greatest effect on Marines marching on foot, the value of carrying the minimum essential supplies and equipment applies for all Marine forces.

a. Fighting Load. The fighting load consists of items of clothing, equipment, weapons, and ammunition that are carried by, and essential to, the effectiveness of the combat Marine and the accomplishment of the immediate mission. The fighting load should not weigh more than 40 pounds. The commander must ensure that his Marines' loads- -and his own- -be stripped to the bare minimum. The addition of small, light-weight items in excess of the limit should not be tolerated, since cumulatively, these items will take a toll in energy. Every extra pound a Marine must carry decreases his combat effectiveness.

b. Existence Load. The existence load consists of items other than those in the fighting load that are required to sustain or protect the combat Marine, that may be necessary for increased personal and environmental protection, and that are not normally carried by the individual. The existence load is normally carried by the Marine's unit. Marching, but not engaged in combat, a Marine may be expected to carry a maximum of 50 pounds of supplies and equipment.

c. Considerations

(1) The primary consideration is not how much a Marine can carry, but how much he can carry without impaired combat effectiveness-moral or physical. The combat strength of a unit is not counted simply in numbers of Marines, but in the numbers of willing and physically able Marines.

(2) It is generally better to risk temporary inconvenience from lack of health and comfort items than to exhaust Marines due to overloading.

(3) It is a fundamental truth that men become physically exhausted more quickly when under the stress of combat. Marines must be required to carry less into battle than they are conditioned to carry in training. Marines should be conditioned for carrying weight, but should be equipped in combat for fleetness of foot.

(4) A common mistake is to base the fighting load on the gear and supplies necessary to meet every contingency. The commander cannot reasonably expect the Marines to carry enough gear for every possible eventuality. The items to be carried must be based on reasonable expectations.

(5) It is the commander's responsibility to procure the transport to carry any additional gear. As a rule of thumb, a rifle company, or unit of similar size, requires one 5-ton truck and trailer in support to carry additional gear. In cold weather, or other conditions in which the necessary personal gear increases, this requirement will increase.

(6) The commander must ensure that the supply system provides, on a dependable and timely basis, the balance of essential supplies and equipment not carried by the unit. Marines must feel confident that they will be supported with the necessary supplies and equipment.

(7) In training, commanders must implant in their Marines the pride in operating under austere conditions. To effectively demonstrate the need for this spartan mentality, the commander must personally set the standard. Marines should be trained in field craft, foraging techniques, and the use of caches and field expedients. Maximum use should be made of captured stores.

3105. MARCH TECHNIQUES AND PROCEDURES

a. Organization. A command executing a march is organized into march units, march serials, and march columns, as necessary for control.

(1) March Units. A march unit is a unit of command which moves and halts at the command of a single commander. The march unit normally corresponds to one of the smaller Marine units such as a platoon or company.

(2) March Serials. A march serial consists of one or more march units organized under the senior officer and given a numerical or alphabetical designation to facilitate control. A serial is usually a battalion, but may be a company if the battalion is marching alone.

(3) March Columns. A march column is composed of elements of a command moving over the same route. It may be composed of one or more serials. To facilitate control, a column commander is designated. A column is normally a regiment or larger unit, but may be a battalion if marching alone.

b. Order of March. The order of march is determined by the mission, terrain, and the commander's desired order of commitment of units into action.

c. Control and Coordination Measures. The commander establishes initial control of the march by designating control measures in his march order. The most frequently used measures include:

o Start point and release point.
o Time at which head or tail of column passes the start point.
o Rate of march.
o Order of march.
o Route of march.
o Location of command post.
o March communications.

d. The Pacesetter. The pace-setter is an experienced individual carrying the same load as the majority of the Marines and marching from 4 to 10 meters at the head of the column. (See fig. 3-3.)The pace-setter's primary duty is to maintain the rate of march ordered by the column commander.

(1) Setting the Pace. The pacesetter does this by establishing the pace (length of step) and cadence (steps per minute) to obtain the prescribed rate of march. He should be of medium height so normal strides will be taken. Overstriding or understriding tends to tire leg muscles quickly and affects the combat efficiency of marching Marines. The officer marching at the head of the column supervises the pacesetter to ensure that he takes normal strides and maintains a uniform cadence.

Figure 3-3 is available in the illustration section at the end of the book

(2) Reducing Accordion Effect. Accordion effect occurs when the column alternately lengthens and shortens, causing the rear elements of the column to double time to maintain distance. It is caused by changes in the rate of march by the lead elements, usually after passing a slowing obstacle. To reduce this effect as much as possible, lead elements should slow their rate of march for a sufficient time after crossing obstacles to allow the rear elements to maintain the prescribed distance without running. The rear elements can signal the head when the last man has cleared the obstacle and the rate can be resumed. Since some accordion effect is nearly inevitable, the order of march should be periodically rotated to prevent units in the rear from becoming physically exhausted at the conclusion of the march.

e. Length, Rate and Pace

(1) **Length of March.** The length of march varies depending on terrain and weather, enemy situation, and physical and mental condition of Marines. The normal length of march for a 24-hour period is from 20 to 32 kilometers (12 to 20 miles) marching from 5 to 8 hours at a rate of 4 kilometers (2.5 miles) per hour (km/hmi/h).A march in excess of 32 kilometers in a 24-hour period is considered a forced march. Well-trained units that have been progressively trained will be able to achieve a unit goal of 60 miles in a 3-day period.

(2) Rate of March. The same factors of terrain, weather, load to be carried, and condition of Marines affect the rate of march. The commander considers these and selects a rate which will place his unit at the destination in the shortest possible time in condition to accomplish the mission. Unit standing operating procedure should state the standard rate of march for that unit over normal terrain for both road and cross-country movement and in both day and night. The commander then modifies this rate if the situation requires. Normally prescribed standards are as follows:

Figure is available in the illustration section at the end of the book

(3) Pace and Cadence. The normal pace is 30 inches. A pace of 30 inches and a cadence of 106 steps per minute result in a speed of 4.8 kilometers per hour or 3 miles per hour and a rate of 4 kilometers per hour or 2.5 miles per hour if a 10-minute rest halt per hour is taken. Since the pace of each man may vary, the cadence may have to be adjusted to maintain the prescribed rate. Slope of ground and surface footing affect pace. A normal cadence is relatively simple to maintain on level or gently rolling terrain on firm footing. Mud, sand, loose gravel, and hills all greatly affect pace.

f. Halts. Halts during a day's march are taken at regular intervals to rest personnel and adjust loads.Halts are specified by standing operating procedures or march order. Such factors as midday heat or enemy action may force the adoption of long midday halts or night marches. At long halts, each unit disperses to a previously selected location near the route of march.

(1) Normal Time and Frequency. Under normal conditions, a 15-minute halt is made after the first 45 minutes of marching. Following the first halt, a 10-minute halt is made after each 50 minutes of marching. Variations of this schedule are made when a scheduled halt time occurs when passing through a built-up area or when cover and concealment are required by the tactical situation and none is available. Observation posts may be established, if required for security of the unit during halts.

(2) Actions at Halts. All units in the column should be halted at the same time. At the halt signal, Marines should move to the side of the road, staying within the immediate vicinity of their unit. Marines should remove or loosen gear and sit or lie with feet elevated. Commanders inspect Marines and equipment, and corpsmen administer medical treatment, as required.

g. Communications. The four primary means of communications used in controlling foot marches are radio, visual, sound, and messenger.

(1) Radio. Radios are normally used for communications between platoons and higher headquarters in the march column. Radio transmissions should be held to the absolute minimum necessary for control and should be at minimum power required.

(2) Visual.. Visual communications may include flashlights, luminous markers, panel sets, pyrotechnics, and hand and arm signals. When contact with the enemy is not expected, the loss of light discipline from some of these signals may be a lesser security risk than the risk of disclosure from radio transmissions.

(3) Sound. Sound communications include voice and such signaling devices as horns and whistles. Some of their uses include assembling Marines at the conclusions of halts and providing warnings of air or chemical attacks.

(4) Messengers. Messengers are particularly effective during periods of radio silence and reduced visibility. Messages should be simple and brief to preclude mistakes by the messenger.

h. Security. Both active and passive measures are used against attack by enemy aircraft and indirect fire weapons during movement. Active measures include the use of air sentries and organic and attached weapons in accordance with the unit air defense standing operating procedure. Passive measures include the use of concealed routes and assembly areas, night marches, and extended interval between elements of the column. With imagination and planning, all of these measures can be practiced in training without detracting from the physical conditioning benefit of the march. In fact, such drills as rapidly taking cover during simulated attacks and then quickly resuming the march can prove physically challenging.

i. Reconnaissance. Reconnaissance determines in advance critical points along the march route such as bridges, fords, and obstacles in order that congestion or delay may be held to a minimum and local security provided. Paragraph 3103.C describes organization of the reconnaissance party.

3106.TACTICAL MARCHES

These marches are movements of Marines and equipment not in direct ground contact with the enemy but expecting early ground contact either en route or upon arrival at the objective. Some characteristics of this type of movement include security elements to the front, flanks, and, if necessary, to the rear of the march column. Additional considerations include the selection of concealed areas on defendable terrain for start and release points and rest stops. In training, infantry units of regimental size or smaller will normally derive the greatest benefit from tactical marches. The security requirements usually dictate a slower rate of march and, therefore, reduce the physical conditioning value for other types of units which would seldom move in combat with the expectation of early ground contact with the enemy. Two key points should be remembered regarding tactical marches in a training program.

a. Security. Although security requirements often slow the rate of march, it is a valid and important training objective for infantry units to work toward reducing march times while still practicing sound tactical security. March standing operating procedures, good land navigation skills, and simple but complete march orders contribute as much to this goal as physical conditioning. The idea that proceeding slowly enhances security is seldom true. Keeping the enemy off balance by rapid movement frequently is the best guarantee of security.

b. Training. In peacetime training, it is critical that support personnel who normally are located with infantry units in combat, participate completely with infantry units in forced march training. Not only must they not slow the infantry rate of march because of inferior conditioning, but they must be thoroughly familiar with the infantry unit's standing operating procedure for security on the march. Personnel from within the ground combat element such as artillery forward observer teams and combat engineer units, can be easily identified for this training and should be habitually associated with the same infantry units. Personnel from other elements, such as low-altitude air defense teams and radio battalion detachments, are more difficult to identify and earmark for support of the same infantry unit. Their normal mode of employment is in vehicles and their units frequently provide general support vice direct support. This does not, however, lessen their requirement to be prepared to provide support to dismounted infantry over terrain or in a tactical situation not conducive to vehicles. Such training needs to take place, and planning and cooperation among the elements of the MAGTF can overcome the obstacles.

3107.ADMINISTRATIVE MARCHES

These marches are movements of Marines and equipment made when no enemy interference is expected except from aircraft or indirect fire. During these marches, units are administratively grouped for ease of control and speed of movement.The term administrative foot marches should not be taken to imply that tactical considerations are disregarded.There are simply less stringent security measures used than when ground contact with the enemy is a distinct possibility.The normal formation is the route column with one file moving on either side of the road and with negligible security. Air sentries are used and dispersion between individuals and units is practiced. Although roads usually provide the fastest route, there may be reasons to conduct these marches cross-country (particularly as part of a training program). Administrative marches are the type of foot marches most practical for training programs for non-infantry units. These marches may also play a key role in the physical conditioning program of infantry units, because of the greater speeds possible due to lessened security requirements.

3108. FORCED MARCHES

A forced march is a march which covers a greater distance than 32 kilometer (20 miles) in a 24-hour period. Normally the extra required distance should be achieved by increasing the number of hours marched in a day beyond the norm (8 hours) rather than increasing the rate of march beyond the norm (4 kilometers per hour/2.5 miles per hour). However, there will be occasions when the situation or mission demands an increase in the rate of march.

a. Maximum Recommended Distances. The maximum recommended distances for forced marches are-56 kilometers (35 miles) in 24 hours; 96 kilometers (60 miles) in 48 hours; or 128 kilometers (80 miles) in 72 hours.

b. Sample Time Breakdown. A sample time breakdown for a forced march of 52 kilometers, beginning at first light is as follows:

Figure is available in the illustration section at the end of the book

3109. NIGHT MARCHES

Night marches are characterized by closed formations, more difficult control and reconnaissance, and a slower rate of march than day marches.

a. Control. Control is increased by reducing the distance between individuals and units, and by using connecting files to maintain contact between platoons and companies. Connecting files are normally constituted from personnel from the rear most march unit.

b. Safety on Roads. Because of reduced visibility, night marches on roads used by vehicles require attention to safety procedures to prevent accidents. Commanders should not think these safety measures apply only to peacetime training. In combat, nearly all vehicles operate with blackout lights at night, making visibility even more difficult than in garrison driving. If consistent with the tactical situation, the following measures can reduce the possibility of marchers being struck by vehicles:

(1) Use off-road trails and routes as much as possible.

(2) Guards to the front and rear of march units should be marked with strips of luminous tape and carry red-filtered flashlights. These measures are generally consistent with light discipline, while still affording reasonable warning to approaching drivers.

Section II. Conditioning Drills (One, Two, and Three)

3201. GENERAL

a. Description and Objective. Conditioning drills are calisthenic exercises. Each drill contains seven exercises organized and numbered in a set pattern. Each drill takes 15 minutes to complete. The objective of conditioning drills is to exercise all major muscle areas in order to develop strength, endurance, coordination, and flexibility.

b. Area and Equipment. Any level area is satisfactory for conduct of the drills. Drills One and Two contain ground exercises. If ground conditions are unsatisfactory, Drill Three should be used as it contains no ground positions. Usually, no equipment is required;however, if the group exceeds a platoon in size, an instructor's stand is necessary.

c. Formation. The extended rectangular formation is used in this drill. (See app. A.)

d. Starting Level and Progression. The starting level is six repetitions of each exercise. An increase of one repetition for each three periods of exercise in which the drill is performed is an acceptable rate of progression. This rate is continued until 12 repetitions can be completed. To maintain, continue the drill at 12 repetitions. To progress, move to 6 repetitions at a more difficult drill which exercises the same muscle group. Progression can also be gained by moving from Drill One to Drill Two, as Drill Two is more demanding.

e. Starting Positions. Starting positions vary with the exercise and are explained in each exercise.

f. Leadership. A principal instructor demonstrates and leads the drills. The instructor must be familiar with leadership techniques peculiar to conditioning drills to include the exercises, commands, counting cadence, cumulative count, formation, method of teaching the exercises, and utilization of assistant leaders.

g. Use With Other Programs. Conditioning Drills One, Two, and Three reach all major muscles of the body. They are easy to learn and to perform, and they are simple to administer and supervise. These features, coupled with the short time required for completion, the fact that no equipment is necessary, and adaptability to most areas of execution, make these drills possible in any programs.

3202. CONDITIONING DRILL ONE

This conditioning drill is similar to the calisthenics drill commonly known throughout the Marine Corps as the "Daily Seven."The "Daily Seven" may be used as a substitute for Conditioning Drill One.

a. Exercise l:High Jumper

(1) Starting Position. Feet separated shoulder width, knees flexed, body bent forward at the waist, arms aligned with the trunk and hips, elbows locked, palms facing, fingers extended and joined, head and eyes to the front.(See fig. 3-4, A.)(Elbows remain locked throughout the exercise.)

Figure 3-4 is available in the illustration section at the end of the book.

(2) Cadence. Moderate.

(3) Movement. A four-count exercise: at the count of—

(a) ONE -- Take a slight jump into the air, swinging the arms forward and up to shoulder level.

(b) TWO -- Take a slight jump into the air and swing the arms downward and back, returning to the starting position.

(c) THREE -- Take a vigorous leap into the air, swinging the arms forward and up to an overhead position, momentarily looking skyward, on returning to the ground the knees are flexed, head and eyes return to the front.

(d) FOUR--Repeat the action of count two.

b. Exercise 2: Bend and Reach

(1) Starting Position. Feet spread more than shoulder width, arms overhead, elbows locked, palms facing, fingers extended and joined, head and eyes to the front.(See fig. 3-4, B.)

(2) Cadence. Moderate.

(3) Movement. A four-count exercise: at the count of—

(a) ONE-Bend at the knees and waist, swing the arms straight downward and reach between the legs. Touch the ground as far to the rear as possible and look to the rear. (Elbows remain locked throughout the exercise).

(b) TWO-- Recover sharply to the starting position.

(c) THREE--Repeat the action of count ONE.

(d) FOUR-- Repeat the action of count TWO.

c. Exercise 3: Pushup

(1) Starting Position. Front leaning rest position. To assume this position there is a silent one-two count:on the silent count of one, assume the squatting position, heels together, elbows locked inside the knees, hands flat on the ground directly beneath the shoulders; on the silent count of two, thrust the legs to the rear, toes and heels together, body straight from head to heels.(Seefig. 3-4, C.)

(2) Cadence. Moderate.

(3) Movement. A four-count exercise: at the count of—

(a) ONE -- Flex the elbows lowering the body until the thick portion of chest touches the ground.

(b) TWO -- Raise the body until elbows are straight and locked.

(c) THREE--Repeat the action of count ONE.

(d) FOUR -- Repeat the action of count TWO. (On returning to position of attention, the silent one-two count is used in reverse).

d. Exercise 4: Trunk Twister

(1) Starting Position. Feet are spread more than shoulder width apart, fingers placed behind neck, thumbs pointing downward, elbows back. (See fig. 3-4, D.)(Elbows remain well back throughout the exercise).

(2) Cadence. Slow.

(3) Movement. A four-count exercise: at the count of-

(a) ONE-- Keeping the knees locked and the back straight, bend forward at the waist sharply, with a slight recovery.

(b) TWO-- Twist the trunk to the left vigorously at the waist, keeping the elbow back. The left elbow is higher than the right.

(c) THREE-- Twist vigorously to the right, so the left elbow comes under the right.

(d) FOUR-- Straighten sharply to the starting position. NOTE: Do not attempt to touch the elbows to the knees on counts two and three.

e. Exercise 5: Squat Bender

(1) Starting Position. Feet are spread less than shoulder width apart, hands on hips, thumbs in small of back, elbows back. (See fig. 3-4, E.)

(2) Cadence. Moderate.

(3) Movement. A four-count exercise: at the count of—

(a) ONE-- Assume the squatting position, maintain balance on the balls of the feet, with trunk erect thrust arms forward to shoulder level, elbows locked, palms down.

(b) TWO-- Recover to starting position. Elbows are well back.

(c) THREE-- Keeping the knees locked, bend forward at the waist, touching the ground in front of the toes.

(d) FOUR-- Vigorously recover to the starting position.

f. Exercise 6: Body Twist

(1) Starting Position. On the back, arms extended sideward on the ground, palms down. The legs are raised to a near vertical position, feet together, knees locked.(See fig. 3-4, F.)

(2) Cadence. Slow-fast.

(3) Movement. A four-count exercise: at the count of-

(a) ONE-- Lower legs slowly to the left until they touch the ground near the left hand, keeping the knees straight and shoulders on the ground.

(b) TWO-- Recover the starting position by quickly raising the legs, keep knees straight and feet together.

(c) THREE-- Repeat movement of count ONE, except the movement is to the right side.

(d) FOUR-- Recover sharply to the starting position.

g. Exercise 7: Stationary Run

(1) Starting Position. Position of attention. (See fig, 3-4, G.)

(2) Cadence. Fast.

(3) Movement.

(a) At the command of execution, start running in place at double-time cadence, lifting the left foot first time cadence. Follow the instructor as he counts two repetitions of cadence;

e.g., 1, 2, 3, 4--1, 2, 3, 4. The instructor then gives informal commands such as FOLLOW ME, running on the toes and balls of the feet, keeping the back straight, speeding up the cadence to a sprint, raising the knees high, leaning forward at the waist, and pumping the arms vigorously.

(b) To halt the exercise, the instructor will count two repetitions of cadence as the left foot strikes the ground: 1, 2, 3, 4 - 1, 2, 3, HALT.

NOTE: When counting cadence, the instructor counts only as the left foot strikes the ground. The duration of the exercise is approximately 1 1/2 minutes.

3203. CONDITIONING DRILL TWO

a. Exercise 1: Jumping Jack

(1) Starting Position. Feet separated more than shoulder width, arms overhead. (See fig. 3-5, A.)

(2) Cadence. Moderate.

(3) Movement. A four-count exercise: at the count of—

(a) ONE-- Jump to position with the feet together and assume the squatting position, swinging the arms sideward and downward, placing the hands palms down on the ground, elbows locked inside the knees.

(b) TWO-- Recover to the starting position by jumping to the side straddle and swinging the arms sideward overhead.

(c) THREE-- Repeat the action of count ONE.

(d) FOUR-- Recover to the starting position.

b. Exercise 2: Turn and Bend

(1) Starting Position. Side straddle, arms overhead. (See fig. 3-5, B.)

(2) Cadence. Moderate.

(3) Movement. A four-count exercise: at the count of-

(a) ONE-- Turn the trunk to the left and bend forward over the left thigh, attempting to touch the fingertips to the ground outside the left foot. Keep the left knee straight. On successive repetitions attempt to touch farther and farther to the side.

(b) TWO-- Recover to the starting position.

(c) THREE-- Turn the trunk to the right and bend forward over the right thigh, trying to touch the hands to the ground outside the right foot. Keep the right knee straight.

Figure 3-5 is available in the illustration section at the end of the book.

(d) FOUR-- Recover to the starting position.

c. Exercise 3: Eight-Count Pushup

(1) Starting Position. Position of attention. (See fig. 3-5, C.)

(2) Cadence. Moderate.

(3) Movement. An eight-count exercise: at the count of—

(a) ONE-- Assume the squatting position, palms on the ground directly beneath the shoulders, elbows locked inside the knees.

(b) TWO-- Thrust the legs to the rear assuming the front leaning rest position.

(c) THREE-- Flex the elbows until the thick portion of the chest touches the ground.

(d) FOUR-- Raise the body on a straight plane until the elbows are locked.

(e) FIVE-- Repeat the action of count THREE.

(f) SIX-- Repeat the action of count FOUR.

(g) SEVEN-- Recover to the squatting position as in count ONE (elbows locked inside the knees).

(h) EIGHT—Return sharply to the position of attention.

d. Exercise 4: Turn and Bounce

(1) Starting Position. Feet spread more than shoulder width apart, arms sideward at shoulder level, palms up. (See fig. 3-5, D.)

(2) Cadence. Slow.

(3) Movement. An eight-count exercise: at the count of—

(a) ONE-- Turn sharply to the left as far as possible, then recover slightly.

(b) TWO-- Again turn to the left as far as possible and recover as in ONE.

(c) THREE-- Repeat the action of count TWO.

(d) FOUR-- Recover sharply to the starting position.

(e) FIVE—Turn sharply to the right as far as possible, then recover slightly.

(f) SIX—Again turn to the right as far as possible and recover as in FIVE.

(g) SEVEN-Repeat the action of count SIX.

(h) EIGHT-Return to the starting position.

NOTE: The head and hips remain to the front throughout the exercise and the knees and elbows are locked at all times.

e. Exercise 5: Squat Stretch

(1) Starting Position. Attention. (See fig. 3-5, E.)

(2) Cadence. Moderate.

(3) Movement. A four-count exercise: at the count of—

(a) ONE-- Squat, placing the hands on the ground about 12 inches in front of the feet.

(b) TWO— Keeping the fingertips on the ground, straighten the knees completely and raise the hips.

(c) THREE-- Recover to position ONE.

(d) FOUR-- Recover to the starting position.

f. Exercise 6: Leg Circular

(1) Starting Position. On the back, arms stretched sideward, palms down, feet raised foot from ground, knees straight. (See fig 3-5,F.)

(2) Cadence. Slow.

(3) Movement A four-count exercise: at the count of—

(a) ONE-- Swing the legs as far as possible to the left, keeping the knees straight and the legs together.

(b) TWO-- Swing the extended legs over head with the thighs as close as possible to the trunk.

(c) THREE-- Swing the legs as far as possible to the right.

(d) FOUR-- Recover to the starting position.

g. Exercise 7: Stationary Run

(1) Starting Position. Position of attention. (See fig. 3-5, G.)

(2) Cadence. Fast.

(3) Movement

(a) At the command of execution, start running in place at double-time cadence, lifting the left foot first. Follow the instructor as he counts two repetitions of cadence; for example: 1, 2, 3, 4--1, 2, 3, 4. The instructor then gives informal commands such as FOLLOW ME. Running on the toes and balls of the feet, keeping the back straight, speeding up the cadence to a sprint, raising the knees high, leaning forward at the waist, and pumping the arm vigorously.

(b) To halt the exercise, the instructor will count two repetitions of cadence as the foot strikes the ground: 1, 2, 3, 4-1,2, 3, HALT.

NOTE: When counting cadence, the instructor counts only as the left foot strikes the ground. The duration of the exercise is approximately 1 1/2 minutes.

3204. CONDITIONING DRILL THREE

a. Exercise 1: Side Straddle Hop

(1) Starting Position. Position of attention. (See fig. 3-6, A.)

(2) Cadence. Moderate.

(3) Movement. A four-count exercise: at count of-

(a) ONE-- Take a slight jump into the air, moving the legs sideward (more than shoulder width apart); at the same time, swing the arms overhead (to an overhead position) clapping the palms together.

Figure 3-6 is available in the illustration section at the end of the book.

(b) TWO-- Take a slight jump into the air, swing the arms sideward and downward returning to the starting position.

(c) THREE-- Repeat the action of count ONE.

(d) FOUR-- Repeat the action of count TWO.

b. Exercise 2: Back Bender

(1) Starting Position. Standing, feet 12 inches apart, fingers placed behind the head. (See fig. 3-6, B.)

(2)Cadence. Slow.

(3) Movement. A four-count exercise: at the count of-

(a) ONE-- Bend the upper trunk backward, raising the chest high, pulling the elbows back, and looking upward. Keep the knees straight.

(b) TWO-- Recover to the starting position.

(c) THREE-- Repeat the action of count ONE.

(d) FOUR-- Recover to the starting position.

c. Exercise 3: Squat Thrust

(1) Starting Position. Position of attention. (See fig. 3-6, C.)

(2) Cadence Moderate.

(3) Movement. A four-count exercise: at the count of—

(a) ONE-- Assume the squatting position; heels together, placing the hands flat on the ground, shoulder width apart, elbows locked and inside the knees.

(b) TWO-- Thrust the legs to the rear, assuming the front leaning rest position, body in line from head to toe, heels and toes together.

(c) THREE-- Return to the squatting position as in ONE.

(d) FOUR-- Return to position of attention.

d. Exercise 4: Side Bender

(1) Starting Position. Feet are spread more than shoulder width apart, arms are raised sideward and overhead, thumbs interlocked palms to front, fingers extended and joined, elbows locked. (See fig. 3-6, D.)

(2) Cadence. Slow.

(3) Movement. An eight-count exercise: at the count of—

(a) ONE-- Bend to left as far as possible, then recover slightly.

(b) TWO-- Again bend to the left as far possible, then recover slightly.

(c) THREE-Repeat the action of count TWO.

(d) FOUR-- Recover sharply to the starting position.

(e) FIVE-- Bend to the right as far as possible, then recover slightly.

(f) SIX-- Again bend to the right as far as possible, then recover slightly.

(g) SEVEN-- Repeat the action of count SIX.

(h) EIGHT-- Recover sharply to the starting position.

NOTE:Keep the elbows and knees locked throughout the exercise. The bend should occur to the side and not the front.

e. Exercise 5: Knee Bender

(1) Starting Position. Feet are spread less than shoulder-width apart, hands on hips, thumbs in small of back, elbows back. (See fig 3-6, E.)

(2) Cadence. Moderate.

(3) Movement. A four-count exercise: on the count of—

(a) ONE-Do a knee bend, lean trunk forward at the waist, thrust arms between legs until the extended fingers touch the ground palms to the ground, hands 6 inches apart.

(b) TWO~Recover sharply to the starting position.

(c) THREE-Repeat the action of count ONE.

(d) FOUR-Repeat the action of count TWO.

f. Exercise 6: Bottoms Up

(1) Starting Position. Front leaning rest position. A silent one-two count is used as in the pushups. (See fig. 3-6, F.)

(2) Cadence. Moderate.

(3) Movement A four-count exercise: at the count of-

(a) ONE-- With the weight on the hands, and knees locked, jump forward bringing the feet as close to the hands as possible; look to the rear.

(b) TWO--Keeping the knees locked, thrust the legs backward assuming the front leaning rest position.

(c) THREE-- Repeat the action of count ONE.

(d) FOUR-- Repeat the action of count TWO.

g. Exercise 7: Stationary Run

(1) Starting Position. Position of attention. (Sec fig. 3-6, G.)

(2) Cadence. Fast.

(3) Movement

(a) At the command of execution, start running in place at double time, lifting the left foot first. Follow the instructor as he counts two repetitions of cadence; for example: 1, 2, 3, 4--1, 2,3, 4. The instructor then gives informal commands such as FOLLOW ME, running on the toes and balls of the feet, keeping the back straight, speeding up the cadence to a sprint, raising the knees high, leaning forward at the waist, and pumping the arms vigorously.

(b) To halt the exercise, the instructor will count two repetitions of cadence as the left foot strikes the ground: 1, 2, 3, 4 - 1 , 2, 3, HALT.

NOTE: When counting cadence, the instructor counts only as the left foot strikes the ground. The duration of the exercise is approximately 1 1/2 minutes.

Section III. Rifle and Log Drills

3301. RIFLE DRILL

a. Description and Objective. Rifle exercises are conditioning exercises performed with a rifle. Each drill contains six exercises (fig. 3-7) and they are numbered in a set pattern. The drill takes 15 minutes to complete. The objective of rifle drills is to exercise the arms, shoulders, and back muscles in order to develop strength and endurance, particularly in the upper body. In units without rifles, log drills may be substituted.

b. Area and Equipment. Any level area is satisfactory for conducting this drill. Each Marine completes these exercises from a standing position and no ground contact is required. Each Marine will need a rifle and, if the group exceeds a platoon in size, then the instructor will need an instructor's stand.

c. Formation. The extended rectangular formation is used in this drill. (See app. A.)

d. Starting Positions. Starting positions vary with the exercises and are explained in each exercise. As in all set conditioning drills, the command used to start the exercise is STARTING POSITION, MOVE. The following directions apply to rifle drill.

(1) In those exercises which start from the rifle downward position, on the command MOVE, execute port arms as prescribed in NAVMC 2691, Drill and Ceremonics Manual, and then assume the starting position. The command to return to the men to the position of attention at the conclusion of the exercise is POSITION OF ATTENTION, MOVE.

(2) In exercises which terminate in the rifle downward position, on the command MOVE, execute the position followed by order arms as prescribed in NAVMC 2691.

(3) In the exercises which terminate in a position other than the rifle downward position, Marines first assume the rifle downward position before executing port arms and order arms.

(4) These movements are executed without command. This procedure promotes uniformity, but precision is not expected. To be effective, rifle exercises must be strenuous enough to tire the arms, but not to the point where the arms cannot move with precision.

e. Leadership. A principal instructor demonstrates and leads the drill. He must be familiar with leadership techniques for conditioning exercises and the peculiar techniques for rifle drill.

3302. EXERCISES PERFORMED WITH RIFLES

The exercises of rifle drill are outlined in the following paragraphs,

a. Exercise 1: Foreup, Behind Back

(1) Starting Position. Rifle downward, feet together.(See fig. 3-7, A.)

Figure 3-7 is available in the illustration section at the end of the book.

(3) Movement. A four-count exercise: at the count of—

(a) ONE-- Swing the arms forward and upward to the overhead position. Inhale.

(b) **TWO**-- Lower the rifle to the back of the shoulders. Exhale.
(c) **THREE**-- Recover to position ONE and inhale.
(d) **FOUR**-- Recover to the starting position and exhale.

b. Exercise 2: Lunge Side, Turn and Bend
(1) Starting Position. Rifle downward, feet together.(See fig. 3-7, B.)
(2) Cadence. Moderate.
(3) Movement. An eight-count exercise:at the count of—
(a) **ONE**-- Lunge sidewards to the left, swing the rifle forward and upward to the overhead position.
(b) **TWO**-- Turn the trunk to the left and bend forward over the left hip. At the same time, swing the rifle to a **low** horizontal in front of the left ankle.
(c) **THREE**-- Recover to position ONE.
(d) **FOUR**-- Recover to the starting position.
(e) **FIVE, SIX, SEVEN, and EIGHT**-- Repeat on the right side.

c. Exercise 3: Foreup, Back Bend
(1) Starting Position. Rifle downward, feet together.(See fig. 3-7, C.)
(2) Cadence. Moderate.
(3) Movement. A four-count exercise:at the count of—
(a) **ONE**-- Swing the arms forward and upward to the overhead position.
(b) **TWO**-- Bend backward, emphasizing the bend in the upper back. The face is up.Keep the knees straight.
(c) **THREE**-- Recover to position ONE.
(d) **FOUR**-- Recover to the starting position.

d. Exercise 4:Up and Forward
(1) Starting PositionRifle downward, feet together.(See fig. 3-7, D.)
(2) Cadence. Fast.
(3) Movement. A four-count exercise:at the count of—
(a) **ONE**-Swing the arms forward and upward to the overhead position.
(b) **TWO**- Swing the arms forward to shoulder level.
(c) **THREE**-Recover to position ONE.
(d) **FOUR**-Recover to the starting position.

e. Exercise 5: Foreup, Full Squat
(1) Starting Position. Rifle downward, feet in narrow stance. (See **fig.** 3-7, E.)
(2) Cadence. Moderate.
(3) Movement. A four-count exercise: at the count **of**—
(a) **ONE**-- Swing the arms forward and upward to the overhead position.
(b) **TWO**-- Swing the arms down to shoulder level and assume the squatting position.
(c) **THREE**-- Recover to position ONE.
(d) **FOUR**-- Recover to the starting position.

f. Exercise 6: Arms Forward, Side Bend

(1) Starting Position. Sidestraddle, regular stance, rifle forward.(See fig. 3-7, F.)
(2) Cadence. Moderate.
(3) Movement. A four-count exercise:at the count of-

(a) ONE-- Bend the trunk to the left. Keep the knees straight.
(b) TWO-- Recover to the starting position.
(c) THREE-- Bend the trunk to the right. Keep the knees straight.
(d) FOUR-- Recover to the starting position.

NOTE:Keep the rifle on the same level as the shoulders throughout the exercise.

3303. LOG DRILL

a. Description and Objective. Log exercises are conditioning exercises performed with a log. Each drill contains six exercises and they are numbered in a set pattern. The drill takes 15 minutes to complete. The objective of log drills is to develop strength and muscular endurance and, in this instance, under maximum loads. Log exercises also develop teamwork. Log may be used in lieu of conditioning drills after the Marines have become somewhat conditioned.

b. Area and Equipment. Any level area is satisfactory for conducting this drill. Each Marine completes these exercises from a standing position and no ground contact is required. If the group exceeds a platoon in size, then the instructor will need instructor's stand. Each six-person group or eight-person group will need a log. The logs should be from 6 to 8 inches in diameter. They may vary in length from 14 feet (for 6 people) to 18 feet (for 8 people).The logs should be skinned, smoother, and dried. The 14-foot logs should weigh approximately 300 pounds and the 18-foot logs, approximately 400 pounds. Rings should be painted on the logs to indicate each person's position. When not in use, the logs should be stored on a rack to keep them off the ground.

c. Formation. All the Marines assigned to the same log team should be about the same height at the shoulders. The recommended method of dividing the platoon is to have the Marines form a single file or column with short people to the front and tall people to the rear. Have the Marine assume their positions in the column according to shoulder height, not head height. When they are in position, they are given the command COUNT OFF BY SIXES (OR EIGHTS), COUNT OFF, to divide them into six- or eight-person log teams. Each team in turn, can then proceed to the log rack, shoulder a log, and carry it to the designated exercise area. The log teams form in columns in front of the instructor. With the Marine holding the log in the chest position, have them face the instructor and ground the log at least 10 yards between columns and 10 yards between log teams within the columns.

d. Starting Positions. The Marines fall in, facing the log, with their toes about 4 inches from it. The basic starting positions and commands are as follows:(See fig. 3-8.)
(1) **RIGHT-HAND STARTING POSITION, MOVE.** It the command MOVE, move the left foot 12 inches to the left, and lower the body into a flat foot squat. Keep the back straight, head up, and arms between the legs. Encircle the far side of the log with the left hand. Place the right hand underneath the log.(See fig. 3-8, A.)

(2) LEFT-HANDED STARTING POSITION, MOVE. These commands are executed in the same manner as in paragraph a. except that the left hand is underneath the log and the right hand encircles its far side. (See fig. 3-8, B.)

Figure 3-8 is available in the illustration section at the end of the book.

(3) RIGHT SHOULDER POSITION, MOVE. At the command MOVE, pull the log upward in one continuous motion to the right shoulder. At the same time, move the left foot to the rear and stand up, facing left. Balance the log on the right shoulder with both hands.(See fig. 3-8, C.)This movement cannot be performed from the left-hand starting position because of the position of the hands.

(4) LEFT SHOULDER POSITION, MOVE. These commands should be given from the left hand starting position. At the command MOVE, pull the log upward in one continuous motion, to the left shoulder. At the same time, move the right foot to the rear and stand facing right. Balance the log on the left shoulder with both hands.(See fig. 3-8, D.)This movement cannot be performed from the right-hand starting position.

(5) WAIST POSITION, MOVE. From the right hand position pull the log waist high. Keep the arms straight and fingers laced underneath the log. The body is inclined slightly to the rear and the chest is lifted and arched. (See fig. 3-8, E.)

(6) CHEST POSITION, MOVE. This command should be given after the waist position has been assumed .On the command MOVE, shift the log to a position high on the chest, bring the left arm under the log and hold the log in the bend of the arms.(See fig. 3-8, F.)Keep the upper arms parallel to the ground.

(7) To move the log from the right shoulder to the left shoulder, the command is: LEFT SHOULDER POSITION, MOVE. On the command MOVE, push the log overhead and lower it to the opposite shoulder.

(8) To return the log to the ground from any of the above positions, the command is: STARTING POSITION, MOVE. At the command MOVE, slowly lower the log to the ground. The hands and fingers must be kept from under the log.

e. Leadership. A principal instructor demonstrates and leads the drill. He must be familiar with the leadership techniques for conditioning exercises and the peculiar techniques for log drill.

3304. EXERCISE PERFORMED WITH LOGS

The exercises of log drill are outlined in the following paragraphs. Figure 3-9 graphically explains log drill.

a. Exercise l:Two-Arm Pushup

(1) Starting Position. Right or left shoulder position. Regular stance. (See fig. 3-9, A.)

(2) Cadence. Moderate.

(3) Movement. A four-count exercise:at the count of—

- **(a) ONE--** Push the log overhead until the elbows lock.
- **(b) TWO--** Lower the log to the opposite shoulder.
- **(c) THREE--** Repeat the action of count ONE.
- **(d) FOUR--** Recover to the starting position.

b. Exercise 2: Forward Bender

(1) Starting Position. Chest position. Regular stance.(See fig. 3-9, B.)

(2) Cadence. Moderate.

(3) Movement. A four-count exercise:at the count of-

(a) ONE-- Bend forward at the waist, keeping the back and legs straight.

(b) TWO-- Recover to the starting position.

(c) THREE-- Repeat the action of count ONE.

(d) FOUR-- Recover to the starting position.

c. Exercise 3: Straddle Jump

(1) Starting Position. Right or left shoulder position, feet together, fingers interlaced on top of the log.(See fig. 3-9, **C.**)

(2) Cadence. Moderate.

(3) Movement. A four-count exercise:at the count of—

(a) ONE-- Jump to a side-straddle. Pull down on the log with both hands to keep it from bouncing on the shoulder.

(b) TWO-- Recover to the starting position.

(c) THREE-- Repeat the action of count ONE.

(d) FOUR-- Recover to the starting position.

Figure 3-9 is available in the illustration section at the end of the book.

d. Exercise 4: Side Bender

(1) Starting Position. Right shoulder position, feet regular stance.(See fig. 3-9, D.)

(2) Cadence. Moderate.

(3) Movement. A four-count exercise:at the count of—

(a) ONE-- Bend sideward to the left as far as possible, bending the left knee.

(b) TWO-- Recover to the starting position.

(c) THREE-- Repeat the action of count ONE.

(d) FOUR-- Recover to the starting position.

(4) After completing the required number of repetitions, change shoulders and execute an equal number of repetitions to the other side.

e. Exercise 5: Knee Bend

(1) Starting Position. Right or left shoulder position. Narrow stance. Fingers interlocked on top of the log.(See fig. 3-9, E.)

(2) Cadence. Slow.

(3) Movement. A four-count exercise:at the count of—

(a) ONE-- Flex the knees to a quarter-squat position.

(b) TWO-- Flex the knees to a half-squat position.

(c) THREE-- Lower the body to a three-quarter squat position.(Lean slightly forward.)

(d) FOUR-- Recover to the starting position.

NOTE:Pull forward and downward on the log throughout the exercise.

f. Exercise 6: Overhead Toss

(1) Starting Position. Right or left shoulder position, regular stance. The knees are bent to a quarter-squat.(See fig. 3-9, F.)

(2) Cadence. Moderate.

(3) Movement. A four-count exercisc:at the count of-

(a) ONE-Straighten the knees and toss the log into the air approximately 12 inches overhead. Catch the log with both hands and lower it toward the opposite shoulder. As the log is caught, lower the body into a quarter-squat.

(b) TWO-- Again toss the log into the air and when caught return it to the original shoulder.

(c) THREE-- Repeat the action of count ONE.

Section IV. Grass Drills

3401. GENERAL

a. Description and Objective. Grass drills are extremely strenuous exercises and are performed at top speed for only short periods of time. No cadence is counted but the Marines continue to execute the multiple repetitions of the command until the next command is given. The grass drill consists of two drills:Drill One and Drill Two. Each drill contains six exercises. The objective of the drills is to decrease reaction time, to develop cardiovascular endurance, and to provide a vigorous workout for all major muscles.

b. Area and Equipment. Any level area suitable for ground contact and of a size to accommodate the group is adequate. No equipment is needed.

c. Formation. All movements are executed in place. The extended, rectangular formation is recommended for a platoon- or company-size unit. The circle formation is suitable for groups of squad or section size. At the beginning of an exercise program, 2 to 3 minutes of grass drills will insure a good workout.

d. Progression. Progression is gained by gradually increasing the length of time devoted to the drills. As the physical condition of the Marines improves, the periods should be gradually lengthened to 5 minutes. As the second drill is more difficult than the first, some progression can be attained by initially executing Drill One; then as the program and the Marines progress, introduce Drill Two. To extend the duration of the drill, it may be necessary to repeat the drill.

e. Starting Position

(1) The drills are started from the GO position. Other basic positions are FRONT, BACK, and STOP. (See fig. 3-10, A.)

(a) GO. Running in place (top speed): on the toes and balls of feet, knees raised high, arms pumping, body bent forward at waist.

(b) FRONT. Prone position: elbows bent (along body), palms flat on ground directly under the shoulders, legs together and straight.

(c) BACK. Supine position (flat on back): arms extended near side on ground with palms down, legs together and straight, feet toward the stand or instructor.

(d) STOP. Football lineman stance: feet spread and staggered, left arm across left thigh, right arm straight, knuckles on ground, head up, back parallel with ground.

(2) To assume the FRONT or BACK position from the STANDING, GO, or STOP position, vigorously get into the prescribed position as quickly as possible. (See fig. 3-10, B.)

(3) To change from the FRONT to the BACK position, quickly do a pushup, move the feet several short steps to the right or left, lift the arm on the side toward which the feet move, and thrust the legs vigorously to the front. (See fig. 3-10, C.)

Figure 3-10 is available in the illustration section at the end of the book.

(4) To move from the BACK to the FRONT position, sit up quickly, place both hands on the ground to the right or the left of the legs. Move the feet several short steps to the rear on the side opposite the hands. When the feet are opposite the hands, thrust the legs vigorously to the rear and lower the body to the ground.(See fig. 3-10, D.)

f. Leadership. A warm up activity of lesser intensity should proceed grass drill. During the instructional phase and conduct of these drills, the following points should be applied.

(1) The instructor executes only GO and STOP with the Marines. This allows the instructor to supervise the drill.

(2) The commands for grass drills are given in rapid succession without the usual preparatory command.

(3) To prevent confusion, the instructor should give the commands sharply to distinguish them from comments or encouragement.

(4) As soon as the Marines know the drill, they should respond to the instructor's commands and perform all exercises vigorously and as rapidly as possible. All exercises are executed continuously until the next command is given. Insist on top speed performance; anything less is not effective.

(5) The commands peculiar to each exercise are identical to the name of the exercise.

(6) Marines are not to be required to assume the position of attention once the drills are started. To halt the drill for instructions or for rest, the command UP is used. At this command, the Marines assume a relaxed standing position. Do not demand formality. At the conclusion of a fast and vigorous 5-minute grassdrill, it is physically impossible for people to stand at attention.

(7) The sequence of commands for the execution of grass drills should occur as follows:Drill One. GO, FRONT, Bouncing Ball; GO, BACK, Bicycle; GO, Full Squatter; GO, BACK, Situps; GO, FRONT, Mountain Climber; GO, FRONT, Roll Left; GO, STOP, UP.

g. Use With Other Programs. Since grass drills can be executed in a short period of time, they may be executed where only a few minutes are available for exercise or in conjunction with another type of activity. Grass drills are an excellent substitute for running when time is a factor.

3402. GRASS DRILL ONE AND TWO

a.Grass Drill One

(1) Bouncing Ball. From the FRONT position, push up, supporting the body on the hands (shoulder-width apart) and feet. (See fig. 3-11.)Keep the back and legs in line and the knees straight. Bounce up and down by a series of short, upward springs from the hands, hips, and feet simultaneously,

(2) Bicycle. From the BACK position, raise the legs and hips. Keep the elbows on the ground and support the hips with the hands. Move the legs vigorously as if pedaling a bicycle.

(3) Full Squatter. From the STOP position, assume a full knee bend, the feet on line, hands on hips. Bounce up and down in place by short, bouncing jumps.

Figure 3-11 is available in the illustration section at the end of the book.

(4) Situps. From the BACK position and with arms stretched overhead, sit up, reach forward, and touch toes. Return to the supine position.

(5) Mountain Climber. From the STOP position, place both hands on the ground directly under the shoulders. Thrust the right leg to the rear, knee straight. The left foot should be close to the left hand, the left knee outside the left arm. Shift the weight to the hands, thrust off with the rear (right) foot and bring that foot up close to the right hand, the right knee outside the right arm. At the same time, thrust the left leg vigorously to the rear, knee straight. Continue at a fast cadence, alternating the legs.

(6) Roll Left. From the BACK or FRONT position, make one complete roll in the direction commanded. On completing the roll, return to the FRONT or BACK position.

b. Grass Drill Two

(1) Legs Over. .From the BACK position and with arms stretched overhead, palms up, raise the legs upward and then swing them backward over the head until the toes touch the ground behind the head. Return legs to the starting position. (See fig. 3-12.)

(2) V-Up and Touch Toes. .From the BACK position, raise the legs with the knees straight, sit up until the trunk and legs form a V, and touch the toes with the hands. Return to the BACK position.

(3) Rocker. .In the FRONT position, clasp the hands behind the back, arch the body, holding the head back. Start rocking, using the front part of the trunk as a rocker.

(4) Bounce and **Clap Hands**. The procedure is the same as for bouncing ball, but while in the air, clap the hands. This requires a more vigorous bounce or spring. (See fig. 3-12.)

(5) Leg Spreader. .From the BACK position, raise the legs so that the heels are 10 to 12 inches from the ground, spread them apart as far as possible, then close them together. Open and close legs as rapidly as possible.

(6) Forward Roll. .For forward roll from the STOP position, place both hands on the ground, tuck the head, and do one complete forward roll, keeping the legs tucked as you roll, and come back to the STOP position.

Figure 3-12 is available in the illustration section at the end of the book.

Section V.Guerrilla Exercise

3501.GENERAL

a. Description and Objective. Guerrilla exercises are individual exercises of various basic skills that are performed rapidly while moving forward in a circle formation. There are two tables of guerrilla exercises. Each table takes 15 minutes to complete. The objective of this exercise is to increase strength and endurance, aid flexibility, and develop coordination.

b. Area and Equipment. Any level area is suitable for conducting guerrilla exercises. No ground contact is required other than the hands. There is no equipment requirement.

c. Formation

(1) The circle formation (app. A) is used for guerrilla exercises. Each platoon forms its own circle and engages in guerrilla exercises under a platoon instructor. If the platoon exceeds 30 Marines, double or concentric circles may be used.

(2) When the circle is formed, the instructor steps into the center of the circle and moves clockwise in a small circle. He commands:QUICK TIME, MARCH, 1-2-3-4. (Rapid cadence of approximately 130 counts per minute. Cadence and step are maintained between exercises.)

(3) To reform the platoon after completing guerrilla exercises, the instructor halts the Marines and places the base man or platoon guide where he wishes and commands:

(a) BASE MAN (or platoon guide),POST.

(b) FALL OUT AND FALL IN ON THE BASE MAN (or platoon guide).

d. Progression. Progression may be attained by moving from table I to table II or by shortening the quick time marching periods between exercises and performing all exercises a second time.

e. Leadership

(1) To execute the exercises, the Marines continue at quick time while the instructor simultaneously explains and demonstrates the exercise to be performed, and then commands the Marines accordingly. In each instance, the preparatory command will be the name of the exercise and, in all instances, the command of execution will ***MARCH**.. To* terminate each exercise, the command *is **QUICK TIME, MARCH.** The* Marines immediately pick up the step as the instructor counts cadence.

(2) Unless specified differently, each exercise should be continued for 20 to 40 seconds depending upon the vigor of the exercise. The leader can determine the duration of each exercise by observing its effect upon the Marines.

(3) To form for double guerrillas, the commands for pairing the Marines (who are in circle formation) are-

(a) PLATOON, HALT.

(b) FROM (designate an individual), BY TWO'S, COUNT OFF(.Examplel-2; 1-2; 1-2; etc.)

(c) EVEN NUMBERS MOVE UP BEHIND ODD NUMBERS(At this time, adjust pairs according to height and weight.)

(d) YOU ARE NOW PAIRED UP FOR DOUBLE GUERRILLAS. To change the Marines position, merely command CHANGE.)
(e) FORWARD, MARCH.

f. Place in the Program. Many Marines have not had the opportunity to perform the simple skills involved in guerrilla exercises. The conduct of these exercises is a simple matter since they can be performed easily and quickly in almost any situation. The tables of exercise are applicable to all personnel. The tables can constitute a station within a 1-hour period or be completed within a separate 15-minute period.

3502.GUERRILLA TABLES

a. Table I
(1) Double Time.(See fig. 3-13, A.)Hold arms in the thrust position. Execute a double-time run, maintaining the circle formation and the prescribed distance between your fellow Marines. Duration-1 minute.
(2) All Fours.(See fig. 3-13, B.) Face downward. Support the body with the hands and feet. Walk forward hands first.
(3) Crab Walk.(See fig. 3-13, C.) Get in the sitting position, face upward, and lift the hips. Support the body with the hands and feet. Walk forward feet first.
(4) Squat Walk.(See fig. 3-13, D.) Assume a full knee bend position. Grasp the ankles (left ankle with the left hand, right ankle with the right hand). Walk forward.
(5)Broad Jump.(See fig. 3-13, E.) Jump forward on both feet in a series of broad jumps. Swing the arms vigorously to assist the jumps.
(6)Toe-Touch Walk.(See fig. 3-13. F.) Walk forward, bending at the waist and touching one hand to the toe of the opposite foot while it is on the ground. Raise the trunk to the vertical position between steps. Keep the knees straight.
(7)Bottoms-Up Walk.(See fig. 3-13, G.) Assume the front leaning rest position and move the feet toward the hands in short steps, keeping the knees locked. When the feet are as close to the hands as possible, walk forward on the hands to the front leaning rest position.
(8) Straddle Run.(See fig. 3-13, H.) Run forward, leaping to the right from the left foot and to the left from the right foot.
(9) Fireman's Carry .(See fig. 3-13,1.)
(lO)Single-Shoulder Carry(See fig. 3-13, J.) Two men execute the carries as indicated by the diagram. No. 1 man executes one type; No. 2 man executes the other.

Figure 3-13 is available in the illustration section at the end of the book.

b. Table II
(1) Double Time.(See fig. 3-14, A.)Hold arms at the thrust position. Execute a double-time run, maintaining the circle formation and the prescribed distance between your fellow Marine.Duration-1 minute.
(2) Toe-Grasp Walk.(See fig. 3-14, B.) Bend forward and grasp toes. With knees slightly bent, walk forward.

(3) Hand-Kick Walk..(See fig. 3-14, C.) Walk forward, kicking the moving foot upward on every step. At the same time, lean forward and touch the elevated toe with the hand of the opposite arm.

(4) Pike Juinping.(See fig. 3-14, D.) Jump forward and upward from both feet, keeping the knees straight, and at the same time, swing the legs forward and touch the toes with the hands at the top of each jump.

(5) Squat Jump.(See fig. 3-14, E.) Leap forward from the squatting position, with the hands on the ground and the arms between the legs. Land on the ground with hands and legs extended. Bring up the legs to the squatting position.

(6) Steam Engine..(See fig. 3-14, F.) Lace the fingers behind the neck and walk forward in the following manner: as the left leg moves forward, raise the knee high, bend the trunk forward, and touch the outside of the right elbow to the outside of the knee. Then lower the left leg and step forward on the left foot and raise the right leg. Repeat with the right leg and left elbow.

(7) Knee-Touch Walk.(Scc fig. 3-14, G.) Walk forward, bending the knees and touching the ground on each step. The knees are bent and straightened on each step.

(8)Hobble Hopping.(See fig. 3-14, H.) Hold foot behind back with opposite hand and hop forward. On the command CHARGE, grasp the opposite foot with opposite hand and hop **forward.**

(9) Cross Carry.(See fig. 3-14,1.)(10)Saddle Back Carry. (See fig. 3-14, J.) Two Marines execute the carries as indicated in the diagram. No. 1 Marine executes onetype;No. 2 Marine executes the other.

Figure 3-14 is available in the illustration section at the end of the book.

Section VI. Running and Orienteering

3601. RUNNING

a. General

(l)Description. The general form and technique for all types of running are fairly constant. (See fig. 3-15.)The head is erect, body slightly forward without bending at the waist, and the arms are at a loose thrust position alternating from front to rear in straight planes. A cross-body arm movement wastes energy .The movement of the legs and feet will be discussed in subsequent paragraphs dealing with the different types of running. Of primary importance is the fact that in all types of running, the toes should be pointed straight ahead. Toeing out is a common error in both running and walking and should be an item of individual correction.

(2) Objective. The objective of running is to develop cardiovascular endurance. Despite the fact that Marines have developed their muscle structure and the strength of their muscle tissue, unless they have developed cardiovascular endurance to a satisfactory degree, they are not entirely physically fit or combat ready. Cardiovascular endurance (wind) depends on the efficiency of the lungs and heart. The efficiency of the lungs and heart depend on the amount of oxygen the lungs can absorb with each breath inhaled and the amount of carbon dioxide the lungs can expel. The process of absorbing oxygen and expelling carbon dioxide (cardiopulmonary process) is performed by the blood that circulates through the lungs. The condition of this process will determine the amount of effort a person can exert over a period of time. Running is one of the best activities to develop this vitally important endurance.

b. Running Skills. In the development of running skills, individuals may require instruction to improve their proficiency. Some important skills to consider are-

Figure 3-15 is available in the illustration section at the end of the book.

(1) Action of the Arms. Arm action is important. Check to see that arms are held loosely and that the action is relaxed. The faster the run, the more rapid the arm action.

(2) Breathing. Allow the individual to breathe through the mouth as the body demands a large supply of oxygen. Oxygen can be inhaled in greater quantities through the mouth.

3602. TYPES OF RUNNING

a. Double Time

(1) Description and Objective. Double timing is marching at the rate of 180 steps per minute, each step being 36 inches in length. It takes practice to double time with precision in formation. The Marines should keep in step, placing their feet flat on the ground. This, however, should not be a stamping motion, but should be done with as slight a jolt as possible. Double timing is like a jog, the difference being that in a jog the feet are lifted well off the ground and the running motion is bouncy. In double timing, the feet skim the ground and there is no bounce to the ran. Double timing is a vehicle for teaching proper running form and for the development of the cardiovascular system.

(2) Area and Equipment. This type of running can be completed over a variety of surfaces. Usually a Marine uses a field or road. There is no equipment requirement.

(3)Progression. There is no set standard for alternating quick time and double time in the early conditioning of Marines. A general rule is to begin with enough quick-time marching to ensure a thorough warming up, then double time about 100 paces. Change again to quick time until the individuals have made a reasonable recovery from the running, then double time another 100 paces. The amount of double time can be increased and the quick time decreased from week to week, until the individuals are double timing about 1800 yards. This type of training should be given at least twice a week, but by no means is it adequate as the sole means of conditioning.

(4) Leadership

(a) The instructor should be to one side of the column or group and toward the rear so there is a full view of all Marines. Inexperienced instructors have a tendency to supervise from a position too far forward.

(b) Select an individual who can maintain the proper cadence to act as the guide during double-time running.

(c) There are several ways for the instructor and group to count cadence while double timing. If not contrary to local policy, learn several methods and use them for variety.

b. Wind Sprints

(1) Description and Objective. This type of running involves a series of 30- or 40-yard dashes, usually conducted in successive waves of squads. Each squad is in line and the squad leader on the right flank. Wind sprints assist indeveloping speed and cardiovascular endurance. Any flat and level area may be used which will permit the squad to form a line and run the required distance.

(2) Progression. One or two 30-yard sprints will be adequate at the beginning. As time passes, sprints can be lengthened and up to six or seven sprints may be used.

(3) Leadership

(a) At the command READY (given by squad leader), each runner assumes the sprinter's starting position. At the command GO, the squad sprints approximately 30 yards, takes 10 yards to stop, and lines up immediately with the squad leader who repeats, READY, GO, and again the squad sprints. At the conclusion of the third sprint, the squad waits until all the squads of the platoon have made three sprints. Then they all line up and the squad leaders conduct three more wind sprints in the opposite direction.

(b) Valuable time is gained by having each squad ready to go when the preceding squad has moved off its second sprint mark.

c. Cross-Country Running

(1) Description and Objective. Cross-country running is a distance run conducted on a course laid out along roads, across fields, over hills, through woods, and on any irregular ground. The cross-country ran may be utilized as a conditioner or as a competitive event; the objective is to cover the distance in the shortest possible time. The course should be 2 to 2.5 miles in length and be laid out to avoid heavy vehicular traffic. The course should be marked by directional arrows until the runners know the course. These runs build leg muscles, increase lung capacity, and develop endurance. Any local area of varied terrain is suitable.

(2) Progression. In the mass training of a large group, leaders should be stationed at the head and the rear of the column and should make every effort to keep the individuals together .After determining the abilities of the individuals in cross-country running, it is advisable to divide the unit into three groups. The poorest conditioned group starts first, and the best conditioned group, last. The starting time of the groups should be staggered so that all of them finish about the same time. In preliminary training, the running is similar to ordinary road work in that it begins with rather slow jogging, alternating with walking. The speed and distance of the run is gradually increased. As the condition of the individual improves, occasional sprints may be introduced. At first the distance run is from 1/2 to 1 mile. It is gradually increased to 2 or 2.5 miles. Well-conditioned personnel can run 2 to 2.5 miles within a 15-minute period.

(3) Leadership. Marines should not be required to take part in distance running until they have been through a progressively scheduled training program which requires a considerable amount of running. Cross-country runs should be scheduled occasionally to provide variety in the program. Cross-country running has the advantage of allowing mass participation. Interest can be stimulated by putting the runs on a competitive basis. As a single activity, short cross-country runs can be scheduled once a week, gradually increasing the distance as the physical conditioning improves.

(4)Use With Other Programs. Cross-country running can be combined with other activities such as conditioning exercises.

d. Fartlek Training

(l)Description and Function. Fartlek training is various running exercises conducted along an intense combat-like course. The running is conducted to overwork the lungs, allowing only partial recovery which is followed by another intense period of overload. This sequence is repeated for the duration of the workout. Fartlek training is a useful combat training method and general cardiovascular conditioner. The entire unit must be in uniformly good to excellent condition to qualify for this training.

(2) Progression. The following is an example of one fartlek session:

(a) Warm up by stretching 3-5 minutes.

(b) Warm up running easily 5-10 minutes.

(c) Run at a fast, steady pace for 3/4 to 1 3/4 miles (dependent on terrain).

(d) Walk/jog at a moderate pace for 5 minutes (recovery).

(e) Run easily sprinting 15 to 20 meters occasionally.

(f) Run full speed uphill for 175 to 200 meters.

(g) Warm down by running easily for 1/2 to 1 mile.

(h) Warm down by stretching 3 to 5 minutes.

(3)Leadership. To add variety, try incorporating combat movements, terrain association, a series of exercises or carrying table of equipment weapons (hit and roll). This will take some initiative in setting the course, but it allows for variety. In this manner, the fartlek course can be used as a training session or part of one. It works extremely well with highly motivated, competitive Marines.

3603.ORIENTEERING

Orienteering is land navigation over a prescribed course as a timed, competitive event. It is an excellent way to integrate land navigation training with physical conditioning and requires Marines to be skilled in both areas. Commanders adjust the training as needed to change the degree of difficulty or to emphasize either navigation or conditioning. Taken to the extreme, an orienteering exercise is conducted in full combat gear over rugged terrain against aggressors. Orienteering requires the following physical skills: endurance, speed, strength, and all-around conditioning.

Section VII. Circuit Training

3701. GENERAL

a. Description and Objective. A strength circuit consists of a series of stations where individuals in small groups exercise vigorously for a short period of time and then move (on signal) to the next station where a different form of exercise is conducted. This rotation continues until all groups move through all stations. Strength circuits contain no set or specific types of exercise stations within the circuit. The objective of circuit training is to develop strength. There are three general types of circuits.

(1) Fixed Circuit. This is a circuit in which apparatus of an immovable type (fixed into the ground) is used.

(2) Movable Circuit. This circuit consists of individual exercise apparatus which is portable and can be moved to and from the training area.

(3) Simplified Circuit. This circuit requires no equipment or apparatus.

b. Formation. The exercises are done at will, but rapid, steady, and continuous work is required of all. Each Marine's nervous and muscular system reacts differently to timed vigorous exercises. His performance should be measured on how many movements per exercise he can complete as an individual. For example, one Marine may be able to complete 5 movements, while another may be able to complete 20, and yet each is receiving the maximum benefit. All three circuits contained in this chapter are designed for platoon-sized groups. Expansion beyond this capacity requires a large amount of equipment, as each Marine in the fixed and movable types of circuits must have an item of equipment available for exercise at each station. A group larger than a platoon could be exercised through use of the simplified type of circuit; however, the group would be unwieldy and control could be a problem.

c. Place in the Program. All circuits illustrated can be completed in a 15-minute period. This feature allows the exercise of a platoonor smaller group on the circuit for a single 15-minute period, or the scheduling of the circuit as a 15-minute period within a longer period. A circuit can thus be utilized within a rotating activity system of scheduling. Choice of a circuit by the unit depends upon area, facilities, and other local factors; however, there is a circuit for every need.

3702. FIXED STRENGTH CIRCUIT

a. Description and Objective. The strength circuit is an arrangement of various types of exercise apparatus which are fixed in position. (See fig. 3-16.)Seven basic exercises are used and each exercise requires an apparatus. All apparatus of one type are positioned together to constitute a station. Each station will accommodate 10 Marines. The circuit is designed to be accomplished in 15 minutes when conducted on a time-rotation basis, as normally executed for unit training. For individual use, a Marine may complete the circuit by executing a specific number of repetitions for each exercise. In this case, the time required for completion of the circuit would vary slightly depending on the number of repetitions the Marine completed. The objective of this circuit is to provide a series of exercises which will improve and maintain the strength of the body's major muscle groups.

Figure 3-16 is available in the illustration section at the end of the book.

b. Starting Level. Marines must be thoroughly warmed up prior to participating in the circuit system. If Marines have not engaged in vigorous exercise immediately prior to starting the circuit, then an instructor should execute the following warm up exercises. These exercises should be conducted in the normal formation for set drills. Seven repetitions of each exercise will normally provide sufficient warm up. The exercises are-

(1) High jumper, Conditioning Drill 1.

(2) Bend and reach, Conditioning Drill 1.

(3) Squat bender, Conditioning Drill 1.

c. Progression. The instructor can adjust the circuit system through several methods to accommodate participants of varying physical ability. One method of adjusting an exercise is to change the method of executing the leg lift or to select a heavier weight for the barbell curls. When it becomes apparent during a unit program that the overall fitness of the group has Improved, then the exercise may be made more strenuous in two additional ways. First, the instructor can have one or more of the supplementary stations added. Second, the time spent exercising at each station can be increased in 5-second increments to a maximum of 60 seconds. Also, as Marines learn the circuit, the 45-second period for movement and instruction between stations can be eliminated, thus allowing only enough time to change stations.

d. Starting Positions. The Marine starts on any station, exercises steadily for a certain period (45 seconds initially), then moves on command to the next type of apparatus where he again exercises steadily for an equal period. The Marine continues until the required exercise is completed at each station. If it is desired to expand the number of stations in the circuit to accommodate more participants at one time, the instructor can provide four additional supplementary exercises, requiring no equipment. This will increase the amount of time required to complete the circuit.

e. Leadership. Close control of all Marines is necessary to ensure that a minimum amount of time is spent in moving them to their initial stations and in moving them between stations. One instructor can control the activity on the strength circuit. A stop watch or wrist watch with a second hand is required. When Marines arrive at the strength circuit, the instructor will form them for exercise and conduct the warm up drill. The group is then reassembled and formed into a number of files equal to the number of stations being used in the circuit. Each file is then directed to a station. As soon as all participants have reached an exercise position at a station, the instructor gives the command READY, GO. After 45 seconds of exercise, the instructor gives the command STOP, CHANGE OVER. The instructor allows Marines 45 seconds for moving to the next station and for preparing for the next exercise before the command EXERCISE is again given. In lieu of verbal commands, a whistle may be used to stop and start the exercises. For large groups, a megaphone or loudspeaker is useful.

f. Individual Conditioning Program. For the Marine working alone on the strength circuit, it would be impractical to time the exercise periods. For individual exercise, the Marine should select a number of repetitions of each exercise to complete, then rotate to the next stations after completing these repetitions. The number of repetitions selected should be at or near the maximum that the Marine is capable of doing without halting for rest.

g. Fixed Circuit Stations

(1) Primary Stations (See fig. 3-16.)

(a) Pullups/Chinups. A horizontal bar placed 8 feet above the ground is necessary for each Marine at this station. The Marine will also need a space on the bar that is 45 inches wide. On the command**EXERCISE,**the Marine grasps the bar with both palms facing either forward or to the rear, arms fully extended, and feet free of the ground and executes the exercise as described in MCO 6100.3_.The exercise is repeated as many times as possible until the command **STOP** is given. Then the Marine moves to the next exercise. If a Marine has done his maximum number of pullups/chinups prior to the command **STOP**, he will remain in the "dead" handing position until the command **STOP** and move to the next station.

(b) Twist Grip. The apparatus is a horizontal bar, free to turn, held between uprights placed 30 inches apart. The bar is 52 inches above the ground. A weight of 20 pounds is attached to the center of the bar by a light rope long enough to permit the weight to rest on the ground. The Marine stands at arm's length from the bar and grasps it with his hands on either side of the rope, palms down, thumbs under the bar. On the command **EXERCISE,** the hands are rotated so that the backs of the hands are rotated away from the body, thus winding the rope on the bar. The elbows are kept straight to ensure that the exercise is performed by the hand and forearm. When the weight is drawn up to the bar, the bar is then rotated in the opposite direction to lower the weight to the ground. This exercise is continued until the command **STOP** is given. The Marine then moves to the next station.

(c) Pulley Weights. The apparatus is a T frame with a system of pulleys that suspends a weight of about 90 pounds. The weight is attached to a light steel cable which has a drawbar attached to the other end. The Marine grasps the drawbar and sits down directly under the bar, legs extended to the front and arms extended overhead. The exercise is executed by pulling the drawbar down behind the head, then extending the arms slowly again until they are fully extended overhead. The exercise is repeated as many times as possible until the command STOP is given. The Marine then will move to the next exercise. Upon completion of the exercise, the weight is lowered slowly to the ground.

(d) Barbell Curls. A barbell is necessary for each Marine at this station. The barbell is constructed of 1 1/4-inch pipe 5 feet long, and two concrete-filled No. 10 cans. Each barbell should weigh about 40 pounds. Variance in the weight of the barbells, up to about55 pounds, will allow appropriate overload to be applied to Marines who are above average in strength or weight. The Marine grasps the bar with the palms forward and assumes a standing position with the barbell held in front of the hips, hands approximately shoulder's width apart. On the command **EXERCISE,** the Marine flexes the elbows and draws the barbell up until it touches the upper chest. The elbows remain at the sides. Breath is inhaled with the upward movement and exhaled as the barbell is lowered to the starting position. The exercise is repeated as many times as possible until the command **STOP** is given. The Marine moves to the next apparatus.

(e) Step-Up. The apparatus is a platform or ledge 18 inches high and of such size. At accommodate 10 Marines. The Marine faces the platform and on the command **EXERCISE,** steps up onto the platform, bringing his trailing foot up beside the leading foot. He then steps back down to the original position, stepping down first with the same foot he initially used in stepping up. After 10 repetitions of the exercise, he changes the order of moving the feet to use the opposite leg for stepping up. Repeat this exercise until the command **STOP** is given. The Marine then moves to the next station.

(f) Leg Lift. The apparatus is a horizontal bar constructed as described in (a). To prevent the body from swaying, a horizontal back support is added 40 inches below the horizontal bar. The arms are kept fully extended. On the command **EXERCISE,** the Marine jumps up, grasps the bar with the palms forward and the back support behind him. The exercise is executed by raising the legs to a horizontal position then slowly lowering them to the vertical position. The Marine does not flex his knees. He does not swing the legs to the rear of a vertical position to gain momentum for raising them in the next repetition of the exercise. The movement is repeated until the command **STOP** is given. The Marine then moves to the next exercise. If unable to raise his legs to a horizontal position without flexing his knees, the Marine flexes his knees and draws them up to the chest, then lowers his legs to the vertical position.

(g) Rope Climb. The rope climb is 20 to 30 feet high with five ropes suspended from a horizontal bar which forms the uppermost part of the framework. To prevent the horizontal bar from sagging and to provide safety, only five ropes are attached to it. There are two frameworks per station. The ropes are 6 feet apart. Any method may be used to climb the rope, and the Marines climb as high as possible. Marines who are proficient should climb therope several times during the time allotted. Inexperienced Marines should be cautioned to take care during descent to avoid rope burns on their hands.

(2) Supplementary Stations. The following exercises are designed to expand the basic circuit by being inserted in specific places within the system. For each supplementary station used, there should be adequate room for 10 Marines to exercise.

(a) Bent Knee Situp or Bottoms-Up. These calisthenics are designed to strengthen the abdominal muscles. These exercises will be inserted between the pullup/chinup and twist grip stations. The primary stomach exercise is the situp. In case of inclement weather or other conditions that make ground contact undesirable, the bottoms up exercise is used. In situps on the command **EXERCISE,** the Marine lies on his back (supine position) with knees flexed and both feet flat on the ground and executes the exercise (minus an assistant) as described in MCO 6100.3_.The Marine then returns to the starting position, repeating the exercise until the command **STOP** is given. The Marine then moves to the next station. In bottoms up on the commandEXERCISE,the Marine assumes the front leaning rest and executes the bottoms-up exercise as described in Conditioning Drill 3. He continues this exercise at a moderate cadence until the command **STOP** is given. The Marine then moves to the next station.

(b) Pushup. This exercise is designed to strengthen the arm- and shoulder-girdle muscles. It should be included between the twist grip and pulley weight stations. Upon the command **EXERCISE,** the Marine executes the pushup as described in Conditioning Drill 1. He continues this exercise at a moderate cadence until the command **STOP** is given. The Marine then moves to the next station.

(c) Knee Bender. This exercise is designed to build leg muscles and is included between the pulley weight and barbell curl stations. On the command **EXERCISE,** the Marine executes the knee bender as described in Conditioning Drill 3.He continues this exercise at a moderate cadence until the command **STOP** is given. The Marine then moves to the next station.

(d) Trunk Twister. This exercise strengthens the major muscles of the trunk and is included between the step-up and pullup/chinup stations. On the command EXERCISE, the Marine executes the trunk twister as described in Conditioning Drill 1. He continues this exercise at a moderate cadence until the command **STOP** is given. The Marine then moves to the next station.

3703. MOVEABLE STRENGTH CIRCUIT

a. Description and Objective. The exercises in this circuit are progressive and the course is planned to gain and hold the interest of the participating groups. The circuit consists of a series of stations, with each station designed to develop a particular group of muscles. Along with muscular development, correct posture and deep rhythmic breathing should be stressed on this circuit at all times.(See figs. 3-17 and 3-18.)

Figure 3-17 is available in the illustration section at the end of the book.

b. Equipment. The equipment is set up in files. Six files of 8 stations will accommodate a platoon

c. Formation. The platoon marches to the area where the equipment is positioned and forms a file within each lane of stations, covering down on a piece of equipment. Movements are made on the double, the important factor being that no time is wasted in getting to work.

d. Progression. Initially 40 to 45 seconds per station is adequate. As individuals become stronger the time should be increased in 5-second increments until a minute to a minute and a half is reached.

e. Leadership

(1) The leader stands in front of the barbell station and controls the rotation from this position. The leader supervises the entire group, with the assistance of several instructors who move about in the platoon correcting and encouraging the Marines.

(2) The leader starts each group but does not count cadence nor lead them through the exercises. Each individual exercises rapidly but individually.

(3) As each Marine finishes his repetitions with the barbell, he places the barbells on the ground. The leader calls, READY, followed by the command, FALL OUT ONE. All will double-time to the station directly in front of them, while the individuals on the barbell stations will do an about face to the rear station in their lane.

f. Movable Circuit Stations. The Marine can obtain best results on the movable circuit if the exercises on the various stations are given in the following manner:

(1) Station 1—The Barbell. (See fig. 3-17, 1.)The exercises at this station stress proper posture; deep, rhythmic breathing; and development of the muscles of the arms, shoulders, and upper body. An instructor teaches the proper methods of lifting before the exercises begin. The methods are to lift with the legs, to keep the back straight, and to merely grip with the hands. Two recommended exercises are given below. The instructor will use only one exercise per period .He may use either exercise. At successive periods, the other exercise is used.

(a) Exercise 1--Two-Hands Military Press. (See fig. 3-18, 1.)Grasp the barbell with both hands, knuckles up at shoulder width, and lift to the chest. Steadily press to arm's length overhead; lower to the chest resisting weight all the way. Inhale as the weight is pressed up and exhale as the weight is brought down.

(b) Exercise 2--Two-Hands Regular Curl. (See fig. 3-18, 1.) Lift the weight to the waist, with the palms of the hands out, heels together, stomach in, chest lifted and arched, shoulders back, elbows in close to the sides; inhale deeply and curl the weight to the shoulders, using the arms only, at the same time keeping the elbows close to the sides; exhale Emphasize posture and the use of the arms only. This is a very valuable exercise for the development of the biceps and the grip and should be repeated from 8 to 16 times, depending on the ability of the Marine.

(2) Station 2--The Jump Rope. (See fig. 3-18, 2.) This exercise develops strength and agility in the legs and stamina of the whole body. It makes the Marine agile on his feet and increases his footwork efficiency and timing. Each Marine should progress until able to jump rope at least 3 minutes at top speed.

(3) Station 3--The Twist Grip. (See fig. 3-18, 3.) The twist grip is an excellent exercise for the hands and forearms, and adds greatly to the Marine's ability in hand-to-hand combat. The handle is gripped and twisted, winding the rope until the weighted can is level with the height of the hands, which are held horizontal. The weight is lowered in the same manner; the Marine resists the weight all the way, occasionally stopping the twisting motion and alternately removing first one hand, then the other, from the handle. A variation of the above exercise is to wind the handle with the palms up and the arms bent and with the elbows held close in to the sides. Each Marine should maintain a good posture and keep the stomach muscles taut throughout this exercise.

(4) Station 4--The Incline Plane. (See fig. 3-18, 4,)The use of the incline plane is a very strenuous exercise and well designed for the development of the abdomen. Although 6 to 10 repetitions are sufficient for the beginners, more can be added as ability increases. Marines with hernias or recent operations will be excused from participation at this station.

(5) Station 5--The War Club. (See fig. 3-18, 5.) The war club is a simple and effective means of exercising the principal muscle groups of the body, especially those of the trunk, back, and shoulders. To gain the maximum benefit from this exercise, the Marine must keep both feet flat on the ground at all times. Throughout the exercise period, the weight is swung from arm's length as follows:

(a) As in chopping wood, first on one side, then on the other.

(b) As a batter warming up with a number of bats.

(c) In large circles, first with one hand and then with the other.

(6) Station 6--The Bicycle Ride. (See fig. 3-18, 6.) The bicycle ride is well suited to exercising many of the muscle groups of the body, particularly those of the abdomen. Vary the speed of the exercise, but keep everyone "riding" the entire period. A variation exercise may be performed by placing the legs together, raising them slowly to a height about 2 feet from the ground, and then lowering them slowly to the ground.

(7) Station 7--The Step-up. (See fig. 3-18, 7.) The step-up exercises the legs. The step-up is performed by initially stepping up with the left foot, followed by the right, then stepping down with the left foot followed by the right. Continue for 20 seconds, then change to the right foot as the lead foot for 20 seconds.

(8) Station 8--The Isometric Pull. (See fig. 3-18, 8.) Two trainees work at this station with a cable pull and perform the following exercises:

(a) Initially start with one Marine in the supine position and one Marine sitting. The sitting Marine lowers the upper body to the ground and pulls the partner up to the sitting position. The partner then performs this same action and this is continued for 20 seconds at a rapid rate.

(b) During the last 20 seconds, the same action takes place but, in this case, the Marine in the supine position resists the pull of the partner for approximately 5 seconds before allowing to be pulled up into the sitting position.

3704. CIRCUIT-INTERVAL TABLE

a. Description and Objective. The circuit-interval table is designed to develop strength and endurance within a short period of time, with no equipment requirement, through a rapid and vigorous routine of exercise. Fifteen minutes is an adequate period to execute all exercises and to secure a vigorous workout with the circuit-interval principle.

Figure 3-18 is available in the illustration section at the end of the book.

b. Formation and Starting Positions

(1) A leader forms platoons or smaller groups in an oval or circular formation with 3- to 5-yard intervals between Marines. The Marines face to the right and move forward at quick time and then double time. (See fig. 3-19, A.)After running several platoon circle laps, the leader calls out the name of an arm and shoulder exercise from the list below, orders quick time and commands, for example, PUSHUPS. On this command, all Marines immediately hit the ground and individually and rapidly begin doing pushups. No cadence is counted. (See fig. 3-19, B.) After 30 seconds of exercise, the leader commands. **QN YOUR FEET, FORWARD, MARCH** The platoon resumes the quick time cadence and the leader, when ready, gives the necessary commands for double time. The double time is continued for one or more laps and the leader calls out the name of the next exercise and the process is repeated. This continues, with running between each exercise, until every body part has been exercised.

(2) The instructor controls the running and quick time to observe the effects of the exercise upon the Marines. Cadence, step, and precision are not important to the objective and the instructor should not use them. What is important is speed and the instructor should stress this. After the exercise period is started, the Marines do not stop. This circuit method emphasizes stress and recovery, the recovery occurring during the quick time periods.

c. Activities. The leader can use the following exercises. These exercises can be repeated if necessary, during a second round. The leader can use other calisthenic exercises in the circuit-interval table.

(1) Arms and shoulders-push-ups.
(2) Stomach situps.
(3) Back-squat thrusts.
(4) Legs-bicycle (on back).

d. Progression. lhe progress is controlled by the leader, who must pace the running, quick-time movement, and exercise in such a way that Marines will receive a vigorous workout yet be able to participate throughout the 15-minute period. Marines who are in the initial stages of physical condition will not be able to double time or exercise as long as those who are better conditioned. The idea is to set a pace which can be increased during each workout, thus progressing gradually to a higher level of physical fitness.

e. Leadership. The platoon leader, platoon sergeant, or section leader can lead the group. The leader must execute the exercise with the unit to feel the effects and thereby adjust the pace.

f. Place in the Program. This activity may be scheduled whenever a short period of time is available. The only requirement is that enough space, indoors or out, be available to form the circle.

Figure 3-19 is available in the illustration section at the end of the book.

Section VIII. Basic Physical Skills and Obstacle Courses

3801. GENERAL

The purpose of this section is to list the basic military skills and the methods for their development. Many of these skills are best developed by obstacle courses but other drills are also discussed. The section explains types of obstacle courses, details of construction, and methods of negotiating the various obstacles.

3802. BASIC PHYSICAL SKILLS

a. Objective. The objective is to develop proficiency in the various military physical skills which are essential to personal safety and effective combat operations. In travel by foot over rugged terrain and in the execution of combat duties, Marines must be trained to perform basic skills, such as running, jumping, climbing, and carrying. During training, Marines will develop agility and coordination in these skills .Fast and skillful execution of these skills may mean the difference between success and failure on the battlefield.

b. Place in the Program. These skills are practiced throughout the entire program in many different activities. Many of these skills are best practiced on obstacle courses.(See par. 3803.)

c. Description. These basic skills are the minimum skills required by all Marines. The basic skills are as follows:

(1) Running. Running is used to strengthen the legs and develop cardiovascular endurance. Marines should be exposed to running in various situations: on roads, over rough ground, up and down hills, cross-country, and running over low obstacles.

(2) Jumping. In broad jumping, the take-off foot is planted firmly and the spring comes from the extension of this leg as the other leg reaches for the far side of a ditch or similar obstacle. (See fig. 3-20.) The arms are forcibly raised forward and upward to assist in propelling the body up and forward. Landing may be on one or both feet depending upon the length of the jump. In vertical jumping downward from a height, the jumper should aim his feet at the desired landing spot and he should also jump with the knees slightly bent and feet together, with the trunk inclined slightly forward. As the feet touch the ground, the shock is absorbed by bending the knees into a full squatting position. If the height is too great or the ground too hard to absorb the shock, then the jumper should forward roll or side roll thus eliminating some of the momentum.

Figure 3-20 is available in the illustration section at the end of the book.

(3) Dodging In combat situations, it is often times necessary to change directions quickly. To execute this movement while running, a lead foot is firmly planted, left foot if the direction is to the right and right foot if the direction is to the left. The opposite foot is moved toward the new direction. The knees are slightly flexed during the movement and the center of gravity is low and balanced. (See fig. 3-21.) At the time of the change of direction, the head and trunk are turned quickly in the new direction.

Figure 3-21 is available in the illustration section at the end of the book.

(4) Climbing and Surmounting. All Marines should know how to climb and surmount various types of obstacles. (See fig. 3-22.)

Figure 3-22 is available in the illustration section at the end of the book.

(a) Vertical Climbing, as in Climbing a Rope or Pole. The hands grasp the rope or pole overhead with the palms toward the face. Gripping the object, the body is pulled upward with the arms and shoulders, assisted by the feet which grip the object and assist by pushing downward. If shoulder girdle strength and body coordination are not adequate to permit alternating the hands, the arms act together in pulling upward.

(b) Climbing as in Surmounting a Wall. Ingoing over a wall, the body should be kept as close to the top as possible, since in combat operations it is important to offer as small a target as possible to the enemy. If an individual climbs a wall while carrying a rifle, both hands should be freed by slinging the rifle over the back. There are two methods commonly used for surmounting a wall of moderate height, but only one for dropping from it.The methods are as follows:

1 Run, Jump, and Vault. Approach the wall at a run, jump forward and upward at the wall and place one foot against it as high up as possible. Use the foot in contact with the wall to help push the body upward while grasping the top of the wall with the hands. Pull the body up with the arms, assisted by pressure of the foot against the wall and swing the legs over, propelling the body weight over the wall.

2 Hook and Swing. Approach the wall at a run and jump forward and upward. Hook one elbow over the wall, locking the arm in place by pulling up until the top of the wall is underneath the armpit. Grasp the top of the wall with the other hand. Draw the leg which is closer to the wall up as far toward the top as possible. Then swing the other leg over the top of the wall.The body is then carried over with a rolling motion. A variation of this leg action can be used by Marines who are unable to draw up the leg as described. While hanging with both legs fully extended, start a swinging motion with the legs together. When the legs have enough momentum, swing the outside leg over the top of the wall with a vigorous kick, then follow with the body.

3 **Dropping.** All drops from a height are executed in the same manner regardless of the method used to gain the top. One hand is placed against the far side of the wall while the other hand grasps the top. From this position, the body is rolled over the wall and vaulted away from it with the legs swinging clear. As the body passes over the wall and drops, it should at all times face the wall. This will keep the rifle and other equipment clear. Break the fall by retaining a grasp on the top of the wall as long as possible.

(c) Climbing Ladders and Cargo Nets. Rope ladders, stationary vertical ladders, and cargo nets employ the same general technique. The important element is to grasp the side supports firmly in the hands about shoulder height and place the feet on a rung which would cause the body to be fully extended. In movement upward, one hand is moved upward and a new grasp is secured and, at the same time, the opposite leg moves up a rung. As the knee straightens, the body is elevated. This process is repeated using the opposite arm and leg. Alternation continues in this manner until the climber reaches the objective.

(5) Traversing Horizontal Objects. The traversing of horizontal objects puts heavy stress on the arms and shoulder girdle area as the feet are usually suspended in the air with all of the body weight on the arms and shoulders.(See fig. 3-23.)

(a) Traversing Horizontal Ropes or Pipes. The hands grasp the horizontal support overhead with the palms facing. To propel the body forward, one hand is released and moved forward to secure a new grasp. At the same time, the opposite side of the body is swung forward (some people are able to "walk" in the air, keeping the body to the front and moving the legs in time with the arms as in walking on the ground). The other hand is then released and moved forward; this alternation is continued until the objective is reached.

Figure 3-23 is available in the illustration section at the end of the book.

(b) Traversing Horizontal Ladders. In this situation, the movement is the same as used in traversing a rope or pipe. The hands, however, are placed on the rung with the palms away from the face. Other than this difference the technique is the same.

(6) Crawling. Crawling in combat situations is an often used skill. Crawling may be high or low.(See fig. 3-24.)

Figure 3-24 is available in the illustration section at the end of the book.

(a) High Crawl.

In the high crawl, the Marine moves on hands and knees, moving one hand and the opposite knee and then continuing to move the hands in alternation with the opposite knee following the companion hand.

(b) Low Crawl. The Marine is in the prone position usually with the forearms and palms of the hand on the ground. He propels forward by bending the knee of one leg and pushing with the inside edge of the shoe. At the same time, the opposite arm moves forward and pulls to the rear. The body remains low and movement is continued by bending the opposite knee and pushing, and at the same time sliding the opposite arm forward and pulling. Alternation of hands and legs continues until the objective is reached.

(7) Throwing. Throwing may be executed from the kneeling or standing positions. The object to be thrown is held in the hand, and the throwing arm is bent at the elbow; the hand is then moved to the rear until the hand is behind the ear. The body is turned so that the lead foot and balance arm on the side toward the target are pointing at the target. (See fig. 3-25.) The balance arm is used to sight over and align the throwing hand and the target. When properly aligned, the elbow is moved rapidly forward until it is at a pointjust in front of the body where the arm is straightened and the wrist snapped. This whip motion propels the object to the target. Underhand throws secure momentum by the thrower bending his knees and swinging the throwing arm to the rear. As the knees are straightened, the arm is forcefully swung forward from the shoulder and the object released.

Figure 3-25 is available in the illustration section at the end of the book.

(8) Vaulting. Vaulting is employed to overcome low barriers or fences. (See fig. 3-26.)The object to be surmounted is approached at an angle. The hand on the side next to the obstacle is placed on the top of the obstacle and, with a straight arm, the body weight is pushed upward. At the same time, the leg on the side next to the obstacle is thrown upward and over the top followed by the other leg (side approach). In landing, the weight comes down on the leading leg first followed by regaining the balance on both legs. The free arm serves as a balance. A direct (front) approach can be used at which time both legs go over the object together.

(9) Man Carrying. There are three basic individual means of carrying personnel in combat situations and one of these methods may be used in carrying objects.

Figure 3-26 is available in the illustration section at the end of the book.

(a) Fireman's Carry."A" stands sideways in front of "B", "A" bends his knees and leans forward, placing one arm through "B's" crotch, grasps the wrist of "B's" arm, which is hanging over the shoulder, and then "A" runs forward.

(b) Saddle-Back Carry. With his back toward "B", "A" stands in front of "B". "B" mounts "A's" hips and clasps his arms in front of "A's" chest. "A" grasps "B's" thighs.

(c) Single-Shoulder Carry. "A" stands facing "B"."A" assumes a semi-squatting position. "B" leans forward until "B" lies across "A's" right shoulder. "A" clasps his arms around "B's" legs and straightens up, lifting "B" from the ground. "A" then runs forward. This method may also be used to carry heavy objects.

(10) Balancing. Balancing the body while walking or running on a narrow object when crossing obstacles is a skill which requires practice and confidence. Balance is required in negotiating a log placed across a stream, in crossing a narrow beam or rail, and in similar situations.(See fig. 3-27.) To perform this skill, place the feet on the object to be crossed, hold the arms to the sides at shoulder level, and fix the eyes on the object approximately 5 yards in front of the feet. Generally, it is not a good practice to look down at the feet. Walk the beam by placing first one foot and then the other in the center of the beam, thereby moving forward, using the arms to aid in maintaining balance.

Figure 3-27 is available in the illustration section at the end of the book.

(11) Falling. Injury can be avoided if Marines are taught to fall properly. They should know how to use the body momentum to their advantage during a fall rather than to try resisting that force. (See fig. 3-28.) If enough force is present, such as occurs during a fall while running or in jumping downward from a height, individuals can extend their hands to catch the weight. At the same time, duck the head and roll forward onto the feet. The key to falling without injury from the standing position is relaxation and rolling on the outside of the leg, hip, and buttocks to take the brunt of the fall.

Figure 3-28 is available in the illustration section at the end of the book.

3803. OBSTACLE COURSES

a. General. Obstacle courses are a valuable part of physical readiness training. The challenge presented by the obstacles assists in developing and testing the basic physical skills. In many combat situations, success will depend upon the Marine's ability to perform one or more of these skills, often while carrying field equipment and when fatigued. In this section, two types of obstacle courses will be discussed-Conditioning Obstacle Course and Confidence Obstacle Course.

b. Course Safety Precautions. Commanders and course instructors should take certain precautions to prevent injury on obstacle courses. A few of the precautions are:
(1) Inspect the course for faulty construction of obstacles, protruding nails, rotten logs, condition of the landing pits, and other hazards to safety.
(2) Conduct warm up exercises before the unit runs the course.
(3) Explain and demonstrate the correct techniques for negotiating all the obstacles before allowing the Marines to try them.
(4) Give Marines at least two weeks of conditioning exercises before scheduling the obstacle and/or confidence courses.
(5) Ensure that negotiation of the higher and more difficult obstacles are under the supervision of an instructor.
(6) Do not permit individuals who have neither practiced the basic skills nor run the conditioning obstacle course to participate in the confidence obstacle course.
(7) Weather conditions may cause footing or handhold surfaces to be slippery. If such is the case, postpone training on the course.

3804.CONDITIONING OBSTACLE COURSES

a. Description and Objective. The Conditioning Obstacle Course is commonly known as the Obstacle Course. This course consists of fairly low obstacles which are designed to be negotiated quickly. The obstacles serve to test various basic skills, and running the course is a test of the Marine's physical condition. After receiving instruction and an opportunity to practice the skills, Marines run the course against time.

b. Area and Equipment
(1) Complete standardization of obstacle courses should not be attempted since topographical conditions always vary. Commanders should use ingenuity in constructing a course, making good use of streams, hills, trees, rocks, and other natural obstacles. Since the course is eventually run at high speed, it should not be dangerous.

(2) The course should be wide enough for six or eight men to run simultaneously, encouraging competition. The lanes for the first several obstacles should be wider and the obstacles themselves easier than those that follow. This avoids congestion until the contestants scatter out over the course. The last two or three obstacles should not be too difficult and should not involve high climbing. This prevents injuries and falls resulting from fatigue.

(3) The total distance of the course should range from 300 to 450 yards and include from 15 to 25 obstacles. Normally the obstacles should be 20 to 30 yards apart and arranged so that those which exercise the same groups of muscles are separated.

(4) The obstacles should be substantially built. Peeled logs, 6 to 8 inches in diameter, are ideal for many of the obstacles. Sharp points and corners should be eliminated. Landing pits for jumps or vaults should be filled with sand or sawdust to prevent injuries.

(5) The course should be constructed and marked so that it is not possible to sidestep or detour obstacles. However, it is desirable to provide alternate obstacles of varying degrees of difficulty.

(6) The course should be in the shape of a horseshoe or figure eight so that the finish is close to the start and signs should be placed to indicate the course route.

c. Leadership. Before Marines run an obstacle course, they should be instructed in the proper technique of negotiating each obstacle. In each case this technique should be explained and demonstrated in detail, with emphasis on avoiding injury. Every individual should be given an opportunity to practice on each obstacle until he becomes reasonably proficient at negotiating it. Before the course is run against time, it is advisable for individuals to make several runs at a slower pace. During such practice or trial runs, the instructor should observe the performances and make appropriate corrections. Marines should never be permitted to run the course for time until they have practiced on all obstacles. The best method of timing the runners is to have the timer stand at the finish and call out the minutes and seconds as each individual finishes. If several watches are available, each wave may be timed separately. If only one watch is available, the different waves should be started at regular intervals, such as every 30 seconds. If an individual fails to negotiate an obstacle, a previously determined penalty should be exacted.

d. Types of Obstacles

(1) Jumping-Type Obstacles. These obstacles may be ditches which are cleared with one leap, trenches which the individuals can jump into, heights which require jumping downward, or hurdles which an individual can jump over. (See fig. 3-29.)

Figure 3-29 is available in the illustration section at the end of the book.

(2) Dodging-Type Obstacles. Obstacles of this type are usually mazes consisting of posts set in the ground at irregular intervals. (See fig. 3-30.)The intervals between posts should be rather narrow so that the Marines must pick their way carefully through and around them. Lane guides may be constructed to guide the Marines to dodge and change direction. Obstacles may be put into a maze pattern to cause the Marines to change direction.

Figure 3-30 is available in the illustration section at the end of the book.

(3) Vertical Climbing and Surmounting Type Obstacles. These obstacles may be climbing ropes, either plain or knotted and 1-5 inches in diameter; cargo nets or walls 7 or 8 feet high; or vertical poles 6 to 8 inches in diameter and 38 feet high.(Seefig. 3-31.)

Figure 3-31 is available in the illustration section at the end of the book.

(4) Horizontal Traversing-Type Obstacles. Horizontal obstacles may be pipes, beams, ladders, or ropes. (See fig. 3-32.)

(5) Crawling Type Obstacles. Obstacles which require crawling may be constructed of large pipe sections, low rails, and wire. (See fig. 3-33.)

Figure 3-32 is available in the illustration section at the end of the book.

(6) Vaulting-Type Obstacles. Obstacles of 3 to 3.5 feet in height such as low walls or fences may be used as a vaulting obstacle.(See fig. 3-34.)

(7)Balancing-Type Obstacles. Beams, logs, and planks may be used as balancing-type obstacles.(See fig. 3-35.)These items may be used to span water obstacles and dry ditches, or raised off the ground somewhat to simulate natural depressions.

Figure 3-33 is available in the illustration section at the end of the book.

3805.CONFIDENCE OBSTACLE COURSES

a. Description and Objective. The Confidence Obstacle Course is commonly known as the Confidence Course. This course is composed of higher and more difficult obstacles than those used in the conditioning course. The confidence obstacle course is designed to give Marines confidence in their mental and physical capacities and to cultivate their spirit of daring. They are encouraged but not compelled to negotiate this course and the course is not run against time. The negotiation of a confidence course, however, is strenuous enough to be an excellent physical conditioner. Marines should NEVER attempt to take the obstacles at high speed and should not compete for speed. The obstacles vary from fairly easy to extremely difficult. Some are of considerable height to accustom Marines to climbing such heights without fear. Considerable emphasis is placed on obstacles that train and test an individual's balance.

Figure 3-34 is available in the illustration section at the end of the book.

b. Area and Equipment

(1) The confidence course accommodates four platoons, one platoon at each group of six obstacles. The course should be made up of about 24 obstacles, numbered and marked as follows:! to 6, white numbers on red background; 7 to 12, black numbers on a white background;

13 to 18, white numbers on a blue background; and 19 to 24, white numbers on a black background.

(2)A few simple pieces of equipment should be provided for individuals who do not have the strength, courage, or ability to negotiate the obstacles.

Figure 3-35 is available in the illustration section at the end of the book.

c. Formation. The obstacles should be divided into groups of six, and each group is designated by a different color. Each platoon starts at a different color. Individuals are separated into groups of 8 to 12 at each obstacle. At the starting signal from the company commander, they proceed numerically through. Anyone may skip an obstacle who is afraid to try. Individuals proceed from obstacle to obstacle until time is called, then assemble as ordered.

d. Leadership. If the Marines are new to the confidence course, an instructor will demonstrate or will give a brief orientation at each obstacle, including an explanation and demonstration of a method of negotiating it. Marines are encouraged to try the various obstacles, but they are not compelled to do so. No compulsion is to be used. The manner of negotiating any obstacle is left to the discretion of the Marine. However, the instructor assists anyone who experiences difficulty. Instructors must supervise closely at all times to prevent injuries, as some of the obstacles are quite high. Also, some of the obstacles should not be used when slippery or wet. The example of instructors and especially selected demonstrators will serve to inspire the individuals to greater effort.

e. Negotiating the Obstacles. Although personnel need not conform to any one method of negotiating the obstacles, there should be some uniformity in the approach to them. A general method of negotiating the obstacles is indicated below.

(1) Red Group This group contains the first six obstacles. (See fig. 3-36.)

(a) The Belly Buster Individuals may vault, jump, or climb over. Warn them that the log is not stationary.

(b) Reverse Climb. Climb the reverse incline and go down the other side to the ground.

(c) The Weaver. Move from one end of the obstacle to the other by weaving the body under one bar and over the next.

(d) Hip-Hip. Step over each bar, either alternating legs or using same lead leg each time.

(e) Balancing Logs. Step up on log, retaining the balance, walk or run along log.

(f) Island Hopper. Jump from one log to another until the obstacle is negotiated.

(2)White Group. This group is composed of the second six obstacles. (See fig. 3-37.)

(a) Tough Nut. Step over each X in the lane.

(b) Slide for Life. Climb the tower, grasp the rope firmly, and swing the legs upward. Hold the rope with the legs to distribute the weight between them and the arms. Braking the slide with the feet and legs, proceed down the rope. Warn Marines that there is danger of getting rope burns on their hands. When the rope is slippery or wet, this can be a dangerous obstacle.

(c) Low Belly Over. .Mount the low log and jump onto the high log, both arms grasping over the top of the log, the stomach area in contact with it. Swing the legs over the log and lower the body to the ground.

(d) Belly Crawl Move forward under the wire, belly down, to the end of the obstacle.

(e) The Dirty Name Mount the low log and jump to or reach the higher logs in succession, then jump or drop to the ground. Warn the Marines about the height of the final log.

(f) The Tarzan Mount the lower log and walk the length of it and each successive, higher log until reaching the horizontal ladder. Grasp two rungs of the ladder and swing the body into the air. Negotiate the length of the ladder by releasing one hand at a time and swing forward, grasping a more distant rung.

(3) Blue Group. This group is formed by the third group of six obstacles. (See fig. 3-38.)

(a) High Step over. Step over each log, alternating the lead foot or using the same lead foot.

(b) Swinger. Climb onto the swinging log and over to the ground on the opposite side.

(c) Low Wire. Move under the wire on the back, using the hands to raise the wire to clear the body.

(d) Swing, Stop and Jump. Gain momentum with a short run, grasp the rope, and swing the body forward to the top of the wall. Release the rope while standing on the wall and jump to the ground.

(e) Six Vaults. Vault over the logs, using one or both hands.

(f) Easy Balancer. Walk up one inclined log and down the one on the other side to the ground.

(4) Black Group. The last group is formed by the final six obstacles. (See fig. 3- 39.)

(a) Incline Wall. Approach the underside of the wall, jump up and grasp the top and pull the body up and over. Slide or jump down the incline to the ground.

(b) Skyscraper. Jump or climb to the first floor, climb up the corner posts or assist each other to any desired floor. Descend to the ground in any desired manner.

(c) Jump and Land. Climb up the ladder to the platform and jump to the ground.

(d) Confidence Climb. Climb the inclined ladder to the vertical ladder. Go to the top of the vertical ladder, then down the other side to the ground.

(e) Belly Robber. Step on the lower log and assume the prone position on the horizontal logs. Crawl over the logs to the opposite end of the obstacle.

(f) The Tough One Climb the rope or pole on the higher end of the obstacle, then go down the ladder and across the log platform. Climb over or between the logs at the end and go down the rope or pole to the ground. Vault over the final log.

Figure 3-36 is available in the illustration section at the end of the book.

Figure 3-37 is available in the illustration section at the end of the book.

Figure 3-38 is available in the illustration section at the end of the book.

Figure 3-39 is available in the illustration section at the end of the book.

Section IX Individual Exercise Programs

3901.GENERAL

Marines at times will be stationed on independent duty and consequently will be responsible for their own physical fitness program. This section will assist them in understanding the need for exercise and will aid in the planning and execution of an individual exercise program. Exercise activities included are the Bench Conditioner, 6-12 Plan, Weight Training, and Isometric Contraction. Keeping physically fit is a problem that faces every Marine. Even though we are frequently engaged in training that requires some physical effort, in many cases, it is not enough to prepare us to meet the intense physical demands of combat. Attaining a satisfactory level of physical readiness is not an insurmountable objective. Available time appears to be the most difficult obstacle to the development of physical readiness. In most cases, regular physical training programs are centralized, requiring the individual to temporarily leave the work area. The problems involved in setting an hour aside two or three times each are numerous. However, most of us can devote a few minutes each day to physical fitness with little, if any, impact on our daily work schedule, especially if it does not require us to leave our work area.

a. Type of Program. There are many good physical fitness programs available to the individual or group. Regardless of the type or duration, to be effective, the program must contain exercises that are strenuous and are challenging to the individual. Space will not permit the inclusion of all available means of individual exercise. The programs selected for this chapter have met the requirement of minimum space and minimum time.

b. Need to Augment Program. These programs are quite strenuous and will develop a satisfactory level of physical readiness. However, if the Marine desires additional development of endurance, it is recommended that he supplement these programs with a 15- to 30-minute period of wind sprints and double timing on an alternating daily basis.

c. Progressive Training. If Marines are performing duties which require little or no physical activity, they must plan a physical conditioning program that assures a moderate beginning, moderate but steady progression, and sufficient warmup before starting the vigorous exercise. To avoid organic of bodily harm, a Marine activity without adequate warmup. He should conduct conditioning programs on a daily basis over an extended period of time, never on an unduly accelerated or crash basis.

3902.THE BENCH CONDITIONER

a. Description and Objective. The bench conditioning program uses a modified bench to employ both isotonic (moving) and isometric (stationary) exercises are as the nucleus of the program. The exercises are designed to develop strength and endurance in all the major muscle groups of the body. The principles of progression, overload, and balance are employed when the exercises are performed properly.

b. Area and Equipment. The conditioning apparatus can be constructed in any unit motor pool with welding equipment found in most salvage yards.(See figs. 3-40, 3-41.)Additionally, there are available any number of commercially produced apparatus available. The important thing is that in utilizing this equipment the following exercise routine be adhered to.

c. Starting Level and Progression.. The program consists of two tables, each with 10 exercises. The Marine controls progression by required repetitions or, in some cases, by application of maximum effort. Each table can be completed within 15 minutes.

d. Starting Position. To start the program, the Marine begins with the Table 1 and executes each exercise for the required number of repetitions as indicated. The Marine controls the starting level and progression. When the Marine executes the maximum repetitions for Table I within a 15-minute exercise period, he progresses to Table II. To maintain this level of development, the Marine should also execute the maximum repetitions for Table II within a 15-minute exercise period. The Marine should keep substitution of exercises to a minimum. However, it he completes a full 15 minutes of strenuous exercise and exercises all muscles, then there should be no appreciable difference in the overall development.

e. Bench Conditioning, Table I

(1) Exercise 1: Side-Straddle Hop. This is a two-count warmup exercise done at moderate cadence. The starting position is the position of attention. On count ONE jump slightly into the air, swinging the arms out to the sides and up to a vertical position, hands touching.(See fig. 3-40, A.)At the same time, spread the feet wider than shoulder-width apart. On count TWO, using a slight flexing of the knees and ankles, jump slightly into the air and return to the starting position by swinging the arms back down to the sides. Twenty repetitions of this exercise is the standard dosage throughout the program.

(2) Exercise 2: Hand Walk. Remove the lower horizontal bar. Adjust the upper horizontal bar so that it is high enough to permit a "dead hanging" position, with the feet off the ground.(See fig. 3-40, B.j From the "dead hanging" position, release one hand and drop the arm to the side of the body. Then raise that arm and regrasp the horizontal bar. Release the bar with the other hand and drop that arm to the side. Repeat this as many times as possible.

(3) Exercise 3: Situps. Lie down with the fingers interlocked and placed behind the head. Hook the toes under the foot braces. Raise the trunk and upper body to an upright sitting position, twisting it to the left and then forward and downward until the right elbow touches the left knee.(See fig. 3-40, C.)Lower the body to the starting position .Sit up again but twist the body to the opposite direction as before, touching the left elbow to the right knee. Again lower the body to the starting position. The starting dosage is 20 situps. The Marine should continue the progression until he has attained 40 situps.

Figure 3-40 is available in the illustration section at the end of the book.

(4) Exercise 4: Double Step-upStarting at one end of the bench, step up onto the bench, and walk across it. Step down from the other end; turn around and repeat the process to return to the starting point.(See fig. 3-40, D.)Each return to the starting point constitutes a repetition. The starting dosage is 20 repetitions. Maximum dosage is 35 repetitions. This exercise should be done at a rapid cadence.

(5) Exercise 5: Isometric Bar Lift. Adjust the lower bar so that it is slightly higher than the beltline. Placing the feet on the footplates at the base of the frame, grasp the lower bar so that the hands are spread shoulder-width apart. Assume a crouched position and lift with maximum effort using the arms, back and legs.(See fig. 3-40, E.)Starting dosage is 4 repetitions of a stress time of 5 seconds followed by a 5-second rest prior to the next repetition. The Marine obtains progression by lengthening stress periods to 6 and later 7 seconds. Do not increase the number of repetitions.

(6) Exercise 6: Knee Lift. Adjust the upper bar to the same height used in Exercise 2.Adjust the lower bar so that is stops rearward movement of the hips when the "dead hanging" position is assumed.(See fig. 3-40, F.) Keeping the arms extended, flex the legs and raise the knees as high as possible. Hold this position for 5 seconds, then return to the starting position. After 2 seconds in the starting position, raise the knees again. Each return to the starting position constitutes one repetition. The dosage is five repetitions. Progression is the same as in Exercise 5.

(7) Exercise 7: Isometric Pull. Adjust the lower horizontal bar to a position where it is slightly higher than the beltline. Grasp the handles and pull outward.(See fig. 3-40, G.)Apply maximum effort and hold for approximately 5 seconds. Relax for 5 seconds between repetitions; perform four repetitions. Moving the body closer to or farther away from the bar will change the stress from the upper arms to the forearms. Progression is the same as in Exercise 5.

(8) Exercise 8: Isometric Compression. Maintain the position as in exercise 7.(See fig. 3-40, H.)Grasping the handles in the same manner, press in **with** maximum effort and hold for approximately 5 seconds. Relax for 5 seconds between repetitions. Progression is the same as in Exercise 5.

(9) Exercise 9: Isometric Press. Remove the lower horizontal bar. Adjust the upper horizontal bar until it is about 6 inches lower than the extended arms can reach. Stepping on the footplates at the bottom of the frame, grasp the bar with both hands and push up.(See fig. 3-40,1.)Keep both the legs and arms slightly flexed and the back straight. Apply maximum effort for 5 seconds then relax for 5 seconds. Complete four repetitions. Progression is the same as in Exercise 5.

(10) Exercise 10: Pushups. Grasping the foot braces with both hands, assume the front leaning rest position.(See fig. 3-40, J.)Keeping the back and legs straight, lower the body until the chest is lower than the **hands,** then return to the starting position. The Marine should complete the maximum possible number of repetitions.

f. Bench Conditioning, Table II. There is no limit on the maximum number of repetitions attainable in Exercises 3, 4, and 6 of Table II. The only limit imposed is that the entire program of 10 exercises outlinted in either table should not exceed 15 minutes.

(1) Exercise 1: Side-Straddle Hop. This is a two-count warm up exercise done at a moderate cadence. The starting position is the position of attention. On count ONE, jump slightly into the air, swinging the arms out to the sides and up to a vertical position, hands touching.(See fig. 3-41, A.)At the same time, spread the feet wider than shoulder-width apart. On count TWO, using a slight flexing of the knees and ankles, jump slightly into the air and return to the starting position by swinging the arms back down to the sides. Twenty repetitions of this exercise is the standard dosage throughout the program.

(2) Exercise 2: Pull-up. Adjust the horizontal bar so that is is high enough to permit a "dead hanging" position with the feet off the ground. Grasp the bar with both hands, palms facing forward. By flexing the arms, raise the body to a position where the chin is higher than the bar.(See fig. 3-41, B.)Then lower the body to the "dead hanging" position. Repeat as many times as possible.

(3) Exercise 3: Bench situps. Sit on the bench and hook the feet under the foot braces. With the fingers interlocked behind the head, lean back until the head touches the floor.(See fig. 3-41, C.)Return to the starting position. The starting dosage is 15 situps.

(4) Exercise 4: Step-Up. Face the bench and step up on it with one foot, bringing the trailing foot up next to the leading foot. Step back down again, leading with the same foot used first in stepping up.(See fig. 3-41, D.)Perlorm half of the total repetitions, then change the sequence of moving the feet to use the other leg in stepping up, and repeat the same amount of exercise. The starting level is a total of 40 step-up. **This exercise should be done at a rapid cadence.**

(5) Exercise 5: Isometric Bar Lift\djust the lower bar so that it is slightly higher than the beltline. Placing the feet on the footplates at the base of the frame, grasp the lower bar so that the hands are spread shoulder-width apart. Assume a crouched position and lift with maximum effort using the arms, back, and legs.(See fig. 3-41, E.)Starting dosage is four repetitions of a stress time of 8 seconds followed by a 5-second rest prior to the next repetition. Progression is obtained by lengthening stress periods to 10 seconds. Do not increase the number of repetitions.

Figure 3-41 is available in the illustration section at the end of the book.

(6)Exercise 6: Leg Lift. Adjust the bars and assume the starting position as shown in figure 3-41, F. Keeping arms and legs extended, raise the legs to a horizontal position and hold in that position for 2 seconds. Then lower the legs slowly to the starting position. Five repetitions is the starting level.

(7) Exercise 7: Isometric Pull. Adjust the lower horizontal bar so that it is slightly higher than the beltline. Grasp the handles and pull outward.(See fig. 3-41, G.)Apply maximum effort and hold for approximately 8 seconds. Relax for 5 seconds between repetitions; perform four repetitions. Moving the body closer to or farther away from the bar will change the stress from the upper arms to the forearms. The Marine obtains progression by lengthening the stress period to 10 seconds.

(8) Exercise 8: Isometric Compression. (See fig. 3-41, H.)Maintain the position as in exercise 7.Grasping the handles in the same manner, press in with maximum effort and hold for approximately 8 seconds. Relax for 5 seconds between repetitions; perform four repetitions. The Marine obtains progression by lengthening the stress period to 10 seconds.

(9) Exercise 9: Isometric Pres. Remove the lower horizontal bar. Adjust the upper horizontal bar until it is about 6 inches lower than the extended arms can reach. Stepping on the footplates at the bottom of the frame, grasp the bar with both hands and push up.(See fig. 3-41,1.)Keep both the legs and arms slightly flexed and the back straight. Apply maximum effort for 8 seconds, then relax for 5 seconds. Complete four repetitions. The Marine obtains progression by lengthening the stress period to 10 seconds.

(10) Exercise 10: Inclined Pushup. Assume the front leaning rest position with the feet on the bench.(See fig. 3-41, J.)Keeping the back and legs straight, lower the body until the nose touches the ground. By extending the arms, raise the body to the starting position. Repeat as many times as possible.

3903. THE 6-12 PLAN

a. Description and objective. The 6-12 plan of physical fitness has been developed to assist in regulating quantity and progression and to provide a convenient set of exercises. This is a basic program and will take 18 weeks to complete if you follow the moderate progression as to the time prescribed for each level of achievement. The time can be shortened as explained below. This plan consists of six basic exercises a day which can be completed in 12 minutes. There are six tables of six exercises each, thus allowing you to progress from table to table. The plan is progressive, fits any age group, contains balance and variety, and applies the principle of overload in a safe and gradual manner. Begin at Table I, Progression Guide, with the number of repetitions as indicated by age.

b. Formation. If just starting an exercise program, do not rush through the first table. Remember, individuals should remain at each level for about a week before moving upward. The time allotment stated for each exercise at the bottom of the tables is a guide; some people may take more and some less time on the individual exercises. At the end of a one-week period, when the individual can comfortable perform the six exercises in 12 minutes, move on to the next level. To a certain degree, the individual must be the judge of his ability to progress form level to level and table to table. If starting with a certain degree of fitness, some of the beginning tables may present little challenge.

c. Starting Level **and Progression.** There are three levels of achievement for each age group, indicated as A, B, and C. Start at the C level for the appropriate age group. At the end of a one-week period, or when the individual can do all exercises at that level within 12 minutes, progress to the B level. At the end of the second week, or when the individual can accomplish that level within 12 minutes, progress to the A level. At the conclusion of the third week or when the individual can achieve the A level within the time limitation, move on to table II.(See Tables I through VI, Progression Guide.)

d. Maintenance Level. Attempt to work through all six tables. If this proves to be too difficult, then maintain exercise at the—

(1) A-level on Table IV, Progression Guide it in the 45 to 49, 50 to 59, or over 60 age group.

(2) A-level on Table V, Progression Guide if in the 17 to 29, 30 to 39, or 40 to 44 age group.

e. Precautions. To achieve the maximum benefit, perform each exercise exactly as specified. Read the descriptions and study the illustrations. Do not slight the movements. Use a sensible approach and follow these four points as they apply before starting or during your exercise program.

(1) If you have the slightest doubt about your ability to participate in this exercise program, consult a physician.

(2) Stop immediately if you notice unusual breathlessness or chest pain while taking part in these exercises. If these conditions persist, consult a physician.

The following Tables of Progression Guides and Exercises are the progression of the 6-12 Plan program.

Table I. Progression Guide is available in the illustration section at the end of the book.

Table I. Exercises.is available in the illustration section at the end of the book.

Table II. Progression Guide.is available in the illustration section at the end of the book.

Table II. Exercises.is available in the illustration section at the end of the book.

Table III. Progression Guide.is available in the illustration section at the end of the book.

Table III. Exercises.is available in the illustration section at the end of the book.

Table IV. Progression Guide.is available in the illustration section at the end of the book.

Table IV. Exercises.is available in the illustration section at the end of the book.

Table V. Progression Guide.is available in the illustration section at the end of the book.

Table V. Exercises.is available in the illustration section at the end of the book.

Table VI. Progression Guide.is available in the illustration section at the end of the book.

Table VI. Exercises.is available in the illustration section at the end of the book.

(3) Unless you have exercised regularly and know yourself to be in good physical condition, start at table I with the C level appropriate to your age.

(4) If you are out of shape, admit that fact to yourself. Hide your pride; after all, you are in the privacy of your own quarters. Set your goal for the longer, steadier pull toward fitness. Resist the urge to pass over the lower numbered tables to find a table that will test your fitness. You are not trying to test, but rather to develop.

3904.WEIGHT TRAINING

Weight or barbell training should not be confused with the more common types of weight lifting used as a competitive sport. Weight lifting is designed to develop specific muscle groups so that the individual is capable of lifting a large amount during a single lift. In contrast, weight training is the systematic development of all the major muscle groups by the use of calisthenics reinforced with weight to provide resistance.

3905.WEIGHT LIFTING

a. Description and Objective. The weight lifting program is progressive and applies the principle of overload in a safe, gradual manner. The exercises of the table can be completed in 15 minutes. The objective of the exercises is to develop strength and muscular endurance, and muscle tone of the five major muscle groups:legs, arms, back, trunk, and shoulder girdle.(See fig. 3-42.)

b. Warmup. A warm up exercise is important to prepare the body for the more vigorous exercises that are to follow. Ten repetitions of the high jumper exercise are excellent for a warm up period. Also, an exercise which will require fast body movement is needed.To provide such exercise, 3 to 5 minutes of rope skipping is recommended to increase the individual's development potential.

c. Formation. Marines should take care in the completion of these weight lifting exercises. They should ensure that their back is straight during the lifting phase of all exercises. When exercises require assuming the standing position with the weight, they should always grasp the weight while in a squatting position and then rise to a standing position.

d. Progression. Each exercise has a starting number of repetitions and specified pounds of weight. After each 4th or 5th day of exercise, the Marine should increase the repetitions by one until he has reached the maximum of 10 repetitions. At this time, the Marine should increase the weight by 5 pounds. Then he should repeat the process again with the initial number or repetitions.

e. Weight Training

(1) Exercise 1:**Squat**(See fig. 3-42, A.)Starting level—six repetitions, 50 pounds (commonly called the flatfoot deep knee bend). Place the bar upon the shoulders. Stand with feet about 18 inches apart. Keeping the feet flat, lower the body into the low squat position. Stand erect and repeat. Exhale while lowering into the squat position and inhale while standing. This constitutes one repetition.

(2) Exercise 2:Waist Bender(.See fig. 3-42, B.)Starting level-6 repetitions, 40 pounds. Assume the standing position with the bar across the shoulders and the feet a shoulder-width apart .Bend forward at the waist until the upper body is parallel to the ground; return to the starting position. Each return to the upright position constitutes one repetition.

(3) Exercise 3:Curl (See fig. 3-42, C.)Starting level—6 repetitions, 40 pounds. Grasp the barbell with the palms facing to the rear and assume the standing position with the feet a should-width apart. With the barbell held in front of the hips, flex the elbows and lift the weight until the bar touches the upper chest. Lower the barbell back to the hip level position. Inhale deeply with the upward movement and exhale on the downward movement. Each time that the bar touches the chest will constitute one repetition.

(4) Exercise 4:Side Bender(.See fig. 3-42, D.)Starting level—6 repetitions per side, 40 pounds. Assume the standing position, with the bar across the shoulders, with feet a shoulder-width apart. Bend to the left as far as possible and return to the starting position. Repeat six times and then execute the same procedure to the right for six repetitions.

(5) Exercise 5:Standing Press(See fig. 3-42, E.)Starting level—6 repetitions, 45 pounds. Grasp the bar with the palms facing forward and assume the starting position. Curl the weight to the upper chest position. Inhale deeply and press the bar upward to an overhead position. Exhale while lowering the bar to the chest position. Each time that the bar is pressed upward constitutes one repetition.

(6) Exercise 6:Upward Row. .see fig. 3-42, F.)Starting level—6 repetitions, 40 pounds. Grasp the bar, hands close together, palms to the rear, and assume the standing position. Starting with the bar held in front of the hips, flexing the elbows and the shoulder girdle muscles, lift the bar straight up to an overhead position. Inhale deeply while lifting the bar. Exhale while lowering the bar to the hip position. Each time that the bar returns to the hips constitutes one repetition.

(7) Exercise 7:Shoulder Curl. See fig. 3-42, G.)Starting level—6 repetitions, 25 pounds. Grasp the bar, palms down, and assume the standing position. Keeping the elbows locked, curl the bar, pivoting the arms at the shoulders until the bar is in an overhead position and as far to the rear as possible. Return the bar in the same manner to the hip position. Each time that the bar returns to the hip position constitutes one repetition.

Figure 3-42 is available in the illustration section at the end of the book.

3906.ISOMETRIC EXERCISING

a. Description and Objective. Isometric exercising is the application of maximum effort during an exercise period. The isometric principle is to apply force gradually over a 5- to 10- second period until the maximum effort is applied. Relaxation follows for approximately 5 seconds and then force is applied again. The Marine continues this process at the prescribed level for each exercise for a period of 15 minutes or less. The objective of isometric exercises is to create muscle growth. Isometric exercises are the fastest means of creating muscle growth. They are founded on the fact that a muscle will grow only so fast regardless of the type or duration of the activity. The principle of overload—that the muscles develop commensurate with demand—reinforces that fact. However, isometrics will not develop cardiovascular or muscular endurance. Consequently, Marines who choose to use isometrics in their exercise routine must also include running or some other type of aerobic activity.

b. Area and Equipment isometric exercises may be designed to be performed with or without equipment.

c. Use With Other Programs. The isometric exercises presented in this section will provide a variety from which to choose. In addition to the exercises contained in this section, the application of isometric force is used in some of the exercises in section II.

d.Door Frame Exercises. The following exercises are designed for use with a standard door frame found in all offices or barracks.(See fig. 3-43.)

(1) Exercise l:Arm Press. Stand in the doorway with the legs straight, knees locked. Using the arm muscles, press hard upward against the top of the door frame. Repeat for three repetitions applying gradual effort to maximum contraction.

(2) Exercise 2:Leg Press. Stand in the doorway with the hands on the top of the door frame, elbows locked .With knees bent, press hard with the leg muscles .Repeat for three repetitions beginning with a gradual effort and increasing to maximum contraction. A low platform may be necessary to reach the top of the door frame and still maintain a bent knee position.

(3) Exercise 3:Side Press. Extend both arms to the side of the doorway. Palms are shoulder high, facing outward. With both arms, press hard against the sides of the door frame. Repeat for three repetitions. Begin gradually and increase to maximum contraction.

(4) Exercise 4:Lateral Raise. Extend both arms to the sides of the doorway, arms down, palms facing inward. With the back of the hands, press hard against the sides of the door frame .Repeat for three repetitions. Begin with a gradual effort and increase to maximum contraction.

Figure 3-43 is available in the illustration section at the end of the book.

(5) Exercise 5:Neck Press?lace the forehead against the door frame, hands clasped behind the back Using the neck muscles, press hard against the door frame. Repeat for three repetitions, then reverse position so that the back of the head is resting on the door frame. Again do three repetitions .Begin gradually with both exercises and increase to maximum contraction.

(6) Exercise 6:Door Pull Stand facing the edge of the open door and grasp the doorknobs. Pull outward with both arms (if doorknobs are not available grasp the edge of the door).Applying outward pressure, move the body toward and away from the door .Repeat the three repetitions. Begin with gradual effort and increase to maximum contraction.

Chapter 4
COMBAT WATER SURVIVAL
Section I. Marine Corps Water Survival Program

4101.MILITARY SWIMMING

Military swimming emphasizes strokes that result in staying power rather than speed. Marines need to swim easily, quietly, and with adequate vision. Strokes should be used that allow the Marine to carry basic combat equipment and to tow or push a wounded buddy while keeping the face out of the water to allow breathing. The best strokes to use are the side stroke and the breast stroke. Marines must be proficient in floating, drown-proofing, and in using their basic combat equipment to make flotation devices. They must be drilled on using the Kapok and Mae West buoyancy compensators and on emergency egress from helicopters, landing craft, and amphibious assault vehicles.

4102.PROGRAM DEVELOPMENT

Commanders of the Landing Force Training Commands (Atlantic/Pacific) are tasked with developing and maintaining the individual water survival and swimming training program. They are responsible for training all Marine Corps water safety and survival instructors and integrating this training in to their combat readiness training program. All Marines are deployable and as soldiers of the sea must constantly practice the skills necessary to achieve mission accomplishment. Commanders must be imaginative in program development, integrating basic skills development which enhances the total mission accomplishment. An example is teaching waterproofing of the Marine's backpack so that it can be used as a flotation device in streams or rivers. Once learned, this skill would be practiced first in a pool and then in a stream or a river. Routine inspections during deployments would ensure that backpacks could be used as flotation devices.

4103.QUAL STDS AND TRAINING GUIDELINES

The inherent nature of Marine Corps operations and training requires that Marines achieve an ability to survive in water. Water survival and swimming training is designed to reduce fear of water, instill self-confidence, and develop a Marine's ability to survive in water. As a minimum, every Marine should be qualified as swimmer, third-class (S3).This qualification should be met during recruit training or Officer Candidate School. Once qualified, a Marine need not be requalified unless requalification at a specific level (S3 or higher) is required by unit mission, military occupational specialty or duty assignment, or for other reasons determined by the commander. The following classifications and standards apply to Marine Corps water survival and swimming training. Abbreviations will be used for service, medical, and training record entries citing MCO 1510.29A, Individual Water Survival and Swimming Training, as authority.

a. Qualification standards and test procedures are described in paragraph 4104.

b. Swimmer classification and abbreviations are:

(1) Unqualified - UQ.

(2) Swimmer, third-class (S3).Minimum aquatic skill level for all Marines.

(3) Swimmer, second-class (S2)-

(a) Minimum standard for all naval aviators, naval flight officers, and aircrewmen with written waiver from the commanding officer.

(b) Requirements for military occupational specialty 1803 (assault amphibian vehicle officer), 1833 (assault amphibian vehicle crewman), 1302 (engineer officer), 1371 (combat engineer), 1381 (shore party specialist), and 1379 (engineer operations chief).

(4) Swimmer, first-class (S1)-

(a) Naval aviator, naval flight officer, and aircrewman qualification requirement.

(b) Requirements for military occupational specialty 0321 (reconnaissance man), 8652 (reconnaissance man, parachute jump qualified), 8653 (reconnaissance man, self-contained underwater breathing apparatus qualified), 8654 (reconnaissance man, parachute/self-contained underwater breathing apparatus qualified), 9952 (self-contained underwater breathing apparatus Marine (officer/enlisted)), and 9953 (parachutist/self-contained underwater breathing apparatus Marine (officer/enlisted)).

(5) Water Survival Qualified (WSQ).Ultimate water survival/swimming goal for Marines.

c.Instructor and assistant instructor classifications are-

(1) **American Red Cross Advanced Lifesaver.** Individuals who are currently certified as American Red Cross advanced lifesavers may assist qualified instructors during Marine Corps water survival/swimming training and testing. Certification is valid for 3 years.

(2) American Red Cross Water Safety Instruction. Individuals who are currently certified as American Red Cross water safety instructors may conduct swimming training and qualify individuals as 3d, 2d, and 1st class swimmers. Certification is valid for 2 years.

(3) United States Marine Corps Water Safety and Survival Instruction. Those personnel certified as water safety and survival instructors by one of the landing force training commands may conduct Marine Corps water survival and swimming training and American Red Cross swimming and water safety training. These instructors may also participate as assistant instructors during the conduct of landing force training command-sponsored water safety and survival instructor courses. Certification is valid for2 years.

(4) American Red Cross Water Safety Instructor Traine. Individuals qualified who have been certified as instructor trainers by the American Red Cross may train American Red Cross water safety and basic swimming instructors. Certification is valid for 2 years.

(5) United States Marine Corps Water Safety and Survival Instructor Traine. Individuals qualified who are certified as instructor trainers through special training courses conducted jointly by the American Red Cross and the Marine Corps, under the direction of the Commandant of the Marine Corps, may train water safety and survival instructors and American Red Cross water safety and basic swimming instructors. Certification is valid for 2 years.

d. Training Guidelines Instructors engaged in training Marines as swimmers will use the current edition of the American Red Cross swimming and water safety courses instructors' manuals as the authoritative textbooks for swimming procedures, strokes, breaks, and holds.

4104.WATER SURV/SWIM QUAL STD AND TEST PROCS

a. The following qualification standards and test procedures are applicable to Marine Corps water survival and swimming training. The utility uniform without cover, boots/oxfords or socks will be worn for all swimming tests through swimmer, first class. The utility uniform with boots/oxfords will be worn as prescribed for the water survival qualification.

(1) Swimmer, Third-Class (S3) To qualify as a swimmer, third-class, a Marine must enter the water feet first from a minimum height of 5 feet and remain afloat for 5 minutes. During this time the Marine must swim 50 yards using any stroke or combination of strokes. This test should be taken by all Marines as early as possible in their initial training. Those unable to pass this test will be classified as UQ and should be given instruction in fundamental swimming skills. Those who barely meet requirements are swimmers who need help and should receive additional instruction in fundamental swimming skills, whenever possible.

(2) Swimmer, Second-Class (S2.) To qualify, a Marine must enter the water feet first from a minimum height of 10 feet and remain afloat for 10 minutes. The Marine must swim 100 yards and use the three basic survival swimming strokes (side stroke, backstroke, and breast stroke) for a minimum distance of 25 yards each.

(3) Swimmer, First-Class (S1)A prerequisite to qualification as swimmer, first-class, is successful completion of the test for swimmer, second-class. To qualify as a swimmer, first-class, a Marine must be able to do each of the following:

(a) Approach a person of approximately the same size while in the water, demonstrate a release, get that person in a carry position, and tow the "distressed swimmer" 25 yards.

(b) Enter water feet first and immediately swim underwater for 25 yards. Swimmer is to break the surface twice for breathing during this distance at intervals of approximately 25 feet.

(c) Remove trousers in water, inflate for support, and remain motionless for a minimum of 1 minute.

(d) Swim 220 yards using any survival swimming stroke or any combination of survival strokes.

(4) Water Survival Qualification (WSQ)A prerequisite to water survival qualification is successful completion of the test for swimmer, first-class. To be water survival-qualified, a Mannemust:

(a) Enter water by jumping from a height of 10 feet.

(b) Stay afloat in full utility uniform (less cover) for 1 hour. Boots/oxfords will be removed after 5 minutes in the water but will be retained by the swimmer.

(c) Traverse 75 yards in deep water in full utility uniform (less cover) with boots/oxfords.

b. All water survival and swimming training will include instruction in the following:

(1) Employment of safety lines or other comparable expedients to assist in swift stream/river crossings.

(2) Adverse physiological effects caused by cold water (hypothermia) and the precautionary measures to be taken prior to exposure to such an environment.

(3) Employment of a standard and expedient flotation devices in a water survival situation.

(4) Removal of individual field equipment after unscheduled water entry. Entry will be made from a minimum height of 10 feet with a complete set of field equipment, properly packed as an expedient flotation device and waterproofed. The ALICE pack will then be removed, the rifle laid across the top of the pack, the helmet remaining or put back on the individual's head, and the individual will traverse 25 yards to the side of the pool or water's edge.

c. The additional water survival qualification (flight physiology training) and requalification required of personnel assigned to flight status will be conducted in accordance with OPNAVINST 3710.7_,Promulgation of NATOPS General Flight and Operating Instructions. Swimming qualification (first-class, SI) is a prerequisite to water survival (flight physiology) training and is a one-time requirement.

Section II. Water Sports and Competitive Activities

4201.GENERAL

The commander, when developing the combat readiness program, needs to look carefully at the mission. If the mission requires Marines who must perform mission-essential water-related activities of long duration, he may want to look to water sports/competitive activities as a method of developing endurance. Endurance and confidence can only be developed by extended time in the water. While defined training can accomplish this goal, the goal can be enhanced through training which is integrated with the physical conditioning program. This approach will provide variety and enthusiasm. Two activities which will accomplish this goal are water relays and water polo. The important consideration is that the unit achieve total participation by all members of the unit.

4202.WATER RELAYS

Development of water relays will generally conform to the guidelines in chapter 5, section IV. When contemplating relays, the commander should review chapter 5 and adapt the principles to the size of the unit and the facility to be used.

4203.WATER POLO

The official rules provided in this paragraph will need to be adapted to the unit's unique situation, its size, and the size of the facility. Commanders should make every effort to have all members of the unit participate.

a. NCAA Rules This style of play is based on two popular sports: swimming and basketball. Playing is done on the surface of the water by teams of seven players each. The size of the playing area is 25 yards by 14 yards, which is the size of the typical high school or college indoor pool. The deeper the water, the better. If the entire playing area is deep-6 feet or more—this is ideal, although most indoor pools have a shallow end. At each end of the pool is a goal. In deep water, the goal is 10 feet across and 3 feet high. In shallow water, where the goalie can stand on and jump from the bottom, the goal is 10 feet across and 5 feet high. One of the seven players on each team is the goalie. He is the only player on the team who can stand on or jump from the bottom (if it is shallow enough to permit this), catch and pass the ball with both hands at the same time, or hit the ball with a clenched fist. However, the goalie may not swim across or pass the ball across the middle of the pool.

b.Equipment. The goalie on the visiting team must wear a white cap with the number 1 clearly marked on it; his teammates, who are designated as guards and forwards, must wear white caps numbered 2 through 7, with the substitutes wearing white caps numbered 8 through 21.The goalie on the home team must wear a dark-colored cap with the number 22 clearly marked on it; his teammates designated as guards and forwards, must wear similar dark-colored caps numbered 29 through 42.The guards and forwards on each team may swim freely up and down the pool as they see fit, interchanging positions as often as they wish. They may not stand on or jump from the bottom, touch the ball with more than one hand at a time when catching or passing or shooting it, or enter inside the opposing team's 2-yard line unless preceded by the ball. The ball is similar to a soccer ball, except that it is yellow and covered with a waterproofing substance that makes it easier to handle with one hand.

c. Playing Time. In intercollegiate competition the game consists of four 7-minule quarters, the teams changing ends after every period of play. In high school competition the game consists of four-,five-, or six-minute quarters, depending on the ages of the participants. Here again the teams change ends after every period. There is a very brief interval between each quarter. At the start of each period the teams line up at their respective ends, and with ablast of the whistle, the head referee throws the ball into the middle of the pool. The fastest swimmers on each team then race to gain possession.

d. Officials. The head referee walks along one side of the pool watching for infractions. He is aided by the assistant referee, who patrols the opposite side of the pool. The referees are both equipped with whistles and two official's flags. These flags should be 12 inches square and should correspond in color to the caps worn by the respective teams. Whenever a referee sees an infraction, he blows his whistle and signals with his flag indicating that the player nearest to the ball with the appropriate colored flag will be given possession of the ball. Whenever the whistle is blown, all players should react by looking immediately to see which color flag is being waived which will indicate which team should be moving into the defense. Other officials are a timer and a scorer. As in all sports, it is important that the officials be skilled and competent.

e. Technical Fouls. The following are some of the common technical fouls: starting before the referee blows his whistle to open play; holding onto or pushing off from the side of the pool during play; taking hold of the ball underwater when tackled by an opposing player; swimming inside an opponent's 2-yard line unless preceded by the ball; touching the ball with both hands at the same time (goalie excepted); standing, walking on or jumping from the bottom when taking an active part in the game. When a technical foul occurs, thereferee blows the whistle and with his flag awards possession. The player on the team awarded possession nearest the point of infraction then has 5 seconds to put the ball into play by passing to a teammate or by dropping the ball into the water and swimming after it.

f. Personal Fouls. The following are some of the more common personal fouls: committing any of the technical fouls mentioned above for the purpose of scoring or preventing a goal; holding, ducking, pulling, pushing off from, swimming over, or impeding the arm or leg movement of any opponent who is not touching the ball; and splashing water in the face of an opponent. When a personal foul occurs, the referee will follow the same procedure as with a technical foul. While doing this, the referee will also call out clearly and loudly the number of the player who was guilty of the infraction and a personal foul will be marked against that player by the official scorer. When a player accumulates five personal fouls, he fouls out of the game and must be replaced by a substitute.

g. Penalty Shot. A penalty shot can be awarded by either of the officials or the scorer when:

o An offensive player inside the opponents' 4-yard line not touching the ball is held, ducked, pulled-back, kicked, or struck.

o A team has accumulated a total of ten personal fouls. When the former occurs, the head or assistant referee should immediately blow the whistle and by holding a two-flagged stick in a vertical position above his head, signal that a penalty shot has been awarded. When the latter occurs, the scorer should use a buzzer located at the scorer's tabic to signal that one team has accumulated ten personal fouls, thereby entitling the other team to a penalty shot. A penalty shot is taken from the 4-yard line in front of the goal. All players except the defending goalie must leave the 4-yard line until the shot is taken. No player can be within 1 yard of the shooter. After ascertaining that the shooting player is on the 4-yard line and the goalie is onThegoal line, the referee will ask the shooter to lift up the ball. When he docs, the referee will give a sharp blast from the whistle. At the whistle, the shooter must shoot without delay and without any faking at the goal. The goalie may try to block the shot. If he does, or the shot is otherwise missed, it is immediately in play and action continues.

h. When a Goal is Scored. When a goal is scored, either from a shot taken by a player out in the field or by penalty shot, the defending goalie must then pass the ball to the nearest referee. The referee will then pass the ball back to the goalie, who puts it into play by a pass to one of his teammates as soon as possible.

Section III. The Battle Swimming Test

4301.GENERAL

The battle fitness test is the evaluation contained in the Guidance for Basic Warrior Training (BWT) as part of the Basic Warrior Training Concept Plan.

4302.THE BATTLE SWIMMING TEST

The battle swim test consist of the following:

a. Be able to climb a three meter high board, while wearing a field uniform, carrying a M-16, web gear, and four canteens. Jump off blindfolded, recover and swim 100 meters with the M-16.

b. Repeat the same process, without the blindfold, and upon entering the water a second time, recover and retrieve all submerged gear.

c. Remain afloat and tread water for five minutes.

d. Demonstrate a knowledge of how to cross a stream using field expedient techniques and a single strand rope bridge.

e. Successfully complete helicopter egress training.

Chapter 5
COMPETITIVE CONDITIONING ACTIVITIES
Section I. Organization of Competitive Activities

5101.LEADERSHIP OF COMPETITIVE ACTIVITIES

a. Place in the Program. Competitive conditioning activities consist of dual combatives, relays, team contests, and team sports in which individuals or teams compete against an opponent to win. Competition is one of the best ways of maintaining interest in the physical training program during the sustaining stage. Organized competition provides enjoyable, vigorous physical activity that has proved to be one of the best supplements to conditioning drill activities. The benefits of competition are the development of aggressiveness, unit pride and identity, teamwork, and the will to win. These activities help develop cardiovascular and muscular endurance, strength, and coordination. Scheduling competitive activities in an orderly and progressive manner is desirable. The progression should be from relays to dual combatives, to team contests, and finally to team athletics. These activities should be conducted as part of the program after a basic level of conditioning has been developed, usually during the slow-improvement stage of conditioning. Muscles and joints should be strengthened by preconditioning to withstand the strain placed upon them by sudden stops and turns, body contact, bearing of weight, and falls. Competitive activities, however, should not be allowed to dominate the physical training program.

b. Area and Equipment. Some of the competitive activities included in this chapter require specific types of areas and equipment; others do not. The area requirement can usually be satisfied on available training fields. When items of equipment are required or specific courts or field layouts are to be marked off, such information is included.

c. Leadership. The principal factor for success as a leader of competitive activity is an energetic, dynamic, enthusiastic approach. The leader's attitude is reflected by the group, so the commander must carry on the activity in a snappy and vigorous manner. Confidence on the part of the commander will create an impression of decisiveness and certainty. Confidence grows out of experience and a thorough knowledge of the activity. Mastery of subject is the first step in developing confidence, assurance, and poise.

(1) The following suggestions are offered for leaders of competitive combatives, contests, and sports:

(a) Get the activity underway quickly by selecting and teaching only the essentials.

(b) Use rules to add to the enjoyment of the activity and do not allow them to interfere with the spirit of competition.

(c) Stop the activity before interest begins to lag.

(d) In team contests, clearly distinguish sides by attempting to maintain unit identity.

(e) Always insist on fair play, enforcing the rules impartially.

(2) The following procedure is recommended for presenting a competitive activity.

(a) Name the activity.

(b) Briefly explain the objective of the activity and give only the pertinent rules.
(c) Have a demonstration at slow speed and answer questions.
(d) Organize groups into teams and appoint captains.
(e) Arrange teams in the proper starting positions.
(f) Conduct the activity.

d. Competitive Units. Units for competition should be the same organizations in which Marines train: battalions, companies, platoons. In most situations, the unit is the squad.

e. Provisions for Instruction. One of the most effective methods of maintaining interest and participation in competitive activities is to provide instruction in activities with which most Marines are unacquainted. Such instruction can be conducted during regular physical training periods. Careful planning is required to keep all individuals continuously engaged in vigorous activity. The materials in the following sections should be used as a guide for instructional purposes.

f. Officiating and Control. Every effort should be made to provide good officials for all competitive activities. Poor officials quickly cause dissatisfaction among participants in team activities and create a situation which can turn an organized activity into a brawl. Each company should have or develop several competent and qualified officials available for games on company and platoon levels.

5102. MILITARY FIELD MEETS

When units reach the latter part of the slow-improvement and sustaining stage of training, interest in the program may lag. A change in course content can arouse the desire to participate. An event which does not require a high degree of skill yet demands strenuous activity is ideal for the military field meet. This is a series of team contests conducted on a station-to-station basis during a given period of time. Team contests carried on simultaneously provide essential training, vigorous exercise, and stimulating competition in an atmosphere that is enjoyable for all. Because of the healthy rivalry that a field meet arouses, it is an excellent form of inter-unit competition.

a. Objective and Advantages. The objective of the military field meet is to provide activity for everyone in the participating units. Activities are chosen that will develop aggressiveness, teamwork, a will to win, and competitive spirit, and that will stimulate interest and build esprit de corps. The military field meet can be included as part of the physical training program or as part of the off-duty recreational program. It is a form of contest that can be conducted in nearly all circumstances because it can be easily modified, requires a minimum amount of equipment, and can be readily organized.

b. Level of Competition. The military field meet is flexible. It can be adjusted to large or small groups. A company-size unit is the most desirable, but it may also be administered within a larger unit. If it is conducted within a company, the participating units will be squads. If it is held within a larger unit, platoons compose the teams. A larger-size unit requires more extensive organization and administration.

c. Selection of Events. In organizing a military field meet, select events that are simple and easy to administer. All rules and regulations should be clearly understood by everyone. No event should require previous practice. In selecting the events, consider the interest and capabilities of the Marines and available equipment and facilities. Select events which will require that all members of the units participate. Events should not be dominated by the athletes; rather, MOS-related events which have significance in producing skills and combat readiness should be used.

d.Equipment and Facilities. The site must be large enough to permit events to be grouped about a central control point. Use available facilities such as parade decks, Softball fields, or tracks. A public address system is desirable at the control point for the initial orientation of teams and for subsequent announcements of time lapses, cumulative scores, and final standings. A tally board is necessary so that all units can constantly monitor how the events are progressing.

e.Administrators. Efficient administration of the military field meet depends on the referees, judges, and scorers. For these positions, choose individuals within the units who have had athletic or officiating experience. Prior to the day of the event, all administrators should be briefed and assigned a specific task in order that they may become familiar with the rules and organization of the contests they will conduct. The following officials should be available:

(1) A primary instructor or supervisor who is in charge of the control center and who is responsible for the successful operation of the athletic carnival. The supervisor must have an assistant to act as a timer and scorer.

(2) One assistant instructor in charge of each event. This person should be-

(a) Familiar with the rules of the event.

(b) Effective as a leader to ensure proper supervision and control over the participating teams.

(c) Enthusiastic to provide proper motivation.

(d) Self-confident of the ability to judge infractions of the rules. The assistant instructor must be fair in judgment and penalize without hesitation when infractions occur.

(3) Runners between event stations and the control point. These individuals collect and deliver scores.

f. Team Organization for Competition. The size of the teams is determined by the level on which the field meet is organized. Maintaining the integrity of the unit promotes esprit de corps, but this does not preclude grouping two squads into one team.

g.Conduct of Events

(1) The assistant instructor at each station takes charge of the group and gives a brief explanation of the major rules of the event. The assistant instructor speaks clearly and distinctly from a position to be seen and heard by all. A short demonstration is desirable if it will help clarify the event.

(2) The assistant instructor should make certain that teams can be clearly distinguished by the use of, for example, T-shirts and fatigue jackets, or caps and no cap, or colored jerseys. The assistant instructor should teach a whistle response (teams stop play immediately upon hearing whistle), get the event started as quickly as possible, and make any necessary corrections as the contest progresses. The rale of good officiating is to use a minimum of calls, yet maintain control of the contest. Penalize when necessary, but refrain from disqualifying contestants or teams.

(3) Keep the activity moving as rapidly as possible. When the central control point sounds the whistle to stop the play, all competition ends immediately. The assistant instructor then assembles the group, forwards the team scores to the central point, and, upon a signal from the central control point, rotates the teams to their next station. It is essential that rotation and orientation of teams be quick and orderly to reduce time spent between contests.

(4) Upon completion of the final event, assistant instructors move their teams to the central control point for the announcement of winning teams, presentation of awards (if any), and final critique.

h. Scoring System. The system for determining the winner of the military field meet should be simple and efficient. At the completion of competition, the scorer totals the points that each team has scored in all contests. The scorer then subtracts the number of points scored against a team (penalty points for nonparticipation,etc.) from the total. The resulting scores are placed in a column with the highest score at the top and the lowest at the bottom. The team with the highest total is the winner. This type of scoring system encourages a team to ensure total unit participation.

Section II.Combatives

5201.DESCRIPTTON AND OBJECTIVE

Combatives are strenuous, short competitive contests in which two individuals attempt to overcome each other in a bout of skill and strength. These contests help to develop Marines' resourcefulness, confidence, strength, agility, coordination, and will to win. Any level ground area can be used. However, extremely hard ground should be avoided as some of the combatives require ground contact. A whistle is needed to control the bouts during the competition since voice commands may go unheeded. The extended rectangular formation is used for dual combatives. Combatives are conducted on an informal basis. Marines are allowed to remain at ease between activities and are allowed to brush themselves off after being on the ground. To get the most out of combatives, individuals must be urged to overcome their opponents as quickly as possible as would be required in combat.

a. Benefits for Marine Training. The possibility of close contact with an enemy in combat faces Marines at all times. They must be trained to react aggressively and violently. Combatives may be used as an introduction to such hand-to-hand contact and should be followed by hand-to-hand combat training. Marines enjoy competition, and this type of activity is a welcome change from formal conditioning activities.

b. Instructor Responsibility

(1) The instructor tells the individuals that all combatives begin and end on the whistle signal. The instructor demonstrates each activity before having it performed, explaining it in simple terms.

(2) After stopping one activity, the instructor gets everyone in place for the next.

(3) The instructor must closely supervise combatives to insure that contestants do not use unfair or unsportsmanlike tactics. To avoid unnecessary injury, instructors must see that the bouts are closely controlled and opponents equally paired. Adjustments should be made in apparent cases of mismatched abilities.

5202.COMBATIVES TABLES

There are three tables of combatives, each of which can be completed in 15 minutes. These tables become progressively more difficult from lower to higher numbers.

a.Combatives Table I

(1) Open-Hand Slap Boxing(See fig. 5-1, A. individuals assume a boxer's stance, palms open, fingers extended and joined. Each contestant tries to slap the opponent about the head and upper body with the open hand. This is a good warm- up activity.

(2) Wrist Tug-of-War(See fig. 5-1, B.)Two individuals sit on the ground with the soles of their feet in contact. Each grasps the opponent's wrists so that the hands are directly over the feet. At the whistle, each individual tries to pull the opponent from sitting to standing position.

(3) Arm-Lock Wrestling.(See fig. 5-1, C.)Two individuals sit back-to-back with legs spread and arms locked at the elbows. Each right arm is inside the opponent's left. Each tries to force the opponent's left arm or shoulder to the ground. The individual who first wins three bouts is the winner.

(4) Bulling(See fig. 5-1, D.)Two individuals assume the Westmoreland wrestling hold, each grasping the opponent's neck with the right hand and the opponent's right elbow with the left hand. Each tries to force the other to move one foot by pushing, pulling, or otherwise manipulating. The individual who first wins two boutsis the winner.

(5) Indian Wrestling(See fig. 5-1, E.)Two individuals lie on the ground, side by side, with their heads in opposite directions. They link right elbows. On the instructor's signal or by mutual agreement, each individual raises the right leg approximately straight and far enough to engage the opponent's heel. To start the contest, each person usually raises the leg three times rhythmically and on the third time engages the opponent's heel. Each tries to roll the other over backward. The right leg is used for three bouts, then the left leg for three bouts.

Figure 5-1 is available in the illustration section at the end of the book.

b. Combatives Table II

(1) Wrist Bending.(See fig. 5-2, A.) Opponents pair off, face each other, raise their arms forward, and, with palms forward, interlock their fingers. At the starting signal, each individual attempts to bend the opponent's wrist. The arms are kept up and forward and are not allowed to swing around and down to the sides. The individual who first wins two bouts is the winner.

(2) Back-to-Back Push.(See fig. 5-2, B.)Two individuals stand back-to-back with elbows locked. Each right arm is inside the opponent's left arm. At the starting signal, each pushes backward, trying to move the opponent. Opponents are not allowed to lift and carry each other; only pushing is permitted. The one who pushes the opponent the farthest wins the bout. The individual who first wins two bouts is the winner.

(3) Hop and Pull Hand.(See fig. 5-2, C.) The individuals are matched in pairs. Each grasps the opponent's right hand and, hopping on the right foot, attempts to pull the opponent off balance. Contestants automatically lose if they touch their free hand or their lifted foot to the ground. For successive bouts, they alternate hands and feet.

(4) Westmoreland Wrestling. (See fig. 5-2, D.)Each contestant grasps the back of the opponent's neck with the right hand and the opponent's right elbow with the left hand. In this position, each attempts to pull, push, or force the opponent to touch the ground with any part of the body other than the feet. The individual who first wins two bouts is the winner.

(5) Crab Fight.(See fig. 5-2, E.)Two individuals sit on the ground facing in opposite directions with their hands on the ground behind them. At the whistle, they raise their hips and push with their shoulders and bodies, each trying to make the other's hips touch the ground. The individual who first wins two bouts is the winner.

Figure 5-2 is available in the illustration section at the end of the book.

c. Combatives Table III

(1) Hand Wrestling.(See fig. 5-3, A.) Two individuals stand facing each other. Their right feet are forward and braced side by side. They grasp right hands for the first bout, left hands for the second bout. Each pulls, pushes, makes sideward movements, and otherwise maneuvers to force the opponent to move one or both feet from the original position. The contestant who first wins two bouts is the winner.

(2) Back-to-Back Tug.(See fig, 5-3, B.) Two individuals stand back-to-back with both arms linked at the elbows. Each has the right arm inside the opponent's left arm. At the starting signal, each attempts to pull the opponent. Lifting and carrying are permitted. The contestants must maintain their original direction and keep their arms linked. After a predetermined time, the player pulled or carried the farthest is the loser.

(3) Wrestling to Lift off Feet.(See fig. 5-3, C.) Contestants face each other. Each places the right arm under the left arm of the opponent and around the body. The left arm is over the opponent's right shoulder. Each tries to lift the other off the ground. The individual who first wins two bouts is the winner.

(4) Arm Pull Between Legs.(See fig. 5-3, D.) Two individuals are paired off, back-to-back. Each bends forward and, extending the right arm between the legs, grasps the opponent's right wrist. At the starting signal, each person attempts to pull the opponent After a predetermined time, the player who has pulled the opponent the farthest is the winner of the bout. The one who first wins two bouts is the winner. Repeat with the left hand and then both hands.

(5) Rooster Fight.(See fig.5-3, E.) Each contestant grasps the left foot with the right hand from behind, and the right arm with left hand. Each hops on the right foot, and, by shoulder-butting the opponent or by feinting and sudden evasions, forces the opponent to let go of the foot or arm. The contestant who first wins two bouts is winner.

Figure 5-3 is available in the illustration section at the end of the book.

Section III .Relays

5301. DESCRIPTION AND OBJECTIVE

Relays are races in which each member of a team runs one leg of the race. The team effort decides the winner. Relays provide stimulating competition and contribute to the conditioning of personnel. They also develop aggressiveness, team spirit, and the will to win. Relays should be dispersed throughout the program for short periods of time to provide a change of activity.

a. Team Organization. Relays are conducted most efficiently in platoon-size groups. Teams of equal size must be organized. Competitive spirit is encouraged and teams are organized better by basing teams on units such as squads, crews, or sections. Team captains should be designated. Extra Marines may be used as officials. The number of individuals on a relay team should be limited to squad size. If larger teams are used, the runners will spend too much time awaiting their turns and too little time actually participating. Two to six teams are ideal for relay competition. It is difficult to keep track of winners when more teams compete.

b. Administration of Relays

(1) The time spent on any one relay should be relatively short. If one team achieves a substantial lead in a long relay, the competitive spirit and enthusiastic participation of the other teams may decrease. Several short relays are generally better than one long relay.

(2) To maintain competitive spirit throughout a number of relays, determine the teams that win, place, and show in each relay and their total points for all relays. This can be done by awarding points to all teams on the basis of position at the finish of each relay. The team with the greatest number of points is the winner of the entire set of relays.

(3) Difficulties commonly encountered in conducting relays may be avoided by the following procedures:

(a) The last player in a relay race should be conspicuously identified. For example, the last runner can tie a handkerchief around the head or arm, take off the shirt, put on a hat or take it off, or use some other means.

(b) Another way to keep track of the progress of the race is to have each player sit or squat as soon as each is finished.

(c) Judges at the starting line can keep the runners from starting too soon.

(d) To prevent contestants from turning before they run the full distance, they should be required to run around a peg, pole, or assistant instructor.

(e) Batons, handkerchiefs, tent pegs, or other objects should be passed from one runner to the next when relays are run on a circular track.

(4) Before a unit's first participation in a relay, inform participants of the rules and scoring system. Violation of the rules should not result in disqualification. Instead, impose point penalties. A point penalty is imposed by subtracting a point from the team total at the conclusion of the relay.

(5) Careful administration will prevent most violations. For efficient conduct of relays, follow this procedure:

(a) Announce the name of the relay.

(b) Form the individuals in relay position.

(c) Briefly explain the relay and the rules for running it.
(d) Demonstrate.
(e) Have a definite finish line, and ensure that all know where it is.
(f) Answer questions.
(g) Conduct the relay.
(h) Determine winner and award points.

5302. RELAY TABLES

a. Events. The following relays are grouped into a table of activities. Each relay table can be completed in 15 minutes. Thus, relays can be used as a sole activity or as a part of a longer period. Each table is planned for a platoon-size group (30-60Marines). Adequate warmup for participants is provided by conducting several repetitions of exercises 1 and 2 of a conditioning drill. The recommended relay tables require an area 40 by 60 meters in size. Each table provides a variety of activity. The tables are progressive in the overload applied and should be scheduled in numerical order although not necessarily on successive days.

b. Relay Table I
(1) **60-Meter Lane Relay** (See fig. 5-4, A.) Each team is assembled in single file behind the starting line. On signal, the first individual of each team runs to the turn-around line 30 meters away, then runs back and touches the next teammate waiting at the starting line. The winning team is the first team to get its last member across the finish line. If an individual starts before being touched by the preceding runner, the team may be penalized.
(2) Wheelbarrow Race.(See fig. 5-4, B.) The players of each team pair off and line up in single file behind the starting line. The first individual walks on the hands while the partner grasps the ankles. They advance to the distance line (25 meters) behind which they exchange positions and return to the starting line. The rear individual must always hold the partner's ankles. After the first pair returns across the starting line, the next pair starts.

Figure 5-4 is available in the illustration section at the end of the book.

(3) Squad Front Relay .(See fig. 5-4, C.) The teams form in a line along the starting line with a 10-foot interval between teams. The members of each team lock elbows so that they are linked together. At the starting signal, the teams run to the distance line (20meters) where the left flank individual acts as a pivot. The team swings around on the pivot and returns to the base line. If a team breaks its links, it must reform before continuing. The first team to completely cross the base line intact is the winner.
(4) Crab-Walk Race.(See fig. 5-4, D.) The players of each team line up in single file. The first person of each team assumes the crab-walk position with feet forward on the starting line, hands on the ground behind it. At the starting signal, contestants move forward to the distance line (10 meters). They touch the line with their feet and then return to the starting line in the reverse position with the head and hands leading. The second person may not start until the first individual touches the finish line.

c.Relay Table II

(1) 100-Meter Lane Relay.(See fig. 5-5, A.)This relay is conducted exactly as the first relay of table 1 except that the start and turn-around line are 50 meters apart. This relay provides progression in sprinting.

(2) Frog-Jump Relay. (See fig. 5-5, B.) Each team lines up in single file. The first individual assumes a squatting position on the starting line. At the starting signal, the contestant progresses to the distance line (15 meters) and back by leaping forward, catching the weight on the hands, and bringing up the legs to the squat position for the next leap forward.

(3) Simple Relay. (See fig. 5-5, C.)Each team lines up in single file. Place a marker on the distance line (20 meters) in front of each team. Each team member, behind the first one, grasps the belt of the individual ahead. At the starting signal, each team runs as a unit to the marker, circles it, and returns to the starting line. The first team to completely cross the starting line intact is thewinner.

(4) Fireman's Carry Relay.(See fig. 5-5, D.)The players of each team line up in pairs, one behind the other. One individual in each pair carries the partner to the distance line (30 meters), using the fireman's carry. At the distance line, individuals exchange places and return to the starting line. As a variation, the person to be carried lies on the ground. The one carrying lifts the partner to the proper position. This relay may be performed with the other carries described in guerrilla exercises in chapter 3.

d.Relay Table III

(1) 200-Meter Circle Relay .(See fig. 5-6, A.)A course is laid out in a circular, rectangular,or oval pattern that is 200 meters around. Each team provides one runner on the starting line. On signal, the runner races around the 200-meter track and touches the next teammate, waiting at the starting line, who runs the same course. Each team member runs one lap of the course.

Figure 5-5 is available in the illustration section at the end of the book.

(2) Bear-and-Crab Race.(See fig. 5-6, B.) Each team lines up in single file. At the signal to start, the first contestant in each column assumes the bear-walk position (hands on ground in front), walks to the distance line (15 meters), and then runs back to the starting line. The returning contestant touches off the second individual and goes to the rear of the line. The second contestant assumes the crab-walk position (hands on ground behind) and crab-walks with feet leading to the distance line. This contestant also runs back to the starting line and touches off the next individual, who walks bear fashion. The rest of the members of each team alternate in this manner. The relay ends when the first person is back at the head of the line.

(3) Pilot Relay.(See fig. 5-6, C.) The players are grouped in threes, arms interlocked at the elbows, and outside players facing backwards. The middle person runs forward; the two outside individuals run backward. They run to the turning point (15meters), where they start back, this time with the middle person running backward and the two outside individuals running forward. The next set of three players starts when the first set crosses the starting line.

(4) Saddle-Back Relay.(See fig. 5-6, D.) Mark two parallel lines 15 meters apart. Each team selects a rider. The remaining members of each team count off. The even-numbered players from each team form in single files behind one line and the odd-numbered players from each team form in single files behind the other line directly across from their teammates. At the starting signal, the rider mounts the back of the first player of the team who carries the rider across the other line. There the rider changes mounts to the first player in the second line without touching the ground. This person carries the rider to the next player waiting in the first line. The relay continues until all of the mounts have carried the rider. A rider who falls off must mount again at the point of the fall. A rider who falls in changing mounts must get back on the original mount before making the change.

e. Relay Table IV

(1) 100-Meter Circle Relay (See fig. 5-7, A.)A course is laid out in either a circular, rectangular, or oval pattern that is 200 meters around. Each team is divided in half with each half positioned at starting lines on opposite sides of the track. Each runner races halfway around the track and touches a teammate who completes the lap. Each runner then waits in file at the halfway point until touched by the runner and then completes the second half of the lap. The first team to return all runners to their original starting line is the winner.

Figure 5-6 is available in the illustration section at the end of the book.

(2) In-and-Out Relay.(See fig. 5-7, B.) Each team lines up in a file with players 2 meters apart. At the starting signal, the first player runs back through the column in a zigzag fashion. He alternates, going to the right of one teammate and to the left of the next. Upon completing the run, the player lines up 2 meters behind the last one in the column. As soon as the first runner has passed the second person, the latter starts to run. This continues until all the players have realigned their original order. The team that finishes first is the winner. It may be desirable to have this relay continue until all individuals have run through their entire team two or three times in succession.

(3) Circle Race.(See fig. 5-7, **C.**) Each team forms a circle and holds hands with all individuals facing out except one who faces in and is the "driver."At the starting signal, the teams race to the distance line (20 meters) and back, keeping the circle intact. All the individuals in the circle must completely cross the distance line. The "driver" gives directions and orders. When the circle breaks, it must be reformed before it can continue. The first team completely over the starting line is the winner.

(4) Horse-and-Rider Relay.(See fig. 5-7, D.)Each team lines up in a single file. At the signal to start, the second player in each column leaps upon the back of the first who carries the rider across the distance line (30 meters). At the distance line, the rider dismounts and runs back to the starting line. There the rider picks up the third individual in the column, and carries this player to the distance line where the first player has remained. This continues until the last person is carried across the distance line.

Figure 5-7 is available in the illustration section at the end of the book.

Section IV. Team Contests and Athletics

5401. DESCRIPTION AND OBJECTIVE

Team contests are competitive activities in which Marines as a team compete with another team to win. They are guided by simple rules and organization. The function of team contests is to provide competition and an opportunity for body contact and to contribute to the development of physical readiness. In competing and working together as a team, individuals develop aggressiveness, the will to win, and teamwork. Team athletics deserve a prominent place in the physical training program because they contribute to increased combat efficiency. Because of the competitive nature of athletics and their natural appeal, individuals take part with enthusiasm. Athletic teams formed at intramural and higher levels are a strong unifying influence and provide one of the best means of developing esprit de corps.

a. Area and Equipment A level training field is sometimes the only area required. Many contests need no equipment. In contests requiring equipment, the need is for standard items such as logs, balls, nets, goals, and similar types of equipment. Specific requirements for area and equipment are listed with each contest description.

b. Progression. Team contests of a strenuous nature should be introduced after a basic period of conditioning has been completed and individuals are in the slow improvement stage of conditioning. Progression can take place from the less active to the more vigorous contests, and then from the non-contact to the contact or combative-type contests.

c. Necessity for Preconditioning Individuals must undergo conditioning prior to participation in athletics. Muscles, organs, joints, and ligaments not accustomed to stress and strain from sudden stops and starts, falls, body contact, rapid turns, prolonged running, and other rigors of athletic competition are subject to injury. Although athletics should not be introduced until players are physically prepared, there is still opportunity to engage in competition through lead-up contests. Individuals learn many of the skills required for athletics while participating in team contests.

d. Benefits. Athletics are beneficial primarily to sustain interest in the program and to maintain an achieved level of physical fitness. Athletics are a supplement, not a substitute for other types of conditioning activities which should continue. All of the desirable traits of physical fitness cannot be developed through athletics, yet their contribution is significant. For athletics to make a proper contribution to physical conditioning, the selected sports must bevigorous. The team contests recommended here are carefully selected for their simplicity, aggressiveness and applicability to squad- teaching teamwork. They are generally strenuous and involve the whole team throughout the entire duration of the contest.

5402. TEAM CONTESTS

a. Pushball

(1) **Players.** 10 to 50 players on a side. (See fig. 5-8.)

(2) Equipment A large pushball 5 to 6 feet in diameter.

(3) Area. A field 240 to 300 feet long, 120 to 150 feet wide. Mark a center line in the middle of the field parallel to the end lines. Mark a line 45 feet on either side of this center line and parallel to it, extending it across the width of the field. Mark another parallel line 15 feet from each end line, extending it across the width of the field.

(4) Game. Four 10-minute quarters are played. Give 2-minute rests between quarters and 5-minute rests between halves. The object of the game is to propel the ball over the opponent's goal line by pushing, rolling, passing, carrying, or any other way except kicking.

(a) The ball is placed on the center line with the opposing captains 3 feet from the ball. The rest of the players are 45 feet from the ball, on their half of the field. On the referee's starting whistle, the captains immediately play the ball with their respective teams coming to their assistance.

(b) At quarter time, the ball remains dead for 2 minutes at the spot where it was when the quarter ended. At half time, the teams exchange goals. The play is then started as it was in the beginning.

Figure 5-8 is available in the illustration section at the end of the book.

(c) Players may use any means of interfering with an opponent's progress except striking and clipping. Clipping is throwing the body across the back of an opponent's leg, while the opponent is running or standing. Legal use of force may be applied to all opponents whether or not they are playing the ball. For striking an opponent, the offender is removed from the game. The team penalty is half the distance to their goal. The penalty for clipping is the same.

(d) When any part of the ball goes out of bounds, it is a dead ball. The teams line up at right angles to the side lines and 3 feet apart at the point where the ball went out. The referee then tosses the ball between the teams.

(e) When, for any reason, the ball remains in one spot for more than 10 seconds, the referee declares the ball dead. The ball is then put into play as it is for an out-of-bounds situation.

(5) Scoring. A goal is scored when the ball, or any part of it, is propelled across the opponent's end line. A goal counts 5 points. The team scoring a goal has the privilege of trying for a point after the goal. To try for this extra point, the ball is placed on the opponent's 5-yard (or 15-foot) line. The teams line up on either side of this line separated by the width of the ball. Before the whistle blows, one player on each team may place hands on the ball. On the referee's signal, the ball is put into play for 1 minute. If any part of the ball is driven across the goal line in this 1-minute period, the offensive team scores 1 point. The defending team may not score during the opponent's try for the extra point.

b. Line Rush

(1) Players. Any number up to 50 on each side. (See fig. 5-9.)

(2) Equipment. None.

(3) Area. A field, 75 by 100 feet.

(4) Game. One team lines up behind one goal line and the other in midfield. On the starting signal, the team standing behind the goal line seeks to cross the goal line at the opposite end of the field within 30 seconds. The team in the center seeks to prevent it by catching and holding the runners. At the end of 30 seconds, the teams change.

(5) Scoring. Count the number of individuals who have crossed the far goal at the end of 30 seconds. After each team has had from three to five tries, the scores are added and the winner declared. A player scores 1 point when any part of the body is across the goal line.

Figure 5-9 is available in the illustration section at the end of the book.

c. Human Tug-of-War

(1) Players. 10 to 20 on a team.(See fig. 5-10.)

(2) Equipment. None.

(3) Area.40 to 60 feet.

(4) Game. Draw a line in the center of the area. Divide the players into two equal teams. Place them in single file on opposite sides of the center line facing each other. Each individual places the arms around the waist of the teammate in front. The two leaders of the opposing teams grasp each other around the waist. On signal, the teams try to pull each other over the center line within 30 seconds.

(5) Scoring. The team pulled across the center line loses. If neither team is pulled over the center line, but one team breaks its file, that team loses the match.

(6) Variation. Use a 3/4- or 1-inch rope. Space the leading players on each team 10 feet apart, each holding one end of the rope. The team pulled across the center line loses.

Figure 5-10 is available in the illustration section at the end of the book.

d. Master of the Ring

(1) Players. Any number. (See fig. 5-11.)

(2) Equipment. None.

(3) Area. A clearly marked circle large enough to contain all the players.

(4) Game. All the players stand inside the circle. At the signal, all players attempt to throw each other out of the circle. All tactics are fair except unnecessary roughness. When any part of the body touches across the line, the player is out and leaves the circle at once. Several officials are needed to spot the players who cross the line.

(5) Scoring The player who remains in the circle when all the others are out is the master of the ring.

(6) Variation l. The players are divided into two equal teams. Each team is clearly marked. On signal, each team tries to throw the opponents out of the circle. The winning team is the team that eliminates all opponents from the circle.

(7) Variation 2. The players are divided into two equal teams. Each team sends only one individual into the circle. When one person has been forced out of the circle, the losing side sends in only one player. The team which eliminates all opponents is the winner.

(8) Variation 3.A pit, approximately 4-feet deep, is used rather than a circle on level ground. The contest may then be an individual or team activity.

Figure 5-11 is available in the illustration section at the end of the book.

e. Log Pivot Circle. Space teams far enough apart so each team can pivot in a circle without colliding with any other team. Each team holds a log in the bend of the arms in front of the chest. (See fig.5-12.) At the command **CIRCLE RIGHT, MOVE,**the left-flank individual holds the pivot while the log is carried around 360 degrees in a counter- clockwise motion, back to the original position. This movement may also be performed to the left by facing in the other direction and pivoting clockwise(**CIRCLE RIGHT, MOVE**).Other commands may be used such as: **CIRCLE HALF-RIGHT, CIRCLE HALF- LEFT** and so on. The first team to complete the prescribed movement is the winner.

Figure 5-12 is available in the illustration section at the end of the book.

f. Log-Rolling Race. Each team tries to roll its log a measured distance by pushing it with the hands and driving the body forward with the legs. (See fig. 5-13.)The first team to get the entire length of the log across the finish line wins.

Figure 5-13 is available in the illustration section at the end of the book.

g. Prone Pushing Contest.. Two teams lie prone, facing each other with a log between them. (See fig. 5-14.)Both teams place their hands against the log, keeping their arms straight. Then, by driving with the legs, each team attempts to push the other a measured distance to the rear.

Figure 5-14 is available in the illustration section at the end of the book.

h. Shuttle Relay Race. Each team in this race divides into two parts, A and B. Members of part A run 50 yards with a log held under their right arms. (See fig 5-15.)At the distance line, they give the log to their teammates in part B who bring it back to the starting line. The team pair finishing first is the winner.

Figure 5-15 is available in the illustration section at the end of the book.

5403.CROSS-COUNTRY AND DISTANCE RUNNING

a. Value of Running. Long-distance running gives some benefits that cannot be obtained in the same degree from any other sport. It builds powerful leg muscles, increases lung capacity, and develops endurance. For these reasons, cross-country and distance running should be included in the physical training program. These sports require only a few miles of open space which is generally available. Short cross-country runs and middle-distance runs can be used to supplement other activities, particularly team sports or the sports that develop precision or agility rather than endurance. Short cross-country runs can be scheduled once a week, gradually increasing the distance as the physical condition of the runners improves.

b. Cross-Country Runs. A cross-country run is a distance run held on a course laid out along roads, across fields, over hills, through woods, and on any irregular ground. A flat cinder or dirt track is not a suitable surface for cross-country running. Some runs are as long as 6 miles while others are as short as 3 or 4 miles. If cross-country running is being used to supplement other activities, the 3-mile course is long enough for most. Cross-country runs should be used only after individuals reach the sustaining stage of conditioning. This type of running should then be scheduled occasionally to provide variety. Cross-country running has the advantage of allowing mass participation. Interest can be stimulated by putting the runs on a competitive basis.

c. Practice Methods. Conditioning is essential to distance and cross-country running. Championship distance running depends on stamina, which can be developed only through constant training. An individual of only average ability can become an outstanding distance runner by steady and careful training. Hiking is an excellent method for getting into condition before the season opens. Long walks build up leg muscles. During the first month of the season, training should be gradual, starting with short distances, and increasing day by day. At first the legs will become stiff, but the stiffness gradually disappears if running is practiced for a while every day. To prevent strain, it is essential to limber up thoroughly each day before running.

(1) In training a large group, leaders should be stationed at the head and the rear of the column. They should make every effort to keep the runners together. After determining the abilities of the unit in cross-country running, it is advisable to divide them into three groups. The poorest-conditioned group is started first, the best-conditioned group, last. The starting time of the groups should be staggered so that all of them come in at about the same time.

(2) In preliminary training, running is similar to ordinary road work in that it begins with rather slow jogging, alternating with walking. The speed and distance of the run is gradually increased. As conditioning improves, occasional sprints may be introduced. Atfirst, the distance run is from one-half to 1 mile. It gradually increases to 2 or 3 miles. On completing runs, require runners to walk for 3 or 4 minutes before stopping to permit a gradual cooling off.

d. Facilities and Equipment. There should be at least one timer with a stopwatch (preferably three) for timing the runners. A course 3 or 5 miles long should be measured and marked by one of the three following methods:

(1) Fasten directional arrows to the top of tall posts and place them at every point where the course turns. Such signs should also be placed at every other point where there may be doubt as to the direction of travel.

(2) Place a lime line on the ground over the entire course.

(3) Place flags, clearly visible to the runners.

- **(a)** A red flag indicates a left turn.
- **(b)** A white flag indicates a right turn.
- **(c)** A blue flag indicates the course is straight ahead.

e. Rules/Scoring. A cross-country team consists of seven individuals, unless otherwise agreed. In dual meets, a maximum of 12 on one team may be entered, but 7 or fewer are scored. First place scores 1 point, second place 2, third place 3, and so on. All who finish the course are ranked and tallied in this manner. The team score is then determined by totaling the points scored by the first five of each team to finish. The team scoring the least number of points is the winner. If fewer than five (or the number determined prior to the race) finish, the places of all members of that team are disregarded. If two or more teams score the same number of points, the event is called a tie.

5404.TEAM ATHLETICS

Team athletics add variety to the combat readiness program. At no time should they be allowed to replace intramural activities or to provide practice time for intramural or base teams. If team athletics are integrated into the program, the goal should be total participation and competition. Therefore, the games selected can be played by the skilled or the unskilled. They require limited equipment, yet will teach competitiveness, aggressiveness, and teamwork while they improve conditioning.

a.Soccer. Soccer is one of the best athletic activities for developing endurance, agility, leg strength, and great skill fulness in using the legs. The most popular sport in the world, soccer is the national game of many European, Asian, and Central and South American countries. In recent years, it has become popular in the United States. A soccer ball is the only equipment needed for the game, and Marines can learn to play it easily. Players do not need much skill to participate, but the amount of skill they can develop is almost unlimited.

(1) Place in the Program. Soccer should be introduced during the latter part of the slow improvement stage of physical training. It can be used as a competitive activity in the sustaining stage. It is primarily a spring or fall sport. Any level field is suitable for competition. The boundaries for the soccer field are similar to the dimensions for a football field. (See fig. 5-16.) Goal posts are essential, but they are easily constructed and usually temporary so that they may be removed when not in use.

Figure 5-16 is available in the illustration section at the end of the book.

(2) Basic Skills

(a) Passing. Passing with the feet is the primary means of moving the ball. Short passes are easier to control. Emphasis should be continually placed on the skill of passing.

(b) Dribbling. The ball is dribbled by a series of kicks with the inside or outside of the foot. Do not kick with the toe. Keep the head over the ball when kicking and propel it only a short distance at a time. Keep it close to the feet. If a player lets a ball move far from the feet, an opposing player can easily take it away.

(c) Instep Kicking. The instep kick is the basic soccer kick. The toe does not come in contact with the ball. The toe is pointed downward, and the instep (the area under the shoe laces) is applied to the ball with a vigorous snap from the knee.

(d) Inside-of-the-Foot Kicking. The ball is kicked with the inside of the foot, and the leg is swung from the hip. The toe is turned outward, and the sole of the foot is parallel with the ground as the foot strikes the ball. This kick is used for short passes and for dribbling.

(e) Foot Trapping. The foot trap is the method of stopping the ball by trapping it between the ground and the foot. Place the sole of the foot on top of the ball at the instant it touches the ground, keeping the foot relaxed. This is an effective way to stop a fast-moving ball.

(f) Shin Trapping. .The shin trap is a method of stopping the ball with the shins. Stand just forward of the spot where the ball should strike the ground and allow it to strike the shins in flight or on the bounce. Use either one or both legs from the knee down so as to absorb the bounce.

(g) Body Trapping. .Intercept the ball with any part of the upper body except the arms and hands. Keep the body relaxed and inclined toward the ball. To keep the ball from bouncing, move backwards from it as it strikes the body. This drops the ball at the feet in position for dribbling or passing.

(h) Heading. Heading is a technique for changing the direction of the flight of a ball by butting it with the head. Tense the neck muscles and jump up to meet the ball. Butt the ball with the forehead at about the hairline to reverse its direction. Use the side of the head to deflect it to the side.

(3) Offensive and Defensive Positions. Forwards usually play on the offensive half of the field. Fullbacks usually play on the defensive half of the field. Halfbacks are the backbone of the team. They move forward on the offense and back on defense. The goalkeeper almost always remains within a few feet of the goal.

(4) Abridged Rules

(a) A soccer team is composed of 11 players. (See fig. 5-16.)

(b) The player propels the ball by kicking it with the feet or any part of the legs, by butting it with the head, and by hitting it with any portion of the body except the arms or hands.

(c) The goalkeeper is the only individual allowed to place hands on the ball, but only in the goalkeeper's area. The term "hands" includes the whole arm from the point of the shoulder down.

(d) A goal is made by causing the ball to cross completely the section of the goal line lying between the uprights and under the cross bar.

(e) Each goal scores 1 point for the team scoring the goal.

(f) The penalty for a foul committed anywhere on the playing field (except by the defensive team in its penalty area) is a free kick awarded to the opposing team.

(g) All opponents must be at least 10 yards from the ball when a free kick is taken.

(h) The penalty for a foul committed by the defensive team in its penalty area is a penalty kick.

(i) A penalty kick is a free kick at the goal from the spot 12 yards directly in front of the goal. The only players allowed within the penalty area at the time of the kick are the kicker and the defending goalkeeper.

(j) An official game consists of two 30-minute halves,

(k) Teams change goals at the end of every quarter.

(l) In the event of a tie, an extra quarter may played. This may be followed by a sudden death period. If the tie still exists, the game can be decided by penalty kick.

(m) After a team has propelled the ball across a side line and out of play, the ball is put back into play by a throw-in from the side line by a member of the opposing team. The ball is thrown in from the point where it crossed the side line as it went out of bounds. The ball must be thrown in with both hands using an over-the-head motion, keeping both feet on the ground.

(n) When the offensive team propels the ball across the defensive team's goal line but not in the goal, the defensive team is awarded a goal kick-a free kick taken from within its own goal area. The ball must come out of the penalty area to be in play.

(o) When the defensive team causes the ball to go behind its own goal line, excluding the portion between the goal posts, the offensive team is awarded a corner kick—a free kick taken by a member of the offensive team at the quarter circle, at the cornerflag post nearest to where the ball went behind the goal line. The flag post must not be removed.

(p) The game is started and, after a goal has been scored, is resumed by placing the ball in the center of the midfield line. Players must be on their side of the line until the ball is kicked. The ball must be kicked forward and must move at least 2 feet to be legal. The first kicker may not touch the ball twice in succession at the kick off. The opposing team must be 10 yards from the ball until it moves.

b.Speedball. Speedball offers vigorous and varied action with plenty of scoring opportunities. It is easy to learn and provides spontaneous fun. Little equipment is needed-a ball is all that is absolutely necessary. Speedball combines the kicking, trapping, and intercepting elements of soccer; the passing game of basketball; and the punting, drop kicking, and scoring pass of football.

(1) Place in the Program.Speedball, like soccer, should be introduced into the physical training program during the latter part of the toughening stage and used as a competitive activity in the sustaining stage. Speedball may be played any time weather permits but is primarily a spring or fall activity.

(2) Method of Play.The game is played as follows:

(a) Two teams of 11 each play the game under official rules, but any number of players may successfully constitute a team.An inflated leather ball, usually a soccer ball, is used. The playing field is a football field with a football goal post at each end. (See fig. 5-17.)

(b) The game starts with a soccer-type kickoff. The kicking team tries to retain possession of the ball and advance it toward the opposite goal by passing or kicking it. Running with the ball is not allowed, with the result that there is no tackling or interference. After the ball touches the ground, it cannot be picked up with the hands or caught on the bounce but must be played as in soccer until it is raised into the air directly from a kick. Then the hands may be used.

(c) When the ball goes out of bounds over the sidelines, it is given to a player of the opposing team and is put into play with a basketball throw- in. When it goes over the end line without a score, it is given to a player of the opposing team who may either pass or kick it onto the field.

(d) When two opposing players are contesting possession of a held ball, the official tosses the ball up between them as in basketball.

(e) Points are scored by kicking the ball under the crossbar of the goal post, drop kicking the ball over the crossbar, or completing a forward pass into the end zone for a touchdown.

Figure 5-17 is available in the illustration section at the end of the book.

(3) Basic Skills

(a) Shared Skills. Skills include kicking, passing, heading, and trapping (from soccer); punting, dropkicking and forward passing (from football); and passing, receiving, and pivoting (from basketball).

(b) Kickups. The kickup is a play in which a player lifts the ball into the air with the feet so as to legally play the ball with the hands. The kickup generally makes the transition from ground play to aerial play. The technique of making the play depends upon whether the ball is rolling or stationary. If the ball is rolling or bouncing toward the player, the foot is held on the ground with the toe down until the ball rolls onto the foot. Then the foot is raised, projecting the ball upward. If the ball is stationary, the player rolls it backward with one foot. Then the player places a foot where the ball will roll onto it. The player then raises the foot, projecting the ball as before. If a ball is rolling away, the player should stop it with a foot and play it as a stationary ball. The player can also raise the ball by standing over it with a foot on either side. The player presses the feet against the ball and jumps into the air, propelling the ball into the hands.

(4) Offensive and Defensive Play. The positions of players in speedball are much the same as in soccer. However, some of the positions have different names. Each team has 11 players. The forward line is composed of five players, the right end, right forward, center, left forward, and left end. The second line consists of right halfback, fullback, and left halfback. In the next line is the right guard and left guard. The player who defends the goal is the goal guard. The strategy employed in speedball during offensive play is very similar to that of soccer. There are two types of defensive formations in speedball: man-for-man and position defense. Man-for-man defense is recommended for beginning players.

(5) Abridged Rules

(a) Field. A speedball field is 100 yards long with additional 10-yard end zones on each end and 55 yards wide, the same as a regulation football field.

(b) Players. Eleven players are on a team. The goal guard has no special privileges.

(c) Time. Ten-minute quarters. Two minutes between quarters. Ten minutes between halves. Five minutes for extra overtime periods.(Begin the first overtime by a jump ball at center, same goals; change goals in the event of a second overtime period.)

(d) Winner of Toss. The winner of the toss has the choice of kicking, receiving, or defending a specific goal.

(e) Kickoff. A kick off starts the first and second half of a game and starts play after a score. The kick off is made from any place on the 50-yard line. The kicking team must be behind the ball when it is kicked. The receiving team must stay back of its restraining line (10 yards distance) until the ball is kicked. The ball must go toward the opposing team's goal before it can be played. Kick off out of bounds goes to the opponent at that spot. A kick off touched by the receiving team before going out of bounds, with no impetus added, still belongs to the receiving team. A kick off, in possession and control of the receiving team and then fumbled out of bounds, belongs to the kicking team at that spot.

(f) Starting Second and Fourth Quarters. The ball is given to the team that had possession at the end of the previous quarter, from out of bounds, as in basketball.

(g) Second Half. The first-half receiving team kicks off at the beginning of the second half.

(h) Fly and Ground Ball. The most characteristic playing rule of speedball is the difference between a fly ball (or aerial ball) and a ground ball. A player is not permitted to touch a ground ball with the hands and must play it as in soccer. A fly ball is one that has risen into the air directly from the foot of a player (for example, a punt, drop kick, place kick, or kickup).Such a ball may be caught with the hands provided the catch is made before the ball strikes the ground again.

(i) Kickup and Overhead Dribble. A kickup is a ball that is so kicked by a player that he can catch it himself. A bounce from the ground may not be touched with the hand because it has touched the ground since being kicked. This rule prohibits the ordinary basketball dribble, but one overhead dribble (throwing the ball into the air and advancing to catch it before it hits the ground) is permitted.

(j) Out of Bounds. If a team causes the ball to go out of bounds over the sidelines, a free throw-in (any style) is given to the opposing team. When the ball goes over the end line without scoring, it is given to the opponents who may pass or kick from out of bounds at that point.

(k) Tie Ball. In case two players are contesting the possession of a held ball, even in the end zone, a tie ball is declared. The ball is tossed up between them.

(6) Scoring

(a) Field Goal(3 points).A soccer-type kick, in which a ground ball is kicked under the crossbar and between the goal post from the field of play or end zone, is a field goal. (A punt going straight through is not a field goal for it is not a ground ball. The ball must hit the ground first.) A drop kick from the field of play that goes under a crossbar does not count as a field goal. A drop kick from the end zone that goes under the crossbar counts as a field goal;if it goes over the crossbar, it is ruled as a touch back, and the other team takes possession of the ball.

(b) Dropkick(2 points). A scoring drop kick must be made from the field of play and go over the crossbar and between the uprights. The ball must hit the ground before it is kicked (usually with the instep).

(c) End goal(1 point). This is a ground ball which receives its impetus (kicked or legally propelled by the body) from any player, offensive or defensive, in the end zone and which passes over the end line but not between the goal posts.

(d) Penalty Kick(1 point).A ball kicked from the penalty mark that goes between the goal posts and under the crossbar is a penalty kick. The penalty mark is placed directly in front of the goal at the center of the goal line.

(e) Touchdown(1 point). A touchdown is a forward pass from the field of play completed in the end zone. The player must be entirely in the end zone. If the player is on the goal line or has one foot in the field of play and the other in the end zone, the ball is declared out of bounds. If a forward pass is missed, the ball continues in play, but it must be returned to the field of play before another forward pass or drop kick may be made.

(7) Substitutions. Substitutions may be made any time when the ball is not in play. A player who is withdrawn may not return during that same period.

(8) Timeout. Three legal timeouts of 2 minutes each are permitted each team during the game.

(9) Fouls

(a) Personal Foul. Kicking, tripping, charging, pushing, holding, blocking, and unnecessary roughness of any kind, such as running into an opponent from behind, are personal fouls. Kicking at a fly ball and thereby kicking an opponent is a personal foul. Four personal fouls disqualify a player from the game.

(b) Technical Foul. Illegal substitution, more than three timeouts in a game, unsportsmanlike conduct, and unnecessarily delaying the game are technical fouls.

(c) Violation. Traveling with the ball, touching a ground ball with the hands or arms, double overhead dribble, violating tie ball, and kicking or kneeing a fly ball before catching it are violations.

(d) Penalties. The offended player shall attempt the kick.

(e) Officiation of Fouls

1 Fouls in the field of play allow no follow-up while fouls in the end zone always allow follow-up.

2 On penalty kicks, with no follow-up, only the kicker and goalie are involved.

3 On penalty kicks, with a follow-up, the kicking side is behind the ball and the defending side behind the end line or in the field of play. No one is allowed in the end zone or between the goal posts except the goal guard. The kicker must make an actual attempt at goal and cannot play the ball again until after another player plays it.

c. Volleyball. Volleyball is a popular sport. The game entails much physical activity, yet it is not strenuous. It is, therefore, a game for young and old alike, for beginners and for skilled players. It may be played indoors or outdoors on any type of terrain. As an informal activity, volleyball can be played by any number; as an organized activity, it provides, as few other sports do, a game for 12 to play in a limited area. While volleyball requires no great skill to play, it permits a high degree of proficiency. Players naturally get more enjoyment when they know the game and play it well. For this reason, instruction in the basic skills should be provided.

(1) Place in the Program Volleyball may be used occasionally as a competitive activity during the sustaining stage. Though it is a year- round sport, it should be included in the physical training program only when a more strenuous activity is impractical. Volleyball is an excellent off-duty activity.

(2) Instruction. Usually during the first three or four classes, a 10- to 15- minute period of instruction, followed by scrimmage, is enough to teach the basic skills, rules, and techniques of volleyball. More time, if available, can be given to teaching basic skills, but emphasize competitive play rather than formal instruction. Divide the unit into 6-person teams, organized on the basis of ability .Teams should be as nearly equal as possible.

(3) Basic Skills

(a) Passing the High Ball. The chest pass is the most effective method of playing the ball. To receive the ball, the knees are fixed, and the body is tilted forward. The elbows are raised sideward to a point in line with the shoulders. The wrists are extended in line with the forearm. The arms, wrists, and hands are rotated inward. To pass the ball, the hands are chest-high, thumbs pointing inward. The fingers are flexed, forming a cup, allowing them to contact the ball. On contact with the ball, the wrists are snapped while the fingers and elbows are pushed upward, sending the ball upward. A high ball is much easier to handle than a low one.

(b) Passing the Low Ball. A ball that is lower than the waist is one of the easiest to hit, but it is also a frequent cause of fouls (holding or carrying the ball). The best position for handling a low ball is to have the knees flexed and arms flexed at the elbows and rotated so the thumbs are pointing outward, the palms up. When the fingers contact the ball, the entire body reacts in a lifting motion. The arms and hands swing upward in a scooping action. It is important that the fingers, not the palms, contact the ball, and that the ball is batted, not thrown.

(c) Underhand Serve. Take a position behind the back line facing the net, left foot forward, holding the ball in the palm of the left hand. The left knee is flexed; the right knee is straight. Swing the right arm back and at the same time move the left hand (holding the ball) across the body in line with the right hip. Then swing the right arm forward hitting the ball off of the left hand with the palm of the right hand, raising the hips and arching the back in the same motion. Be certain to swing the right arm in a straight line, or the ball will be difficult to control. When the opposition is in formation, the server should try to place the ball in the right or left back area, and not near the net.

(d) Setup. A setup is a ball hit into the air near the net by one player, so that a teammate may hit or "spike" it sharply downward into the opponent's court. The chest pass is the best pass to use. The ball is sent approximately 10 feet into the air toward the spiker so it will descend from 4 to 20 inches from the net.

(e) Spiking. The spike is a leap into the air and a sharp downward hitting of the ball into the opponent's court. A spiker must be able to spring easily from the floor, judge the movement of the ball, and strike it with a downward movement of the arm. To jump from the floor, step off with one foot and jump with the other. Stand with the right or left side to the net, facing the setup player. Much depends upon the setup player to place the ball in the proper position. The spiker jumps into the air and strikes the ball above its center so as to drive it downward. A snapping movement of the arm and wrist will drive the ball forward and downward with power and control. Aim for a weak spot in the opponent's defense.

(f) Blocking. The block is a defense technique used to prevent a spiker from driving the ball across the net. It is an attempt by one or more defensive players at the net to block a hard-hit shot by using the force of the ball to send it immediately back into the opponent's court. In an effective block, forwards on the defensive team spring into the air at the time of the spike, placing both hands and arms in the expected path of the ball. An effective block tends to upset the offense and presents another element for the spiker to worry about. To be effective, the blocker must anticipate the path of the ball and time the block with the spike.

(4) Abridged Rules

(a) The volleyball court is 30 feet wide by 60 feet long. (See fig. 5-18.)

(b) The top of the net is 8 feet high.

(c) A volleyball team consists of 6 players.

(d) A match consists of the best two out of three games.

(e) The first team scoring 15 points wins the game, provided that they have 2 points more than their opponents.

(f) A deuce game is a game in which both teams score 14 points. The game is continued until one team obtains a 2-point advantage over the other.

(g) Only the serving team can score. If the serving team commits a fault, it loses the serve to the opposing team.

(h) The team members who receive the served ball rotate one position in a clockwise direction.

(i) The ball is put into play by serving from behind the back line.

(j) A served ball touching the net results in the loss of the serve. At any other time during play, a ball touching the net is still in play.

Figure 5-18 is available in the illustration section at the end of the book.

(k) The ball is out of play when it touches the ground or goes outside one of the boundary lines.

(l) All line balls are good.

(m) The players must hit or bat the ball; they may not throw, lift, or grasp it.

(n) A player may not touch the ball with any part of the body below the knees.

(o) A player may not play the ball twice in succession. In receiving a hard-driven spike, a defensive player may make several contacts with the ball even if they are not simultaneous. All such contacts, however, must constitute one continuous play, and all must be above the knees.

(p) The ball may be touched no more than three times on one side of the net before being returned across the net to the opposing team.

(q) A player must not touch or reach across the net.

(r) A player may touch the line under the net, but may not cross it.

Chapter 6
EVALUATION OF PERFORMANCE DURING TRAINING

6001.OBJECTIVES OF COMBAT READINESS

The objectives of the Marine Corps physical fitness program are-

a. To contribute to the health and well-being of every Marine through regular exercise and health education.

b. To develop Marines who are physically capable of performing their duties in garrison and in combat.

c. To develop Marines with a reserve level of physical fitness that will help them win in combat.

d. To provide a medium for developing the self-confidence of the individual Marine and thereby enhancing overall discipline, morale, esprit de corps, unit efficiency, and the desire to excel within the Marine Corps.

6002. THE PHYSICAL FITNESS TEST

The physical fitness test (PFT) has distracted attention from the proper goal of the physical fitness program. Simply stated, the goal of the physical fitness program is the success of Marines in combat. Unit commanders must not allow the PFT to become an end in itself. The PFT should be used within the commander's program of combat readiness training as a standard, an index of the physical fitness of individual Marines at a given time and place. The PFT is not difficult, and it must be kept in perspective. All athletes should be first class, and first class is attainable for all Marines. The conduct of the PFT is described in Marine Corps Order 6100.3_.

6003. OTHER METHODS OF EVALUATION

The commander should use other ways of evaluation to provide variety and imagination. Intangible objectives, such as confidence and aggressiveness, are hard to measure. The use of either inspection or observation is necessary to evaluate objectives of this nature. The commander has several other methods available when evaluating the physical condition of the command. The additional methods of inspection, observation, medical examination, and testing are summarized as follows:

a. Formal Inspections. Formal inspections, using inspection officers and standardized rating criteria, may help evaluate unit physical fitness.

b. Daily Observation. Routine observation of physical performance and appearance can serve as an indicator of an individual's or unit's physical readiness. However, mere observation is not a totally reliable or accurate means of evaluation.

c. **Physical Examination.** Medical examination may detect any individual disability or detrimental physical condition. It may guide in application of remedial, therapeutic, or limited exercise programs.

d. Other Physical Fitness Tests. Some other tests are described in detail in this manual for information purposes. Commanders are cautioned, however, not to place too great an emphasis on testing.

6004. RESPONSIBILITY

The commander is responsible for the physical fitness of the command and for the measurement and evaluation of its physical readiness. The goal of combat readiness training remains to prepare Marines and units for the physical demands of combat. The bulk of the commander's combat readiness training program should be such activities as obstacle and confidence courses, endurance courses, and progressive load-bearing marches which will prepare Marines for combat. Group activities which are competitive and combative in nature are advantageous and add imagination and enthusiasm to the program. Activities such as orienteering, which is not only physically demanding but also competitive and which teaches essential subject skills, are excellent and motivate all concerned. A commander's goal should be to provide an integrated training program that is so aggressive and demanding of Marines and leaders alike that a PFT administered at any time will reflect that each and every Marine is both physically and mentally prepared for the demands of combat.

6005. COMMANDER'S PHYSICAL FITNESS PROGRAM

The Commandant of the Marine Corps has directed that the physical fitness test shall be the universal measure of adequate individual physical fitness. The focus of a commander's physical fitness training program, however, should not be to prepare Marines to pass the physical fitness test, but to prepare them for the physical rigors of combat.

a. Administration. The physical fitness program requires each Marine to spend a minimum of 3 hours per week in physical fitness training, to be tested semiannually, and to obtain a minimum standard of third class. The attainment of a higher score is a laudable individual goal which should be encouraged but should not become a unit objective. Overemphasis could be detrimental to training required to develop the complete Marine. All commanders should exercise sound judgment as well as positive, aggressive leadership in striving toward the training objective.

b. Variety of Activities. The program must include the following elements:

(1) Physical Readiness Training. physical readiness training is a complete physical training program which develops and maintains the strength, endurance, and physical skills needed to sustain the individual during combat.

(2) Remedial Physical Conditioning. Remedial physical conditioning is a process by which physically substandard Marines are conditioned to meet prescribed standards.

(3) Competitive Conditioning Activities. Competitive conditioning activities consist of teams or individuals competing against an opponent to win. This involves a combination of sports and military skills designed to foster competitive attitudes and develop unit pride and esprit de corps.

(4) Occupational Conditioning. Marching long distances, dry net training, etc., should not be ignored as a means to fulfill the time requirement.

c.Time. In order to develop the desired level of physical fitness, unit commanders are advised that a minimum program of 3 hours per week is mandatory. The program must encompass at least three exercise periods weekly. Each period should include calisthenics, running, or other forms of vigorous activity .The minimum desirable length for each exercise period is 60 minutes. Shorter periods may be authorized by the unit commander if dictated by the local training situation. Unit commanders are encouraged to use the normal working day to satisfy this training requirement. However, commanders are authorized to conduct required physical fitness training during off-duty hours when the mission, workload, personnel status or other significant factors preclude fulfilling it during duty hours.

d.Action. Marine Corps Order 6100.3_ requires that commanders-

o Establish and maintain an effective physical fitness program which ensures that all Marines maintain an acceptable level of physical fitness.

o Ensure that all individuals are medically qualified prior to participating in the physical fitness program.

o Establish a minimum physical fitness program of at least 3 hours per week. Physical training may be authorized on an individual basis at the discretion of the unit commander.

o Ensure that all Marines participate in physical fitness conditioning activities commensurate with their medical qualifications and limitations.

o Conduct physical fitness testing for all Marines.

o Place all Marines who fail to pass the PFT on a daily command-supervised remedial physical conditioning program until they pass the test.

o Ensure that results of physical fitness testing are entered on section A of the fitness report. Comments are placed in section C in connection with attainment of superior physical performance or a medical excuse from the PFT.

6006. OTHER FORMAL TESTS

Commanders must realize that the best index of combat readiness is personal observation of performance on conditioning marches of 15 miles while under load. This manual presents three additional indicators of unit fitness. These indices are the physical readiness test, which was formerly the Marine Corps standard and is now routinely administered at Officer Candidate School; the airborne trainee physical fitness test which many Marines who attend U.S. Army formal schools are required to take; and the battle fitness test, the ultimate test of the combat readiness of Marines. Commanders may wish to include these tests or portions thereof within their combat readiness programs.

6007. PHYSICAL READINESS TEST

a. Purpose. The purpose of this test is to measure the Marine's ability to meet the minimum standard of physical readiness for duty in the field. This test is currently administered to all male officer candidates at Officer Candidate School, MCCDC, Quantico, VA. Female candidates do not take this test.

b. Test Area. Administration of this test requires an area suitable for crawling, jumping, and running 160 yards to include a trench that is 8 feet wide.

c. Equipment. The equipment required for this test includes—
o A platform, rail on a ledge, or a box 18 inches high,
o A 20-foot climbing rope,
o A stopwatch.

d. Conduct of Test. Adequate timers and other supervisory personnel must be provided by the company staff. All events should be conducted in a single session of one morning or afternoon. The events may be run in any sequence.

e. Uniform. The uniform for the physical readiness test is as follows: full utilities, cartridge belt, belt suspender straps, two canteens (full of water), rifle, and helmet. Depending on the heat condition, the PT shirt may be worn in place of the utility jacket. All equipment will remain secure throughout the test. Except for emergencies, canteens are not to be emptied.

f. Events
(1) Event 1: Climbing Uphill (Step-ups)This event simulates marching uphill at a rapid and steady pace. The Marine stands in front of the step. On the command GO, the Marine places either foot on the step and steps up. Hand pressure on the knee may be used if desired. The Marine then stands erect and steps backward and down, on one foot at a time. The same leg may be used for each repetition or the legs may be alternated. For 100 points, Marines will complete 60 repetitions in the maximum allotted time of 1 minute and 20 seconds (80 seconds).
*(2) Event 2: Rope Climb. T*his event resembles entering and leaving a hovering helicopter, using ropes in house-to-house fighting, and ascending and descending landing nets. The Marine assumes a sitting position on the ground at the bottom of the rope with hands at the highest point that can be reached. On the command GO, the Marine jumps up and climbs to the top. The Marine must touch the 20-foot mark with one hand, then descend without sliding or dropping. Marines must reach the 20-foot mark in the maximum allotted time of 30 seconds or less.
(3) Event 3: Evacuation. This event simulates reaching and evacuating a wounded Marine under fire. The Marine assumes the prone position at a distance of 50 yards from a casualty who is of approximately equal weight. On the command GO, the Marine springs up and covers the distance in a sprint. The Marine lifts the casualty to a fireman's carry and returns to the starting point. Assistance by the casualty is permitted. The tested individual must carry all equipment; the casualty has none. Marines must negotiate the course in a maximum time of 47 seconds or less.
(4) Event 4.'Advance by Fire and Maneuven. This event simulates advancing and assaulting during an attack. The Marine assumes the prone position at the starting point. On the command GO, the Marine then alternately crawls, runs in a zigzag manner, rolls, jumps, and assumes different firing positions for the next 130yards. Before reaching the objective, the Marine must "hit the deck," roll over, and assume the firing position three times. While assuming the firing position, the Marine must pull back the charging handle, take aim, and squeeze the trigger before rising from the ground. After rising from the ground for the third time,the Marine continues running and jumps to clear the 8-foot trench in one leap. The Marine then finishes the course in

the maximum allotted time of 2 minutes. The Marine is allowed multiple attempts only within the specified time limit.

(5) Event 5: Forced March. This event is a combat run of 3 miles without halts and against time. The Marine must double-time. Marines must complete the 3-mile course in a maximum allotted time of 30 minutes or less.

g. Event Standards

(1) To successfully pass the physical readiness test, the Marine must pass each event with a minimum of 80 points for each. Points are then assessed for each event in accordance with the scoring tables. (See fig. 6-1.)

(2) If a Marine fails one event, the highest possible score that the Marine may receive for the entire test is 79 percent. (See example 1 in fig. 6-2.) If two events are failed, the highest possible score for the entire test is 69 percent, regardless of the total points accumulated. (See example 2 in fig. 6-2.) If three events are failed, the highest possible score is 59 percent. Failure in four or more events will result in a score of zero for the entire test.

(3) To convert the total points for all five satisfactorily passed events to a final percentage score, total the accumulated points for all five events and divide by five. (See example 3 in fig. 6-2.)

Figure 6-1 is available in the illustration section at the end of the book.

Figure 6-2 is available in the illustration section at the end of the book.

6008. AIRBORNE TRAINEE PHYSICAL FITNESS TEST

a. Use and Composition of the Test.

(1) Use. The airborne trainee physical fitness test is a means of determining the physical ability of the applicant for acceptance to and retention in the airborne course of instruction.

(2) Test Events The test battery consists of five events as follows: chinups, knee bender, pushups, situps, and an endurance run.

b. Method of Scoring and Standards

(1) Scoring. The Marine will be scored by a trained scorer who is thoroughly familiar with the minimum standards for the test events. The Marine will be scored on a pass or fail basis. The performance on each event may be recorded on the reverse side of DA Form 705 (Physical Fitness Testing Record).The examinee records the personal information on the face of the card as specified on the first, second, and third lines. The face of the card is identified by writing diagonally across the lower half of the card "Airborne Trainee PPT—See reverse side."Using line 7 on the reverse side of the scorecard, the scorer enters the test title in the block entitled "(Other (Specify))" and completes the required information on the remainder of that line. The "Remarks" section can be used to record the test event titles and the applicant's performance on each test event.

(2) Standards. To successfully pass the test, the examinee must reach the standard in each test event. The standards follow:

(a) Chinups

(b) Knee bender 80(2-minute period)

(c) Pushups 45(2-minute period)

(d) Situps45(2-minute period)

(e) Endurance runs:2 miles in 15:54 minutes or less in athletic gear and 4 miles in 32 minutes in utilities and running shoes.

c. Uniform for Testing

(1) Examinees. The prescribed uniform for test participation is boots and the work uniform of the season. No headdress is worn. When climatic conditions permit, jackets or outer shirts may be removed.

(2) Officials. Scorers and other test officials should be uniformly and distinctively dressed for contrast with Marines being tested.

d. Test Administration Procedure

(1) Preparation for Test. The administration of the test to a large group makes it mandatory that the test effort be organized and efficiently operated. All testing is not completed with large test groups; on certain occasions, individuals and small groups are tested. Care must be exercised to administer the test uniformly and to standardize the conduct of all elements of the test. Regardless of the size of the test group, the following elements of sound test administration should be included:

(a) An orientation to include the purpose, method of administration, scoring of the test, preparation of the score card, and required standards.

(b) A correct demonstration of each event to ensure that there is no misunderstanding of the proper form and required standards.

(c) Completion of all five test events in one test period with all Marines taking the events in the same sequence.

(d) Adequate rest periods between the test events to allow for recovery before the next event.

(2) Method of Administration with a Large Group. With a minimum of 12 lanes per test event, 14 officials can administer the test battery to 150 or 200 Marines in 2 hours. The officials are designated as follows: one officer in charge, one demonstrator,and 12 scorers. If more or fewer Marines are to be tested, a greater or lesser number of officials will be required. The following procedure is recommended:

(a) Conduct an orientation and ensure the examinees have properly completed their score cards.

(b) Assign Marines to lanes and caution them to remain in the same lane order throughout the test.

(c) Explain and demonstrate the chinup event, administer it, and score it. Then proceed to the knee bender and pushup events, and administer them in the same manner.

(d) Grant a 5- to 10-minute rest period after the pushup event. Advise against excessive consumption of water during the break period.

(e) Explain and demonstrate the situp event, administer it, and score it as prescribed. Then move to the run area, explain the running event, and complete it.

(f) Retain the scorecards at the completion of the running event.

(3) Method of Administration with a Small Group. A similar procedure is followed for the testing of individuals and small groups. The informality usually associated with small groups must not conflict with sound test administration. With fewer examinees, fewer officials are required.

e. Description and Explanation of Test Events

(1) Test Event 1:Chinups

(a) Purpose. This event is devised to test arm and shoulder flexor strength.

(b)Equipment There is one horizontal bar per lane, made of plumber's pipe or a gymnasium horizontal bar 1 1/2 inches in outside diameter. The bar must be rigidly supported at a height of 8 feet above the ground. The upright supports must be 5 feet apart. There must be a movable stand at each bar for short Marines to stand on to reach the bar.

(c) Officials There is one scorer per lane.

(d) Organization. The examinees, holding their scorecards, stand in order behind the restraining line in their respective lanes. The scorers take each scorecard when the Marine is called forward for the test.

(e) Starting Position The bar is grasped with the palms turned toward the face, the thumbs underneath the bar. The body is fully extended in a "dead" hanging position with the arms straight and the feet above the ground.

(f) Movement. Pull the body directly upward until the chin is placed over the bar. Lower the body until the elbows are completely straight and the body is again in the "dead" hanging position. Repeat as many times as required.

(g) Instructions. Explain and demonstrate the fully extended "dead" hanging position with the proper grasp. Show that the chin is placed over the bar at the top of the movement and that the arms are fully extended, the elbows completely straight, at the bottom of the movement (the hanging position). Explain that the body must be kept from swinging and that it is permissible to raise the legs and flex the hips when pulling up, but that any kicking, bicycling, or jerking motion with the trunk or legs is not acceptable. Inform the examinees that no penalty is exacted for hanging on the bar to rest in the bottom position but that this is not to their advantage. Tell them that half-completed chinups are not counted, and that the scorer will repeat the number of the last correct chinup when incorrect execution is detected.

(h) Administration and Scoring. Caution the examinees to assume the "dead" hanging position and wait for the scorer's command to begin. The scorer is at the examinee's left with a clear view of the bar. If the examinee begins to swing widely, the scorer should stop the swinging by extending the left arm across the front of the examinee's body, being sure not to hinder the execution of the chinup. The scorer counts aloud the number of chinups correctly executed. When a chinup is not correctly executed, the scorer repeats the number of the last correct one. The scorer records the number of correct chinups on the scorecard and returns the card to the examinee.

(2) Test Event 2:Knee Bender

(a) Purpose. This event measures the strength and endurance of the leg muscles.

(b) Equipment. None.

(c) Officials There is one scorer per lane.

(d) Organization. Marines stand in numerical order behind the restraining line in their respective lanes. The scorer takes each scorecard when the Marine is called forward for the test.

(e) Starting Position. The feet are spread less than shoulder width apart, hands on hips, thumbs in the small of the back, elbows back.

(f) Movement. Do a knee bend and at the same time bend slightly forward at the waist and thrust the arms between the legs until the extended fingers touch the ground. The hands are about 6 inches apart. The bend is approximately a three-quarters bend. From this knee bend position, recover to the starting position by moving the body upward, straightening the knees, and returning the hands to the hips. Repeat as many times as required.

(g) Instructions. Explain and demonstrate the correct starting position. Be certain examinees understand the correct knee bend and that only the tips of the fingers touch the ground. Tell them the scorer will repeat the number of the last correct knee bender when incorrect execution is detected. Some of the common errors are failure to correctly bend the knees, failure to touch the ground, and failure to assume the completely erect position after the bend has been executed.

(h) Administration and Scoring. The scorer stands to one side so as to see that the knees are properly bent and the fingers touch the ground as prescribed. From this position, the scorer can view the examinee to see that a properly erect position is assumed after each knee bend. The scorer counts aloud the number of correctly executed knee bends. When a knee bend is done incorrectly, the scorer repeats the number of the last correct one. The scorer records the number of correct knee bends on the scorecard and returns it to the examinee.

(3) Test Event 3:Pushups

(a) Purpose. Pushups measure arm and shoulder extensor strength.

(b) Equipment. None.

(c) Official. There is one scorer per lane.

(d) Organization. Marines stand behind the restraining line in their respective lanes until the scorer calls on them to perform. The scorer takes each scorecard when the Marine comes forward.

(e) Starting Position. The front-leaning rest position is the starting position. The body is straight from head to heels, palms are flat on the ground directly underneath the shoulders, and elbows are straight and locked. The body weight is supported on the hands and toes throughout the event.

(f) Movement. Bending only the elbows, lower the body in one straight plane until the chest touches the scorer's hand. Straightening and locking the elbows, raise the body in one straight plane, returning to the original front-leaning rest position. Repeat as many times as required, keeping the body in a straight line from head to heels.

(g) Instructions. Explain and demonstrate that the arms are straight at the beginning and completion of the movement and that the chest must touch the scorer's hand, but the stomach and thighs must not touch the ground. Also explain that the whole body must be maintained in a straight line as it is lowered and raised; that is, there is to be no breaking at the hips or shoulders so that any body part is lowered or raised in advance of the other or as a separate segment .Dipping or rolling through the shoulders is illegal, as is lowering or raising the body with one arm or shoulder at any time. Resting is not permitted during repetitions. Instruct

Marines that the scorer will repeat the number of the last correct pushup when incorrect execution is detected.

(h) Administration and Scoring. It is recommended that Marines assume a prone position while placing their feet and hands in the proper positions. This permits them to rest while the scorer gets into position and, at the same time, provides a feel of the body in a straight plane from head to heels. The scorer lies on the right hip and side to the right of the examinee. This gives the scorer a clear view of the examinee's body to see any errors. The palm of the scorer's right hand rests flat on the ground underneath the lowest part of the examinee's chest. By keeping the right forearm flat on the ground at an angle in front of the examinee's right arm, the scorer's position will not prevent the examinee from lowering the body completely. The scorer's left hand is free to test the straightening of the elbow at the completion of the movements and to point out body segments being lowered or raised separately. When in position and ready, the scorer has the examinee assume the starting position and begin doing pushups. The scorer counts aloud the repetitions done correctly and repeats the number of the last correct pushup if one is done incorrectly. There is no penalty if the contour of the examinee's body causes the hips to protrude slightly out of line, provided that the whole body is raised and lowered simultaneously. The scorer enters the number of repetitions on the scorecard and returns it to the examinee.

(4) Test Event 4: Situps

(a) Purpose. This event primarily measures abdominal strength.

(b) Equipment. None.

(c) Official. There is one scorer per lane.

(d) Organization Marines stand behind the restraining line in their respective lanes until the scorer calls on them to perform. The scorer takes each scorecard when the Marine comes forward.

(e) Starting Position. The examinee lies flat on the back with knees flexed, both feet flat on the ground. The correct angle of the thighs to the ground is 45 degrees. If the heels are too near the buttocks, the applicant will not be able to sit up. Legs are spread shoulder width apart. The examinee interlaces fingers and places them behind the head in contact with the ground. The feet are not held by another person.

(f) Movement. Bend forward at the waist and raise the upper body until the head is directly over the knees. Heels are not to leave the ground. Elbows remain in the same plane to the head and body throughout the event. The upper body is slowly lowered to the starting position until the head touches the ground. Repetitions are done at a slow cadence with no rest periods.

(g) Instructions. Explain and demonstrate the correct starting position and the proper execution of the situps to be sure that Marines understand the movement. Warn them that their knees must remain flexed during each situp, the heels cannot leave the ground at any time, and they may not roll up on one side and push up with one elbow. Tell them they must do the repetitions at a slow cadence with no rest periods. Instruct Marines that the scorer will repeat the number of the last correct situp when incorrect execution is detected.

(h) Administration and Scoring. When the performer is in position and ready, the scorer has the performer assume the starting position and begin doing situps. The scorer counts aloud the correct executions. When a situp is improperly done, the scorer repeats the number of the last correct one. No situp is credited if the hands are unclasped from behind the head, if the back is used to bounce up from the ground (which means the shoulders would not touch the ground), or if one shoulder or elbow is used to push up. The examinee is not penalized if the heels slide forward slightly so long as the knees remain flexed and the heels maintain contact with the ground. The scorer enters the number of repetitions on the scorecard and returns it to the examinee.

(5) Test Event 5:Endurance Run

(a) Purpose. This event measures cardiovascular endurance.

(b) Equipment One stopwatch or watch with a sweep second hand.

(c) Area. A large training field on which a quarter-mile track has been staked out or a level road over flat terrain may be used as a running surface. A 1-mile route is designated with wooden stakes marking the start point, finish point, and one-quarter mile intervals.

(d) Officials For large groups, a scorer times the event and controls the conduct of the run, and a guide runs with the group and sets the pace.

(e) Organization The run is conducted with groups of Marines in a column formation. Company-sized units may run at the same time with platoons serving as running groups. The scorer issues the command to assume double-time.

(f) Starting Position. Marines are assembled in proper column formation (column of twos, threes, or fours, as appropriate to the size of the group) with short men to the front. When all is ready, the column is moved forward a short distance before the running period is started.

(g) Movement. At the command DOUBLE-TIME, MARCH, the examinees retain their places in the column formation and execute the command. Length of steps is about 40 inches. The scorer has the group execute the run. The formation is maintained during therun.

(h) Instructions. Marines are instructed to maintain formation while running and are informed that the guide will set the proper pace. They will be instructed in the command used to control the column in the execution of the test. The scorer should announce the 4-minute, 2-minute, 1-minute, and 30-second remaining time intervals.

(i) Administration and Scoring. The event may be administered as previously prescribed to a large group, to several Marines, or to an individual. An individual examinee usually does not require a guide or pacer. If the event is administered on a training field, the scorer may stand in the center as the runner(s) circles about the field. This method of administration relieves the scorer of running with each group to be tested. Scoring is based on successful completion of the run as prescribed. The scorer should announce the remaining times as prescribed in (h).

6009. BATTLE FITNESS TEST

The battle fitness test is a physical fitness evaluation which is currently under development as part of the basic warrior training concept plan.

Chapter 7
THE HUMAN BODY

7001.GENERAL

To implement an effective physical conditioning program, a leader must understand how the human body functions and how exercise affects the body .This chapter provides only an introductory discussion of anatomy and functioning.

a. Body Functioning During the Stages of Conditioning. For more on the stages of conditioning, see chapter 1.

(1) During the toughening stage, the waste products of muscle activity (lactic acids) collect more rapidly than the blood can remove them. This acid waste builds up in the muscle tissue and irritates the nerves in the muscle fiber, causing pain and stiffness. As the exercise program continues, more blood is carried through the muscle, removing the waste materials more rapidly and eventually causing the soreness to disappear.

(2) During the slow improvement stage, the blood circulation in the muscles increases, and the body as a whole becomes a more efficient machine. The improvement is rapid in the first few weeks, but as a higher level of skill and conditioning is reached, the improvement becomes less noticeable. The body reaches its maximum level of performance after 6 to 10 weeks and should then be maintained at this peak.

(3) During the sustaining stage, it is possible to maintain this state of conditioning through 15 to 20 minutes of exercise a day, but the exercise must be quite strenuous.

b. Diet. Regular exercise increases the appetite. If the desire for greater amounts of food is satisfied by a balanced diet, the body benefits.

(1) There are two main types of foods: body-building and energy- producing.

(a) Body-building foods consist of proteins, which build up tissue and repair wear and tear.

(b) Energy-producing foods include carbohydrates and fats. Carbohydrates provide a quick source of energy, while fats act as a reserve store of energy.

(2) In addition, food contains vitamins, mineral salts, and water. During hot weather and strenuous training periods, the body requires greater fluid intake.

(3) Diet should be supplemented with proper rest to allow the digestive system to digest the lood.

(4) Occasionally, especially during early stages of conditioning, strenuous exercise may cause vomiting. Although vomiting is not a frequent occurrence, it is not usually a cause for concern.

7002. SYSTEMS OF THE BODY

The systems of the body include the skeletal, muscular, circulatory, respiratory, endocrine, digestive, genitourinary, and nervous systems. Each has a different function, but all must work in cooperation with one another to insure a sound body. Of these systems, the first four are the most affected by exercise and are discussed in succeeding paragraphs.

7003. THE SKELETON

a. Bones. The skeleton (see fig. 7-1) is composed of about 206 bones, which-
o Provide structure for the body and support for the attachment of muscles,
o Protect vital organs, such as the brain, lungs, and heart,
o Manufacture red blood cells, which carry oxygen through the body.

Figure 7-1 is available in the illustration section at the end of the book.

b. Joints. A joint is a place of union between two or more bones. Joints can be-
o Immovable, such as in the face and head.
o Slightly movable, such as in the vertebrae or sternum.
o Freely movable, such as in the shoulder, hip, knee, ankle, and elbow. The bones in a freely movable joint are held in place by ligaments. Freely movable joints are of the greatest importance in physical training because exercise improves their mobility and stability.

c. Cartilage and Ligaments. The joints in the body, where bones connect, are supported by cartilage and ligaments.
o Cartilage is a tough, elastic, translucent tissue that acts as a shock absorber or buffer between bones. Examples are the discs between the vertebrae, the tissue attaching the ribs to the sternum, and buffers in the knee joints. Damaged cartilage does not heal.
o Ligaments are connective tissue that bind bones together. They contract and extend but are not elastic. Consequently, ligaments can be torn or strained. Damaged ligaments will heal to some extent.

7004. THE MUSCLES

a. General. Muscles are tissue, or an organ composed of tissue, which contract or extend to produce movement. Muscles are of three classifications: involuntary, voluntary, and cardiac. The voluntary and cardiac muscles are the most affected by physical training; physical training has little effect on involuntary muscles.(Seefigs. 7-2, 7-3, 7-4.)
o People have no control over the involuntary muscles, an example of which is the diaphragm,
o Voluntary muscles are the larger skeletal muscles which cause movement and which the individual can control.
o Cardiac muscle is found only in the heart and for all practical purposes is an involuntary muscle. Physical training has a significant effect on the fitness of cardiac muscles.

b. Muscle Structure. Muscle is composed of cells which are grouped into bundles called fibers. Groups of fibers form muscles. These bundles of fibers are held in place by sarcolemma, a thin, sheath-like material that surrounds the muscle bundles and secretes a fluid that lubricates the muscle tissue. The fused ends of the sarcolemma form the tendons which attach the muscles to the bones.

c. Attachment of Muscles. The arrangement of muscles on the skeleton provides the proper angle of pull to make movement possible. Voluntary muscles are usually attached to the skeleton in two places.

d.Action of Muscles. To produce motion and to do work a muscle usually shortens its fibers. The movement may be flexion such as bending the arm at the elbow, or extension, such as the straightening the arm. Some muscles raise the arms or legs, others depress the raised limb. Some muscles have the primary function of rotating the trunk from side to side. In addition, muscles act as stabilizers as well as prime movers.

Figure 7-2 is available in the illustration section at the end of the book.

Figure 7-3 is available in the illustration section at the end of the book.

Figure 7-4 is available in the illustration section at the end of the book.

7005. FUNCTIONING OF THE SKELETON AND MUSCLES

a. General. The skeleton and muscles function in close coordination. Muscles move the body, and bones support the muscles.

b. Effect of Exercise on Bones. Continuous exercise, particularly among younger people, usually strengthens the bones, allowing them to withstand greater stress and strain. Bones which are not used lose much of their strength, a factor which should be considered when individuals return to physical training after a prolonged period of inactivity. People in this category should refrain from activities which might result in bone breakage before the bone is exercised back to normal condition. The condition known as "march fracture" is usually a result of this condition.

c. Muscular Strength. When exercised vigorously to improve strength, muscle grows in size. In general, the larger the muscle, the stronger the muscle. Furthermore, trained muscle functions more efficiently than untrained muscle. Trained muscles contract more vigorously and with less effort. Regular and strenuous exercise also toughens muscle, making it firmer and able to stand more strain.

d. **Muscular Endurance.** Muscles naturally become fatigued during continuous, repeated work. Through specific exercises, such as chinups or situps, local muscle groups become exhausted long before an individual fatigues. Training to develop muscular endurance enables people to continue a relatively heavy load of exercise over a long period of time. Lengthened exercise periods improve muscular endurance.

e. Muscular Coordination. Improved speed and strength result in part from improved muscular coordination. An unskilled individual may use irrelevant muscles to perform a particular activity, increasing the amount of work without increasing the mechanical output. This increase in skill is highly desirable, but it should be offset by greater effort in duration or intensity to compensate for the loss in overload due to increased skill.

f. Muscular Fatigue. When the rate of work is increased, the energy required is proportionately much greater than the increase in rate. For example, if an individual doubles running speed, the amount of power demanded to do this is increased eight times.

g. Circulation in Muscles Regular, strenuous exercise causes the creation of new capillaries and the opening of inactive, latent capillaries, increasing blood circulation within the muscle by as much as 400 percent. This increases the supply of food materials and oxygen to the muscle, improving its endurance. This process takes about 8 to 12 weeks of regular conditioning in young adults and longer as age increases.

7006. THE CIRCULATORY AND RESPIRATORY SYSTEMS

a. The Circulatory System. The circulatory system transports blood to all parts of the body, removes waste products for disposal, and delivers protecting and repairing substances where needed. The circulatory system includes the heart and blood vessels.

(1) The Heart. The heart is a pump which forces blood through the blood vessels to the parts of the body. The heart is a little larger than a fist and is located in the left center of the chest between the lungs.

(2) Blood Vessels. The vessels carrying blood away from the heart are the arteries. They eventually divide into capillaries, the very small vessels through which diffusion and osmosis take place. The capillaries gradually increase in size, forming the veins which carry blood back to the heart.

b. The Respiratory System. The respiratory system performs the function of breathing, whereby oxygen is inhaled and carbon dioxide is exhaled. The respiratory system consists of the mouth, nose, trachea, lungs, and diaphragm.

o Trachea. The trachea, or windpipe, is a hollow, tube-like structure that carries air from the mouth to the lungs.

o Lungs. The lungs are elastic bags in the chest where the exchange of oxygen and carbon dioxide takes place.

o Diaphragm. The diaphragm is a thin, sheet-like muscle just below the lungs. During inhalation, the diaphragm flattens out and lowers, allowing the lungs to expand and fill with air. During exhalation, the diaphragm raises into a dome shape, helping to expel air from the lungs.

7007. CARDIOVASCULAR FUNCTIONING

a. General. Cardiovascular functioning is the combined functioning of the circulatory and respiratory systems. The chief organs of these systems, the heart and lungs, function together to provide oxygen to the body.(See fig. 7-5.)

Figure 7-5 is available in the illustration section at the end of the book.

b. Heart Action. The heart, the organ which propels blood through the blood vessels, is the chief organ of cardiovascular endurance. A person tires quickly if the heart lacks the capacity to circulate the blood through the body. The heart is a muscular organ and is developed by exercise just like any other muscle. The heart cannot be exercised alone; any exercise that puts a sufficient load on the heart, such as running or cycling, will also exercise other body parts. A conditioned heart tends to beat more slowly but pump more blood when at rest. This is known as an increase in "stroke volume,"a desirable condition because it enables the heart to pump more blood with a slower contraction rate.

c. Functions of the Lungs. Strenuous and regular exercise can improve the efficiency with which the lungs transmit oxygen to the blood by as much as 25 percent In a poorly conditioned person, some of the alveoli (air sacs) within the lungs are closed or collapsed. The forced breathing created by exercise over a period of weeks causes the air sacs to slowly expand, increasing the ability to absorb oxygen.

d. Relationship of Heart and Lungs. The heart pumps carbon dioxide-laden blood through the pulmonary artery to the lungs. In the lungs the carbon dioxide is exchanged for oxygen and the purified blood is returned to the heart by way of the pulmonary vein. The heart then pumps the blood through the aorta for circulation throughout the body. Blood moving into the muscles exchanges oxygen for carbon dioxide, after which time it is ready for the return trip to the heart.

e. Cardiovascular Functioning in High Altitudes. Marines to be employed in high altitudes should be acclimated in a similar area for 10 to 14 days prior to employment. Persons not accustomed to the rarefied air of higher altitudes tire more quickly and may collapse after rapid physical exertion. Air is much less dense at high altitudes than at sea level. Persons accustomed to sea level inhale only about 80 percent of the oxygen they are used to no matter how hard they breathe. Further, people accustomed to sea level or moderate altitude simply do not have enough red corpuscles in the blood to fulfill their needs at high altitudes. The red blood cells will increase over time.

f. Symptoms of Cardiovascular Malfunctioning. The brain is the first organ to react to a lack of oxygen. Unconsciousness results when the brain is denied sufficient oxygen. "Blacking out" is actually a defense mechanism to keep the body alive.

APPENDIX A
INSTRUCTOR HINTS AND INSTRUCTOR TRAINING

1. General. This appendix informs instructors on administration of exercise periods, commands, the extended rectangular and circle formations, methods of instruction and conduct of exercise activities, explanation of basic positions associated with the various activities, and instructor training.

2. Aim of the Physical Fitness Program. The aim of the physical fitness program is to prepare Marines physically for war. This can be conveniently split into two phases.

a. Preparation Phase

(1) Development of all-around physical fitness.

(2) Development of purposeful physical skills.

(3) Development of mental alertness.

(4) Development of character and leadership.

b. Maintenance Phase. Once the preparation is completed, the program must progress to a maintenance program. This phase revolves around the need to maintain the trained Marine's physical fitness for a particular role. This will include—

(1) Maintenance of a high standard of all-around physical fitness.

(2) Application of purposeful physical skills to combat skills.

(3) Development of leadership and self-confidence.

(4) Development of mental toughness and endurance.

3. Physical and Mental Development

a. Marines need to be developed physically and mentally. The physical training (PT) instructor must ensure that the program accomplishes this.

b. The system of the body functions best in response to progressive exercise. This means that Marines should master the easier exercises before they attempt the more difficult ones.

c. The body and mind are linked inseparably and must be considered as one in physical fitness training. Physical exercises assist in developing the mind.

d. The confidence in oneself and in one's comrades attained by physical achievement helps to promote the indomitable will to win. The Marine must be taught to realize this. It will help during arduous training, which at all times must be imaginative, realistic, and filled with enthusiasm. All physical fitness training must be associated with developing the qualities required of a Marine.

4. Planning and Preparation of ProgramsThe following factors must be considered when compiling a program (see also ch. 2)~

a. Aim or desired skill level.

b. Duration of the complete training program and number of physical training lessons.

c. Purpose of included exercises,drills, or practical applications.

d. Beginning/inventory fitness levels.

e. Medical facilities, medical personnel, and Marine Corps or local command regulations governing their proximity to the training.
f. Number of Marines to be trained.
g. Availability of instructors, assistants, and facilities.
h. Apparatus and equipment available.
i. Alternative training areas for varying weather conditions,
j. Stages of progression and tests of achievement.

5. Five Basic Principles of Program Design
a. Regularity. Regularity of exercise is far more important than the amount of exercise performed on a given day. A Marine should exercise a minimum of three times per week. Human muscle begins to atrophy after 3 days without exercise.
b. Balance. The program must be balanced in terms of total physical fitness. Imbalance is created when, for example, an individual merely runs. While the cardiovascular system is improved, general upper body strength may be lacking. Conversely, a person who just does weight training may develop strength, but not cardiovascular endurance.
c. Overload. If total body fitness is to be developed, then the muscular and cardiovascular systems must be made to perform more work than previously. This is done by increasing the frequency, intensity, and duration of the exercise. The body adapts to the amount of stress placed upon it and becomes stronger and more efficient.
d. Progression. To be successful, any type of training program must be progressive. This is an area where untrained persons conducting PT can cause injury to Marines resulting in failure of the PT program. Close supervision must be made of untrained instructors. The result of progressing too fast is that the class begins to reject PT because it is no longer enjoyable or motivating. The program should start at a slow pace, and the pace should increase as proficiency improves. Average persons will encounter three stages of progressiom the toughening stage, the slow improvement stage, and the sustaining stage. (See ch. 1.)
e. Variety. Some programs fail because the training becomes boring. Variety is essential and perhaps presents the greatest challenge to the commander and staff.

6. Grouping. Providing for different levels of physical fitness is particularly recommended in the early stages of conditioning. Older Marines and those in poorer physical condition should be expected to attain a group level of fitness, but they should be given more time to do it.
a. Homogeneous Grouping One simple method of providing for the difference in levels is to group them according to their condition. A two-group classification would divide Marines into highly conditioned and average groups. A finer classification could be obtained by dividing them into three groups—a highly conditioned, a moderately conditioned, and an unconditioned group.
b. Performance or Age Grouping. The segregation of Marines into different exercise groups should be based on physical fitness test scores or on the level of fitness they demonstrate. They may also be grouped at first according to age. A common classification by ages is under 30, 30 to 34, and 35 and above.

7. Preparatory Commands and Commands of Execution. The prepreparatory command describes and specifies what is required. The command of execution calls into action what has been prescribed. All preparatory commands are given with a rising inflection. The interval between commands is long enough to permit the average Marine to understand the first one before the second one is given.

8. Extended Rectangular Formation. The traditional formation for carrying on many physical training activities is the extended rectangular formation (fig. A-1). This formation is more compact than any other. It is the best type to use for large numbers of Marines because it is simple and easy to assume.

a. For the formation of oneplatoon, the base man paces off five paces from the stand, faces left and moves five paces, halts, and again faces left. With the base man positioned facing the stand, the platoon leader then commands. **FALL OUT AND FALL IN ON THE BASE MAN**. At this command, all Marines run to the designated area and reform. This procedure is preferred to marching the unit into position. If more control is desired, the unit may march at double time to the vicinity of the base man and then be directed to fall out and fall in on him. Much valuable time is wasted in the field by needless maneuvering of Marines at quick time in an effort to position the platoon or unit on the exact spot for the exercises.

b. A company-size unit assumes the extended rectangular formation from a column of threes or fours at normal intervals between squads. This extension can also be executed from a company mass without interval between platoons. In extending either a platoon or company-size unit, take your place at the head of the column and command:
(1) TAKE INTERVAL TO THE LEFT, MARCH. At this command, the Marines in the right flank file stand fast with arms extended sideward. All others turn to the left and run forward at double time. After taking a sufficient number of steps, all Marines face the front with both arms extended sideward. The distance between finger-tips is about 12 inches, and dress is right.

Figure A-1 is available in the illustration section at the end of the book.

(2) ARMS DOWNWARD, MOVE. At this command, the arms are lowered smartly to the sides.
(3) LEFT, FACE
(4) EXTEND TO THE LEFT, MARCH. At this command, those in the right flank file stand fast with arms extended sideward. All others turn to the left and run forward at double time. Spacing is the same as in (1) above, and dress is right.
(5) ARMS DOWNWARD, MOVE. Same as in (2) above.
(6) RIGHT, FACE.
(7) FROM FRONT TO REAR, COUNT OFF. At this command, the leading Marine in each column turns the head to the right rear, calls of **ONE** and faces the front. Successive Marines in each column call off in turn **TWO, THREE, FOUR, FIVE** in the same manner.
(8) EVEN NUMBERS TO THE LEFT, UNCOVER. At this command, each even-numbered Marine stride jumps sideward to the left, squarely in the center of the interval. In doing this, each one swings the left leg sideward, jumps from the right foot to the left foot, and smartly brings the right into position against the left.

c. To assemble the unit, you commandASSEMBLE TO THE RIGHT, MARCHAt this command, all return to their original position in the column at double time and reform on the base man.

d.It is recommended that the area for grounding equipment and arms be at the edge of, or well away from, the area to be used for exercising. To conserve time and ensure proper position of the unit, the base man or, if the unit is composed of several platoon-sizegroups, the various base men may precede the unit and establish their positions in relation to the instructor's stand.

9. Circle Formation.The circle formation is effective for conducting various exercise activities (fig. A-2). This formation has an advantage in that supervision of all Marines is facilitated and a moving formation is available which provides control. Guerrilla exercises, grass drills, and some forms of running are examples of activities which are more easily conducted in the circle formation than in the extended rectangular formation.

a. It is not advisable to have more than 60 Marines in a circle. When more must be accommodated, separate circles should be used.Concentric circles may be employed to reduce the size of the circle or to accommodate more persons. If concentric circles are formed, the different circles are made by designating squads for each circle.Each additional circle requires more Marines than the one inside it.For example, one squad of a platoon may form the inner circle and theremaimng three squads the outer circle.

b. When a platoon is to form a circle, the commands ai€IRCLEFORMATION, MARCH, FOLLOW MEUpon this command, the right flank squad of the column moves forward at double time with the leader of the platoon group gradually forming a circle in a counterclockwise direction. Each succeeding file falls in behind that on the right. After the rough outline of the circle is formed, the leader commandsPICK UP A 5-YARD INTERVAL.

Figure A-2 is available in the illustration section at the end of the book.

c.The group may be halted and faced toward the center, or, if instruction is not necessary, the exercise activity may be executed without stopping the platoon.

lO.Leadership Techniques

a. Unless you experience all the exercises, you cannot appreciate how arduous they are, what movements are most strenuous and difficult, where the errors in performance are likely to occur, and what the proper cadence should be.

b. You must give everyone careful supervision and participate in the exercises to show that you can do them. When you participate, your assistant instructors should supervise because it is difficult for you to supervise and exercise at the same time.

c. Marines should never be kept too long in one position, especially a constrained one. They should never have to perform an exercise more times than they can do it without losing the proper form. You should insist upon proper form in the execution of all exercise activities.Even slight deviations from the proper form reduce the value of the exercise.

d. Avoid long explanations.As a rule, it should be necessary to give a full explanation of new exercises only. Explain the most essential features of an exercise first; add details later. Too many details at one time are more likely to confuse than to assist them.Minor corrections should be made to the entire class while the exercise is in progress (for exampHEADS **UP,** KNEES STRAIGHT). If necessary, follow this direction by the name of the person who is particularly at fault. If a Marine requires special attention, givethat person separate instruction by an assistant instructor to avoidwasting the time of the group.

e. The heavy demand on your voice can be lightened by training assistant instructors to assume some of the instruction. Using mass cadence is also an effective method of lessening the demand on your voice.

“Illustrations”

DAY	TIME	ACTIVITY
MONDAY	60 MIN	RUNNING ACTIVITY
TUESDAY	60 MIN	COMBATIVES
WEDNESDAY	60 MIN	CONDITIONING & LOG DRILLS
THURSDAY	60 MIN	CONDITIONING & RELAYS
FRIDAY	60 MIN	RUNNING & CONDITIONING
SATURDAY	60 MIN	RUNNING ACTIVITY

Figure 2-1. Sample A- -Single Activity.

DAY	TIME	ACTIVITY	REMARKS
MONDAY	20 MIN 20 MIN 20 MIN	RIFLE DRILL COMBATIVES RUNNING	PROGRESSIVE ACTIVITY SYSTEM
TUESDAY	30 MIN	GUERRILLA EXERCISES TABLE I	SINGLE ACTIVITY
WEDNESDAY	30 MIN	RIFLE DRILL	SINGLE ACTIVITY
THURSDAY	30 MIN	CIRCUIT INTERVAL TRAINING	SINGLE ACTIVITY
FRIDAY	20 MIN 20 MIN 20 MIN	STRENGTH CIRCUIT CONDITIONING DRILL CROSS-COUNTRY RUN	ROTATION ACTIVITY
SATURDAY	60 MIN	WARM UP RUN AND GRASS DRILL	SINGLE ACTIVITY

Figure 2-2. Sample B- -Various Time Periods, Activities, and Systems.

DAY	TIME	ACTIVITY
MONDAY	45 MIN	RUNNING – CROSS-COUNTRY
TUESDAY	75 MIN	TEAM ATHLETICS – TOUCH FOOTBALL
WEDNESDAY	45 MIN	RUNNING – CROSS-COUNTRY
THURSDAY	20 MIN 40 MIN	LOG DRILL DUAL COMBATIVES
FRIDAY	90 MIN	SPEED MARCH
SATURDAY	90 MIN	TEAM CONTESTS

Figure 2-3. Sample C--Sustaining Activities.

Route	Kilometers from SP	Recommended rate of march (kmph)	Remarks
SP: RJ 8th Div Rd—Superhighway	......	4	Bridge; hard surface; two guides.
RJ Jamestown—Hersey Rd	2,25	4	Traffic heavy; two guides; Jamestown Rd bears to right.
RJ Jamestown—Yankee Rd	7.10	4	Two guides
RJ Jamestown—Lightning Rd	10.50	4	Two guides
RJ Jamestown—Sunshine Rd	12.40	4	Light traffic; one guide Sunshine Rd; gravel, poor traction when wet.
RP; RJ Sunshine—Sedan Rd	14.80	4	Two guides

Figure 3-1.Example of Route Reconnaissance Report.

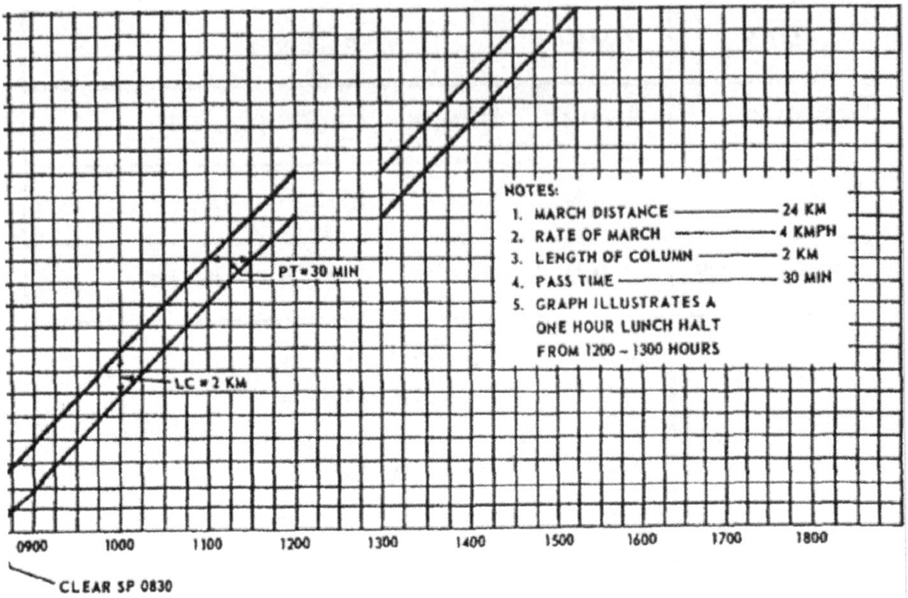

Figure 3-2.Road Movement Graph.

Figure 3-3.The Pace-Setter.

	SINGLE FILE	COLUMN OF TWO
2 m/MAN APART	2 4	1 2
5 m/MAN APART	5 4	2 7

GAP DISTANCES			
SERIAL:	(3-1) × 100 m	=	200 m
MARCH UNIT:	(12-1-2-) × 50 m	=	450 m
	TOTAL COLUMN GAP		650 m

Figure 3-4.Conditioning Drill One.

Figure 3-5.Conditioning Drill Two.

Figure 3-6.Conditioning Drill Three.

Figure 3-7. Rifle Drill.

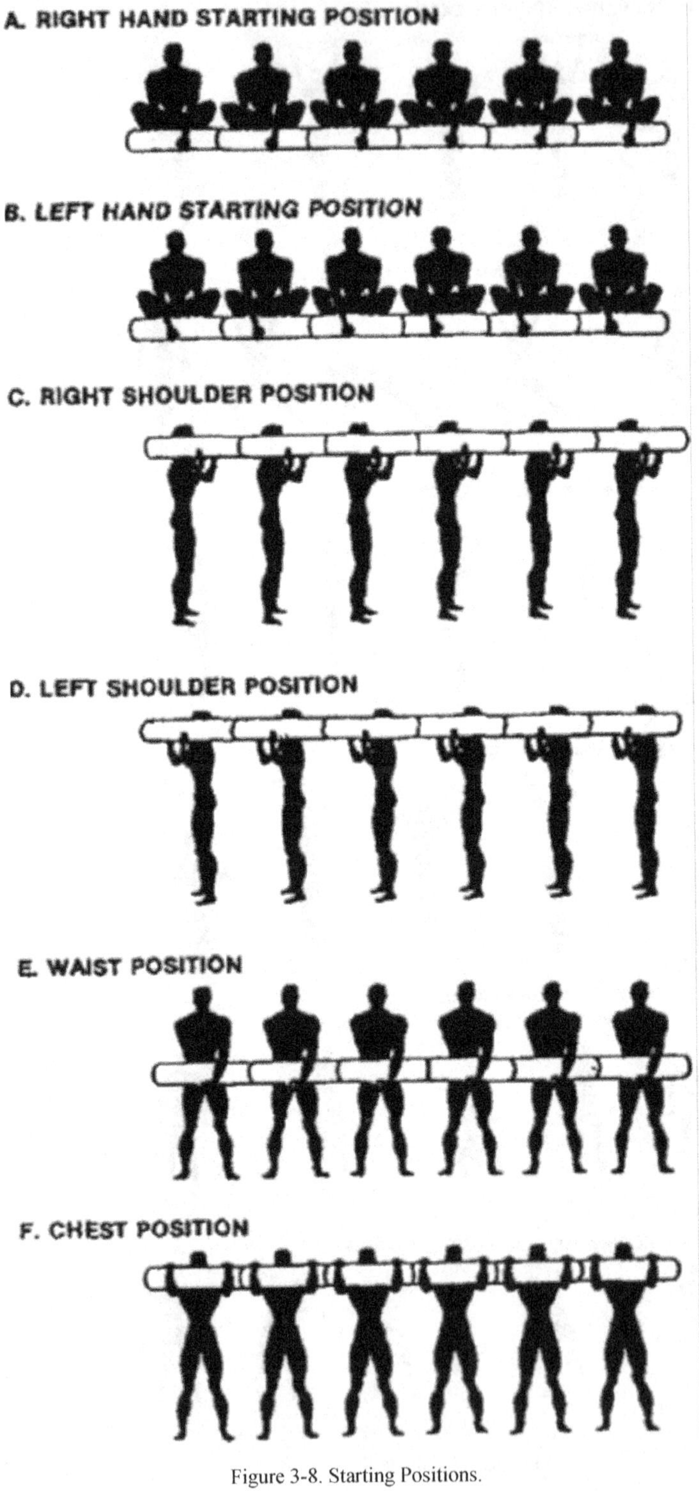

Figure 3-8. Starting Positions.

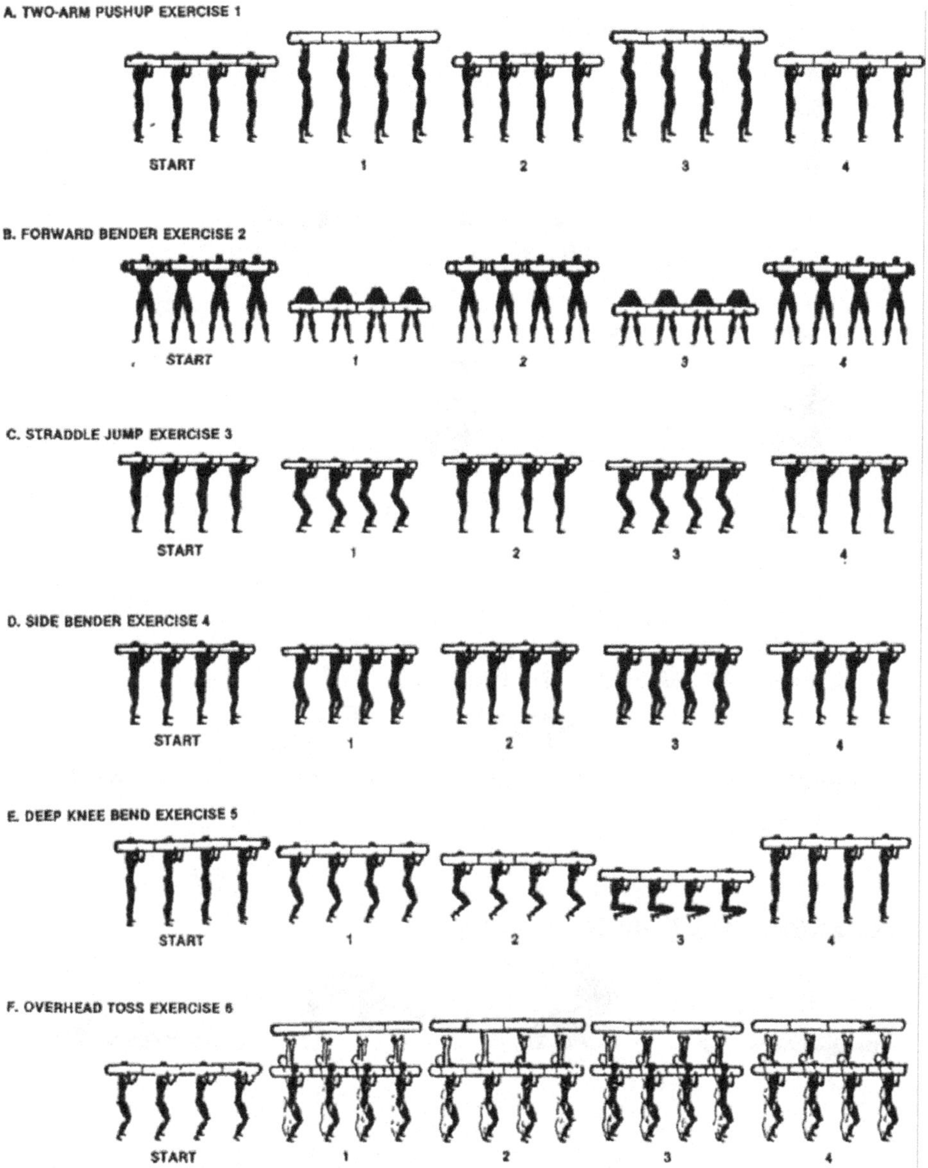

Figure 3-9. Log Drill

Figure 3-10.Basic Positions for Grass Drill.

Figure 3-11.Grass Drill One.

Figure 3-12.Grass Drill Two.

Figure 3-13.Guerrilla Table I.

Figure 3-14.Guerrilla Table II.

Figure 3-15.Proper Running Form.

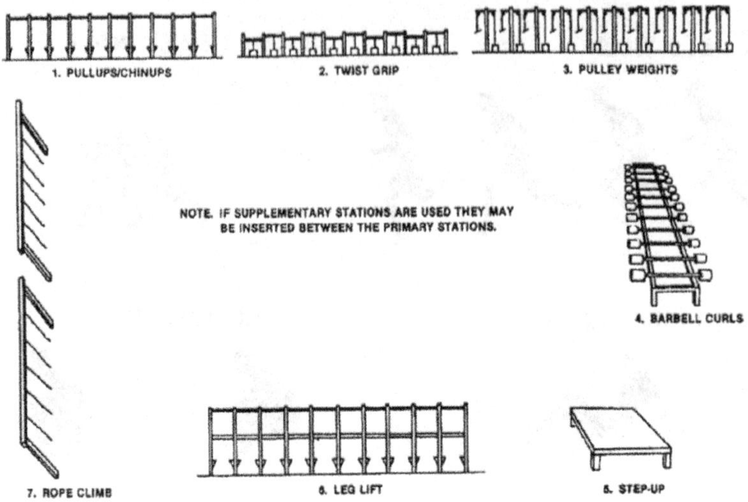

Figure 3-16. Fixed Strength Circuit.

STATION	ITEM	NO.	SPECIFICATION
1	BARBELL		1¼-INCH PIPE 5 FEET LONG WITH CONCRETE FILLED NO. 10 CANS
2	JUMP ROPE		¼- OR ⅜-INCH ROPE, 10 FEET LONG
3	TWIST GRIP		HANDLE 12 INCHES LONG, ROPE 4 FEET LONG, NO. 10 CAN CONCRETE FILLED.
4	INCLINE PLANE		¾-INCH PLYWOOD PLATFORM 2 FEET WIDE AND 6 FEET, 6 INCHES LONG, ELEVATED 10 INCHES AT ONE END. STRAP TO HOLD FEET DOWN.
5	WAR CLUB		HEAD IS 6 BY 12 INCHES, HANDLE IS 14 INCHES LONG BY 1¼ INCHES IN DIAMETER, ABOUT 20 POUNDS.
6	BICYCLE RIDE		PLYWOOD BOARD OR PLATFORM 2 BY 3 FEET WITH 2 BY 2 RUNNERS.
7	STEP—UP		A BOX OR STURDY PLATFORM 18 INCHES HIGH, 18 INCHES WIDE, 24 INCHES LONG.
8	ISOMETRIC PULL		TWO HANDLES 12 INCHES LONG WITH 4 FEET (BETWEEN HANDLES) OR LIGHT WIRE CABLE OR ¼-INCH ROPE

Figure 3-17.Movable Strength Circuit.

Figure. 3-18.Stations of Movable Strength Circuit.

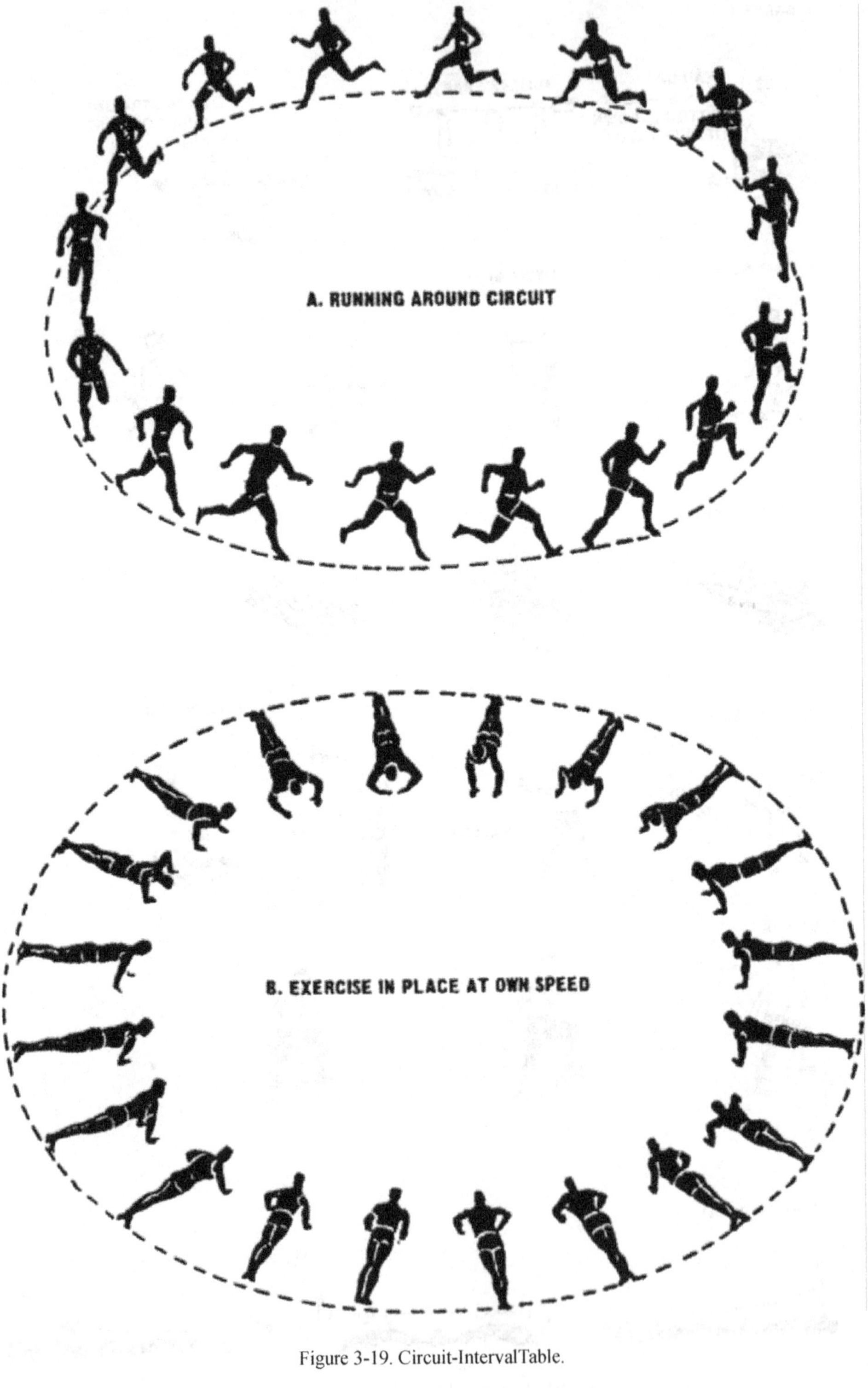

Figure 3-19. Circuit-IntervalTable.

Figure 3-20. Jumping.

Figure 3-21.Dodging.

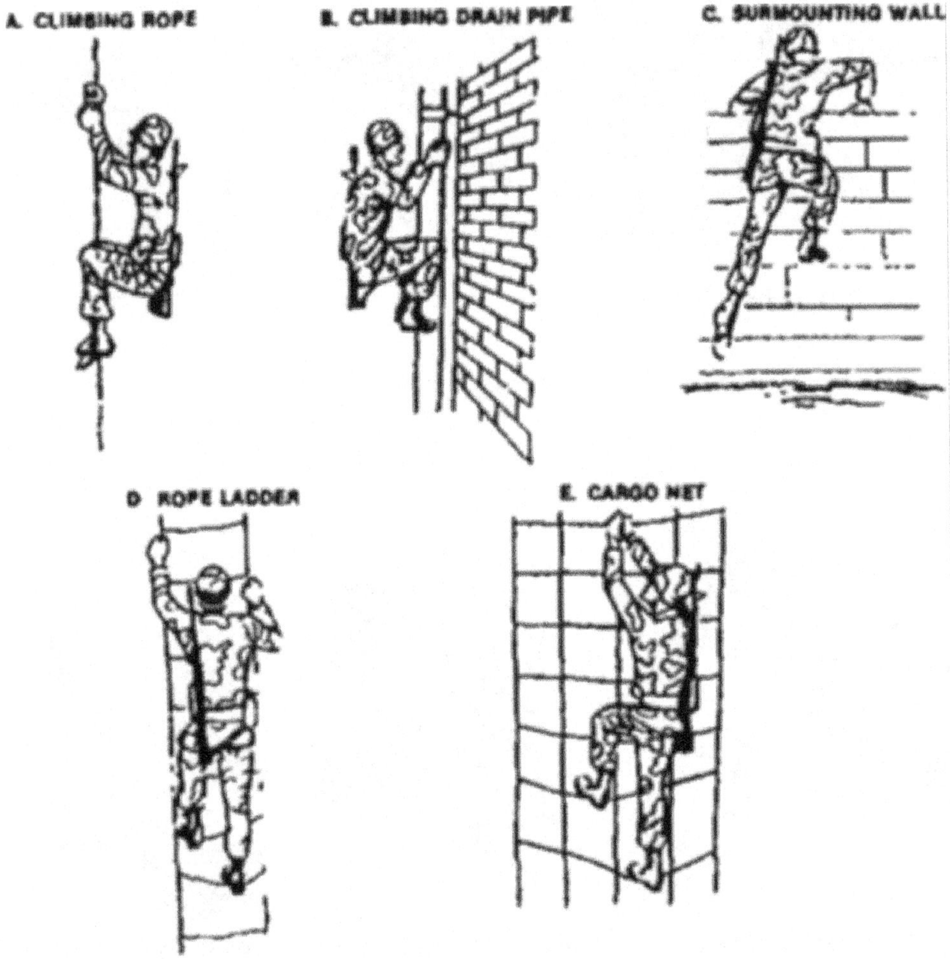

Figure 3-22.Climbing and Surmounting Vertical Objects.

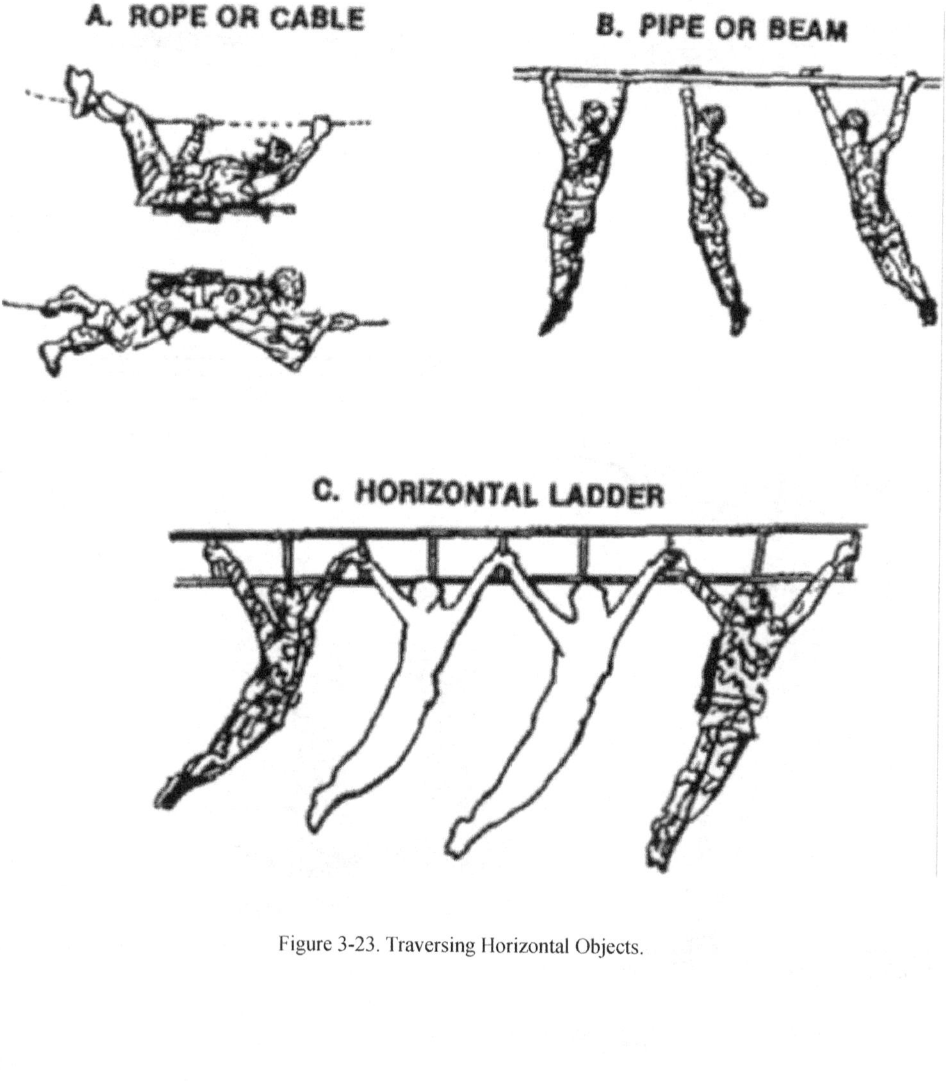

Figure 3-23. Traversing Horizontal Objects.

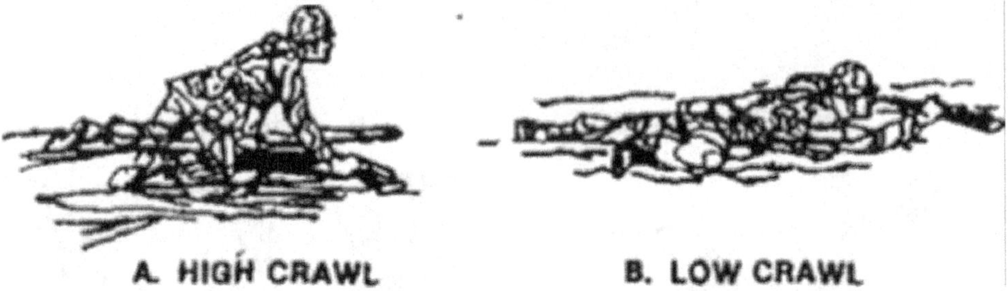

Figure 3-24.Crawling.

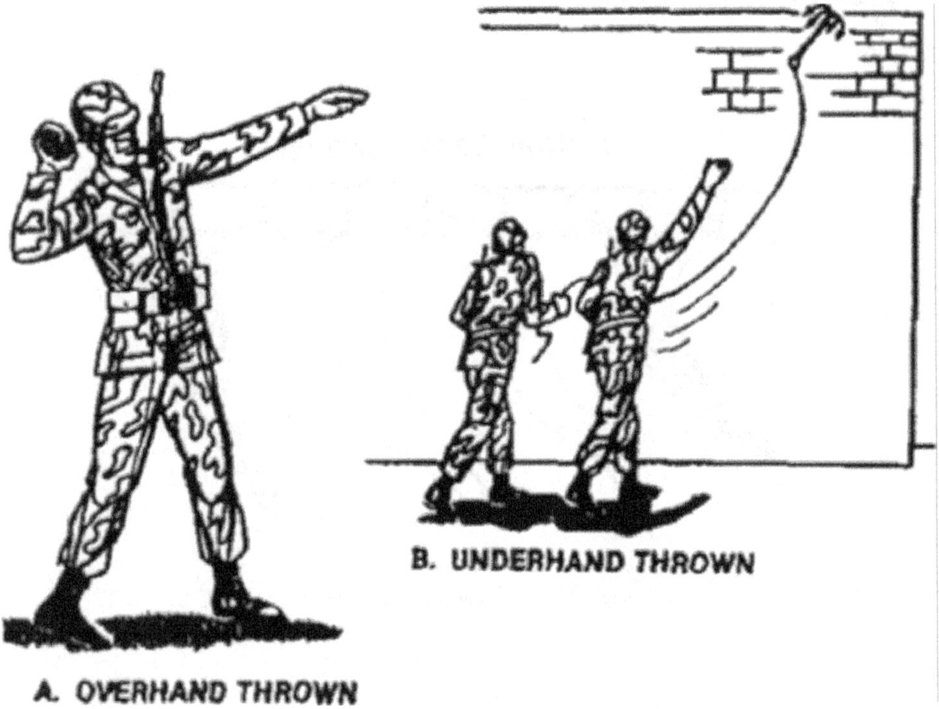

Figure 3-25.Throwing.

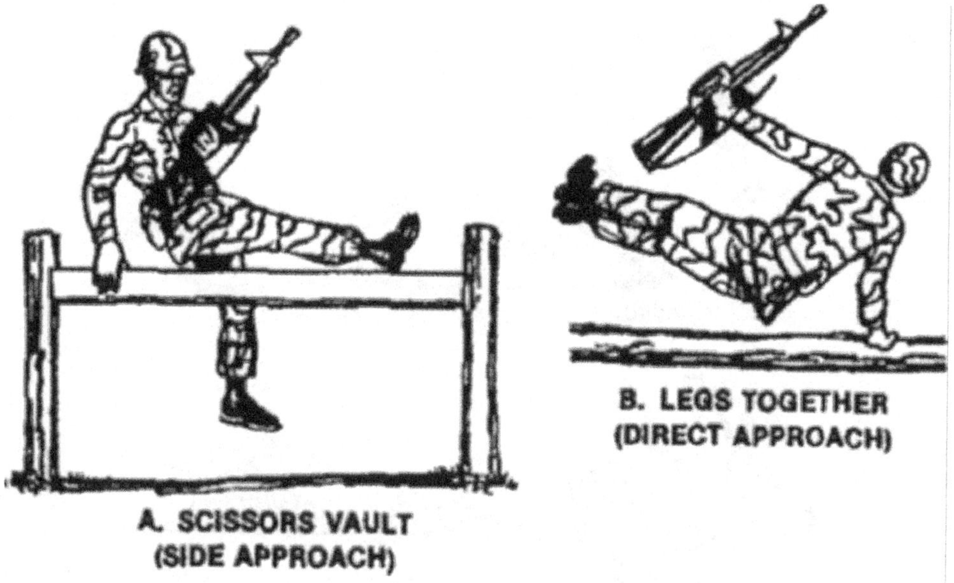

Figure 3-26. Vaulting

Figure 3-27.Balancing.

Figure 3-28.Falling.

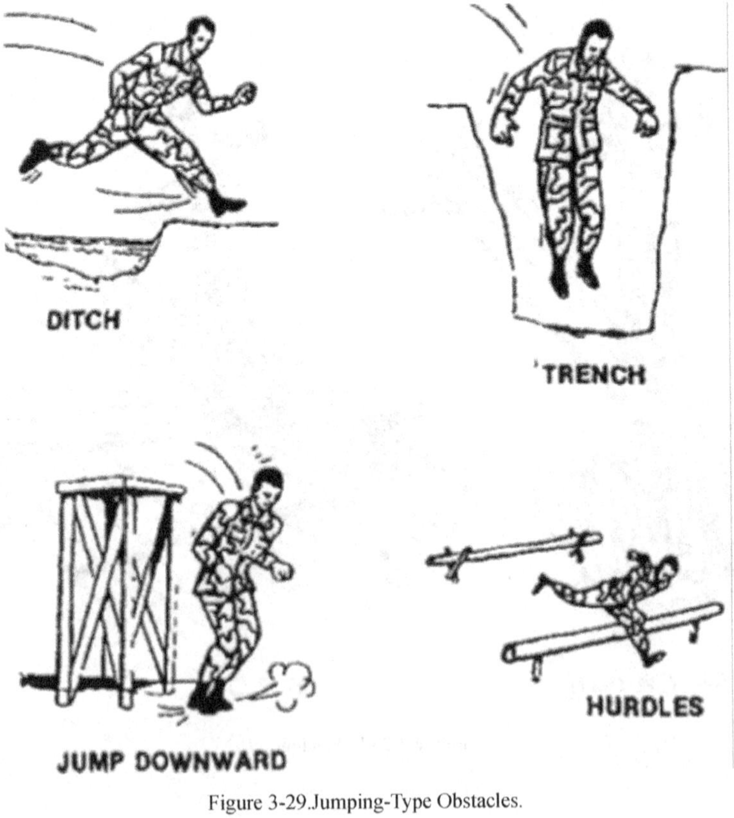

Figure 3-29.Jumping-Type Obstacles.

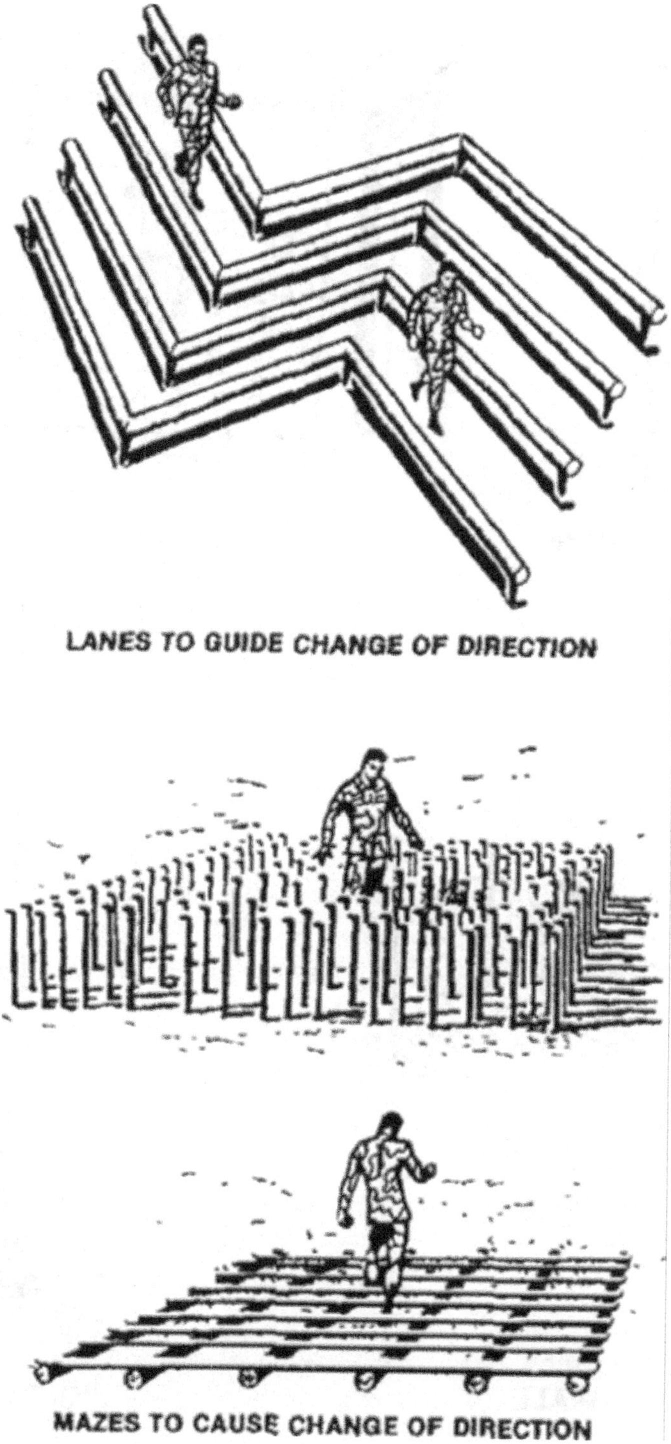

Figure 3-30.Dodging-Type Obstacles.

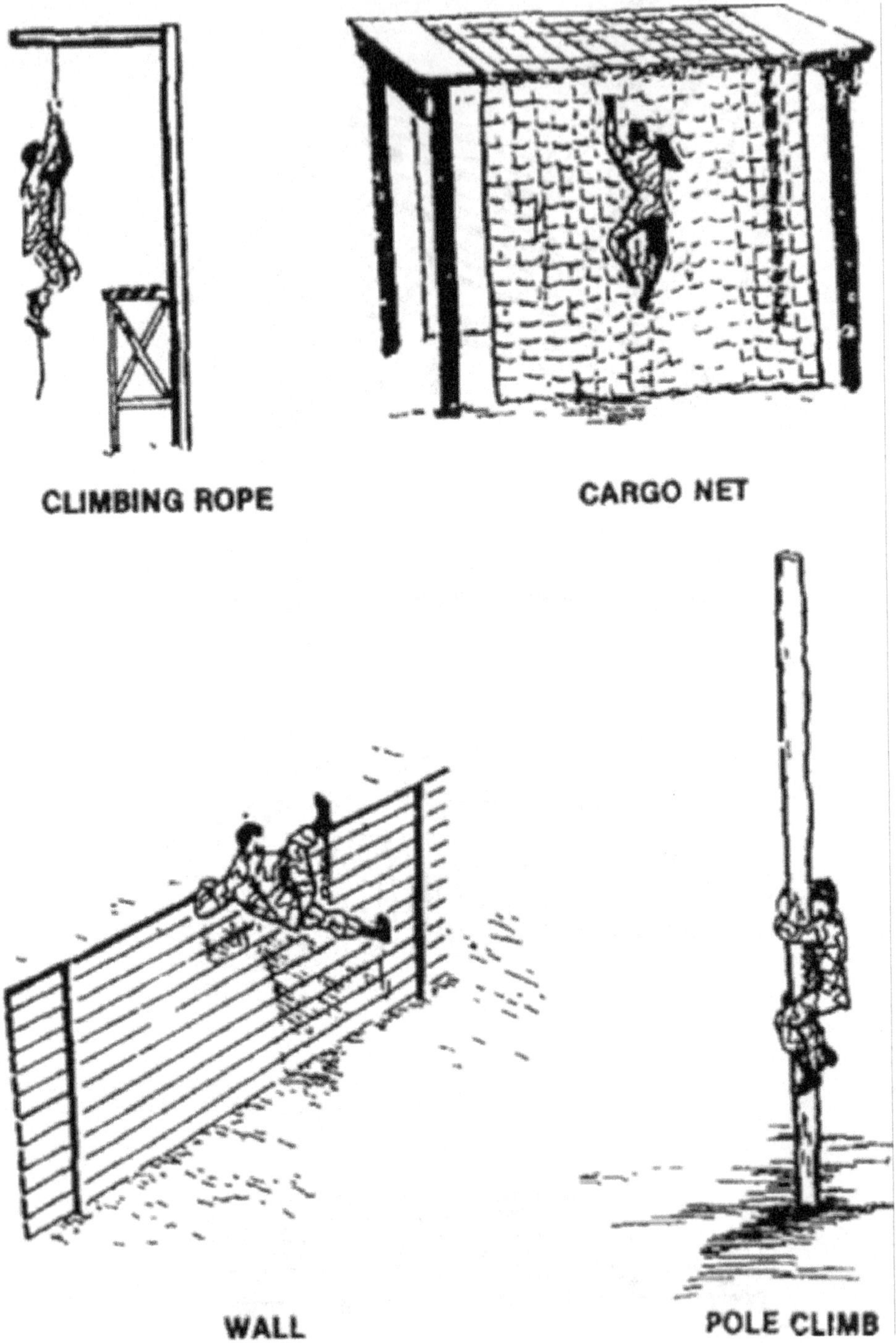

Figure 3-31. Vertical Climbing Obstacles.

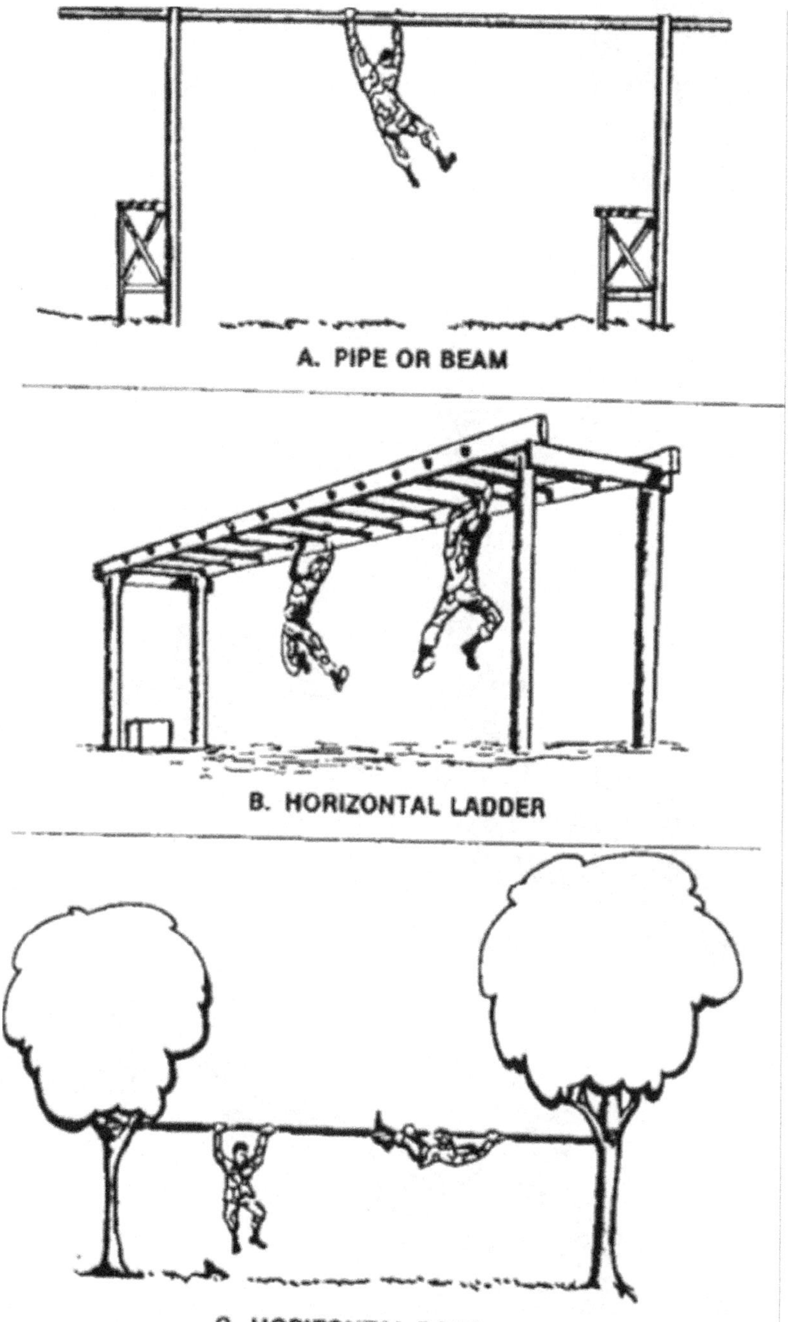

Figure 3-32.Horizontal Type Obstacles.

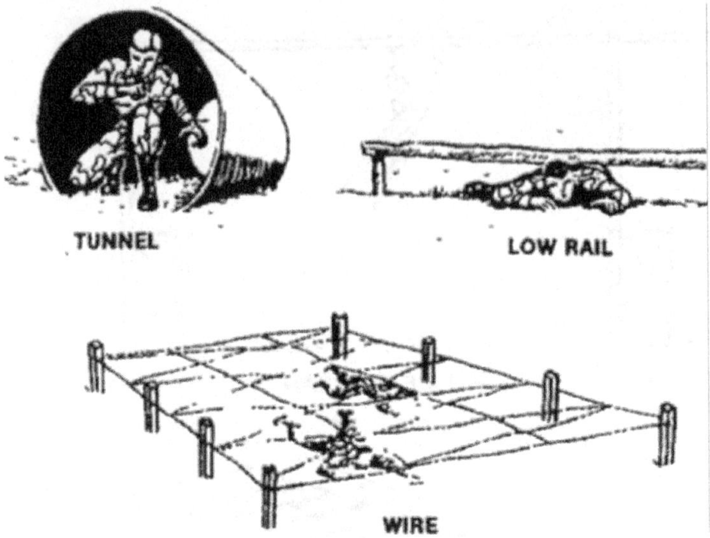

Figure 3-33.Crawling-Type Obstacles.

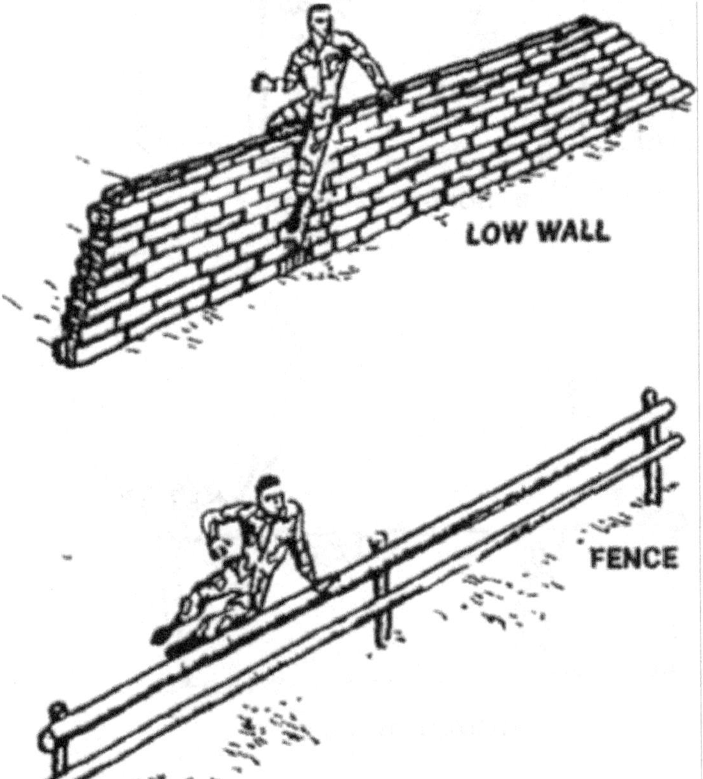

Figure 3-34.Vaulting Obstacles.

Figure 3-35.Balance-Type Obstacles.

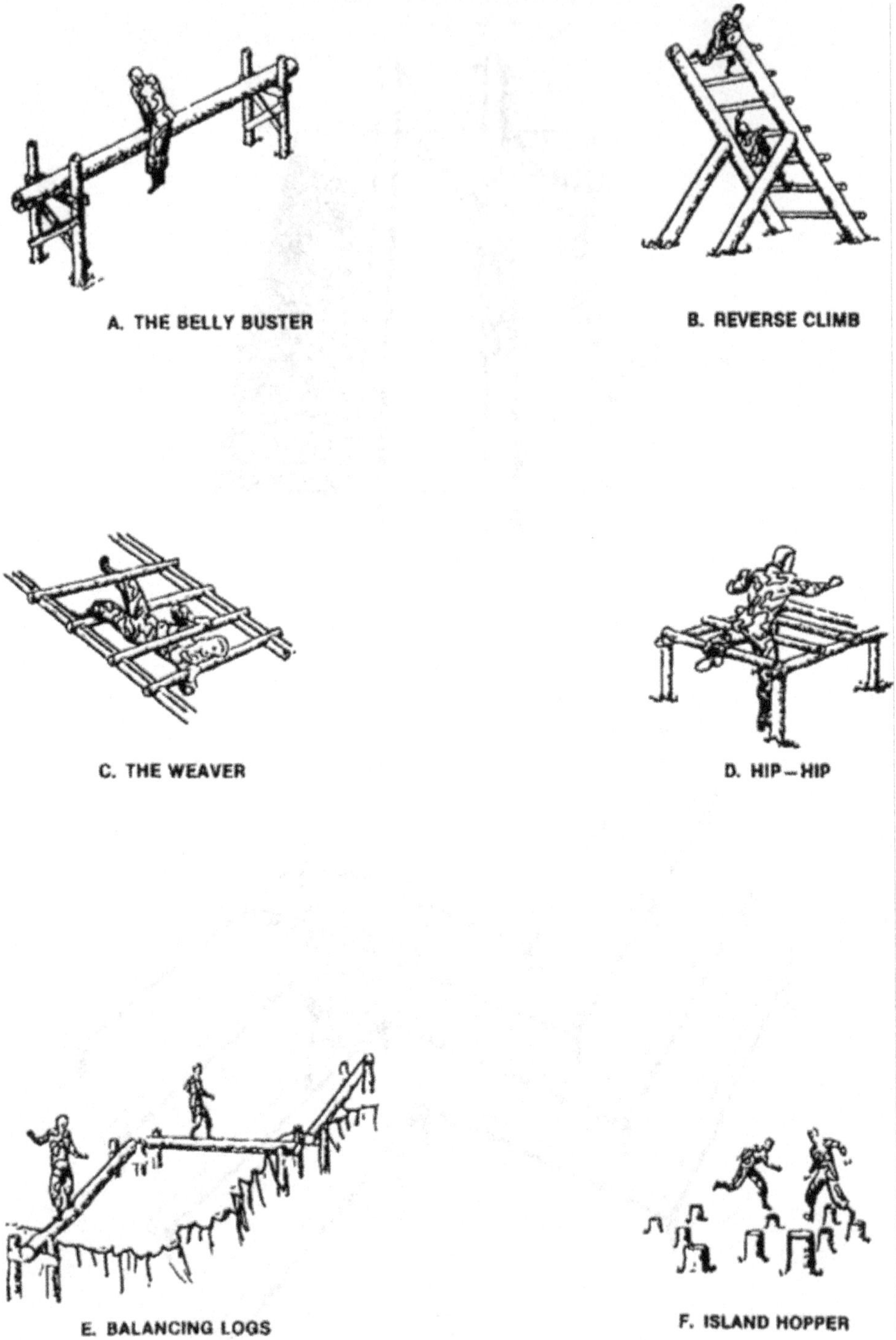

Figure 3-36.Red Group.

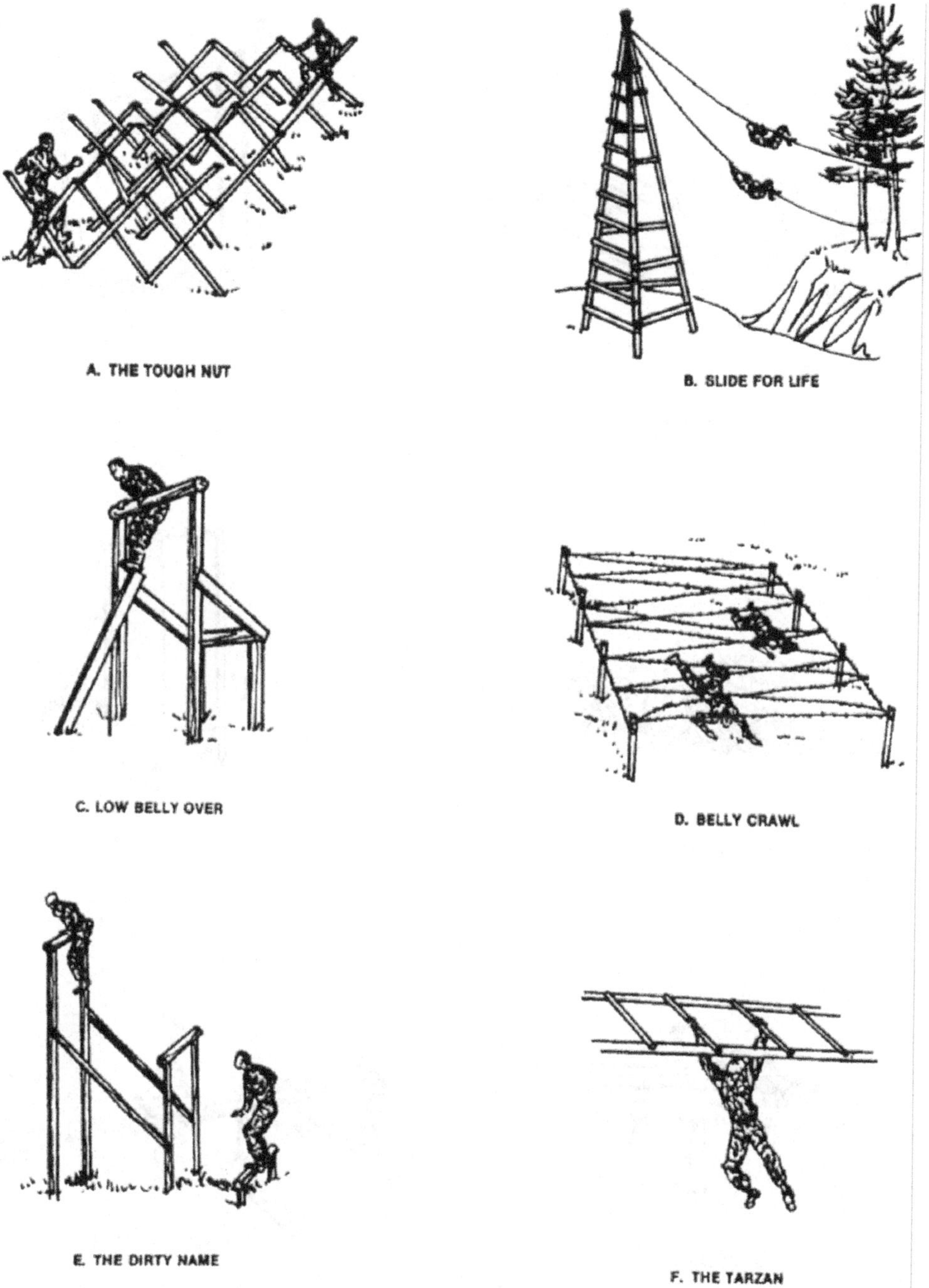

Figure 3-37.White Group.

Figure 3-38.Blue Group.

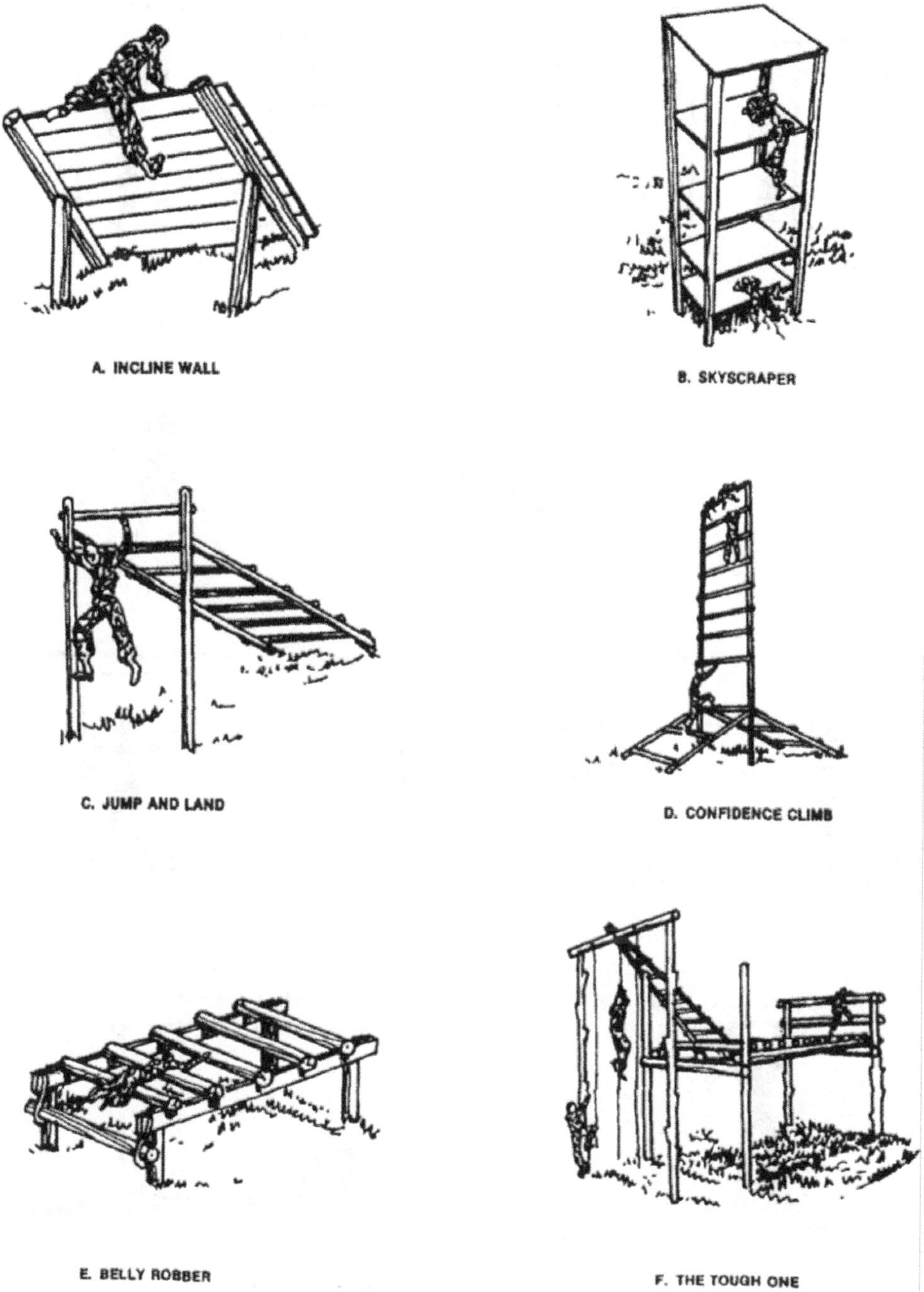

Figure 3-39.Black Group.

Figure 3-40. Bench Conditioning, Table I.

Figure 3-41. Bench Conditioning, Table II.

PROGRESSION GUIDE

AGE GROUP	LEVEL	EXERCISES 1	2	3	4	5	6
17	A	15	18	14	15	15	250
to	B	13	16	13	13	13	235
29	C	11	14	12	11	11	[illegible]
30	A	13	14	12	13	13	200
to	B	11	13	11	11	11	185
39	C	9	12	10	9	9	165
40	A	11	11	10	11	11	150
to	B	9	10	9	9	9	135
44	C	7	9	8	7	7	120
45	A	9	8	8	9	9	100
to	B	7	7	7	7	7	90
49	C	5	6	6	5	5	80
50	A	7	6	6	7	7	75
to	B	5	5	5	5	5	70
59	C	3	4	4	3	3	60
60	A	4	5	4	4	4	50
and	B	3	4	3	3	3	40
over	C	2	3	2	2	2	30
Minutes for each exercise		2	1	1	1	2	5

Table I. Progression Guide.

Table I. Exercises.

PROGRESSION GUIDE

AGE GROUP	LEVEL	EXERCISES 1	2	3	4	5	6
17	A	17	17	17	9	19	300
to	B	15	15	15	8	17	270
29	C	13	13	13	7	15	245
30	A	15	15	15	8	17	235
to	B	13	13	13	7	15	210
39	C	11	11	11	6	13	190
40	A	13	13	13	7	15	175
to	B	11	11	11	6	13	155
44	C	9	10	9	5	11	135
45	A	11	11	11	6	13	125
to	B	9	9	9	5	11	110
49	C	7	7	7	4	9	100
50	A	9	9	9	5	11	95
to	B	7	7	7	4	9	85
59	C	5	5	5	3	7	75
60	A	6	7	7	4	9	70
and	B	5	5	5	3	7	60
over	C	4	4	4	2	5	50
Minutes for each exercise		1	1	1	1 1/2	1 1/2	6

Table II. Progression Guide.

Table II. Exercises.

PROGRESSION GUIDE

AGE GROUP	LEVEL	EXERCISES 1	2	3	4	5	6
17	A	10	19	19	16	10	350
to	B	9	17	17	15	9	315
29	C	8	15	15	14	8	280
30	A	9	17	17	14	9	270
to	B	8	15	15	13	8	240
39	C	7	13	13	12	7	210
40	A	8	15	15	12	8	200
to	B	7	13	13	11	7	180
44	C	6	11	11	10	6	160
45	A	7	13	13	10	7	150
to	B	6	11	11	9	6	135
49	C	5	9	9	8	5	120
50	A	6	11	11	8	6	115
to	B	5	9	9	7	5	105
59	C	4	7	7	6	4	95
60	A	5	9	9	7	5	90
and	B	4	7	7	6	4	80
over	C	3	5	5	4	3	70
Minutes for each exercise		1 1/2	1	1	1 1/2	1	6

Table III. Progression Guide.

Table III. Exercises.

PROGRESSION GUIDE

AGE GROUP	LEVEL	EXERCISES 1	2	3	4	5	6
17	A	12	9	12	24	25	400
to	B	11	8	11	22	23	380
29	C	10	7	10	21	21	360
30	A	11	8	11	23	23	305
to	B	10	7	10	21	21	290
39	C	9	6	9	20	20	275
40	A	10	7	10	20	21	225
to	B	9	6	9	18	18	215
44	C	8	5	8	16	16	205
45	A	8	6	8	16	16	175
to	B	7	5	7	14	14	165
49	C	6	4	6	12	12	155
50	A	6	5	6	13	13	135
to	B	5	4	5	11	11	130
59	C	4	3	4	10	10	120
60	A	5	4	5	10	10	100
and	B	4	3	4	9	9	95
over	C	3	2	3	8	8	90
Minutes for each exercise		1	2	1	1	1	6

Table IV. Progression Guide.

Table IV. Exercises.

PROGRESSION GUIDE

AGE GROUP	LEVEL	EXERCISES 1	2	3	4	5	6
17	A	14	13	28	14	30	450
to	B	13	12	27	13	28	430
29	C	12	11	26	12	26	410
30	A	12	12	25	12	26	350
to	B	11	11	24	11	24	330
39	C	10	10	23	10	22	310
40	A	11	11	23	11	23	250
to	B	10	10	21	10	21	240
44	C	9	9	19	9	19	230
45	A	9	9	20	9	20	200
to	B	8	8	18	8	18	190
49	C	7	7	16	7	16	180
50	A	7	7	16	7	16	170
to	B	6	6	14	6	14	155
59	C	5	5	12	5	12	140
60	A	6	6	12	6	12	115
and	B	5	5	11	5	10	110
over	C	4	4	9	4	9	105
Minutes for each exercise		2	1	1	2	1	5

Table V. Progression Guide.

Table V. Exercises.

PROGRESSION GUIDE

AGE GROUP	LEVEL	EXERCISES 1	2	3	4	5	6
17	A	17	15	32	32	35	500
to	B	16	14	30	30	33	480
29	C	15	13	28	28	31	460
30	A	15	13	30	30	31	400
to	B	14	12	28	28	29	380
39	C	13	11	26	26	27	360
40	A	13	10	27	27	27	310
to	B	12	9	25	25	25	285
44	C	11	8	23	23	23	265
45	A	11	9	23	23	23	250
to	B	10	8	21	21	21	230
49	C	9	7	19	19	19	210
50	A	9	8	19	19	19	200
to	B	8	7	17	17	17	190
59	C	7	6	15	15	15	175
60	A	8	7	15	15	17	140
and	B	7	6	13	13	15	130
over	C	5	5	10	10	12	120
Minutes for each exercise		2	1	1	1	1	6

Table VI. Progression Guide.

Table VI. Exercises.

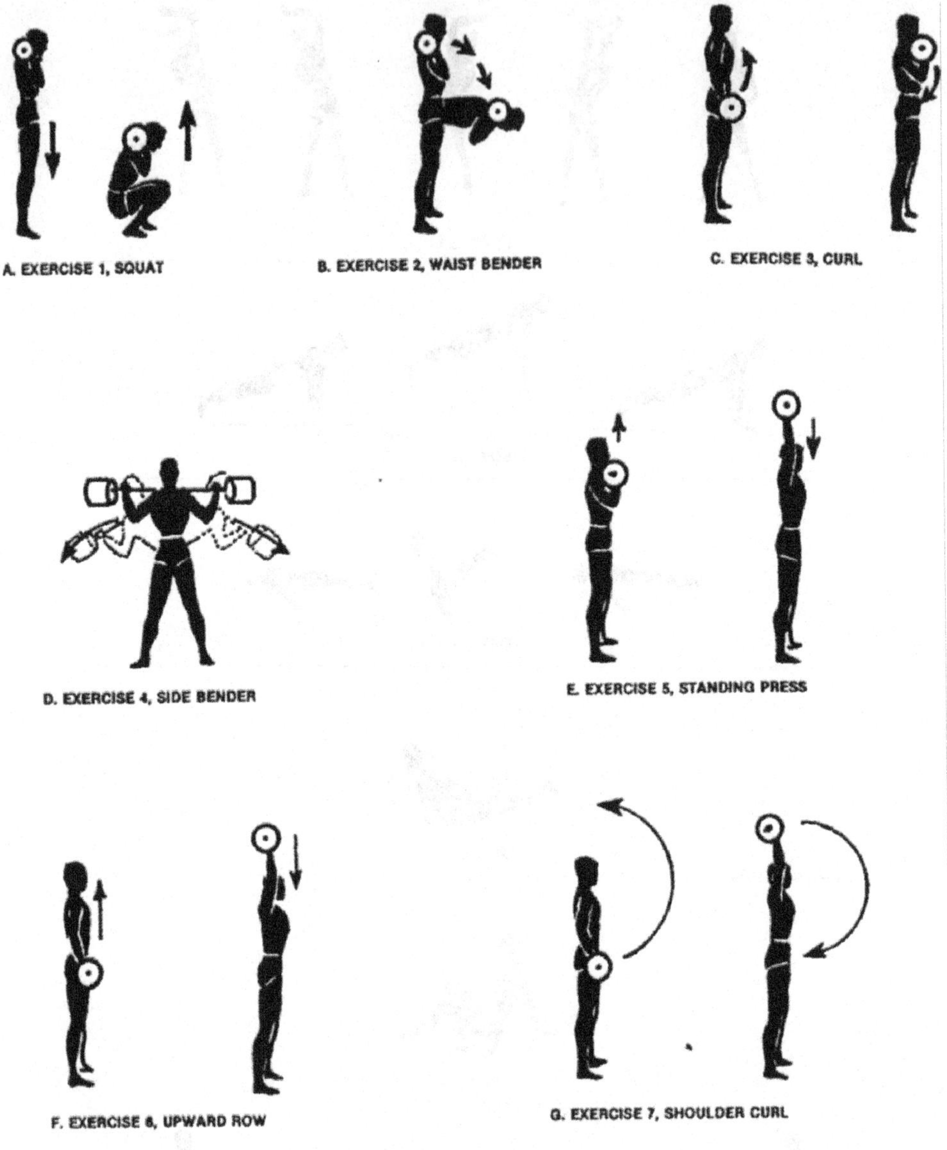

Figure 3-42. Weight Training.

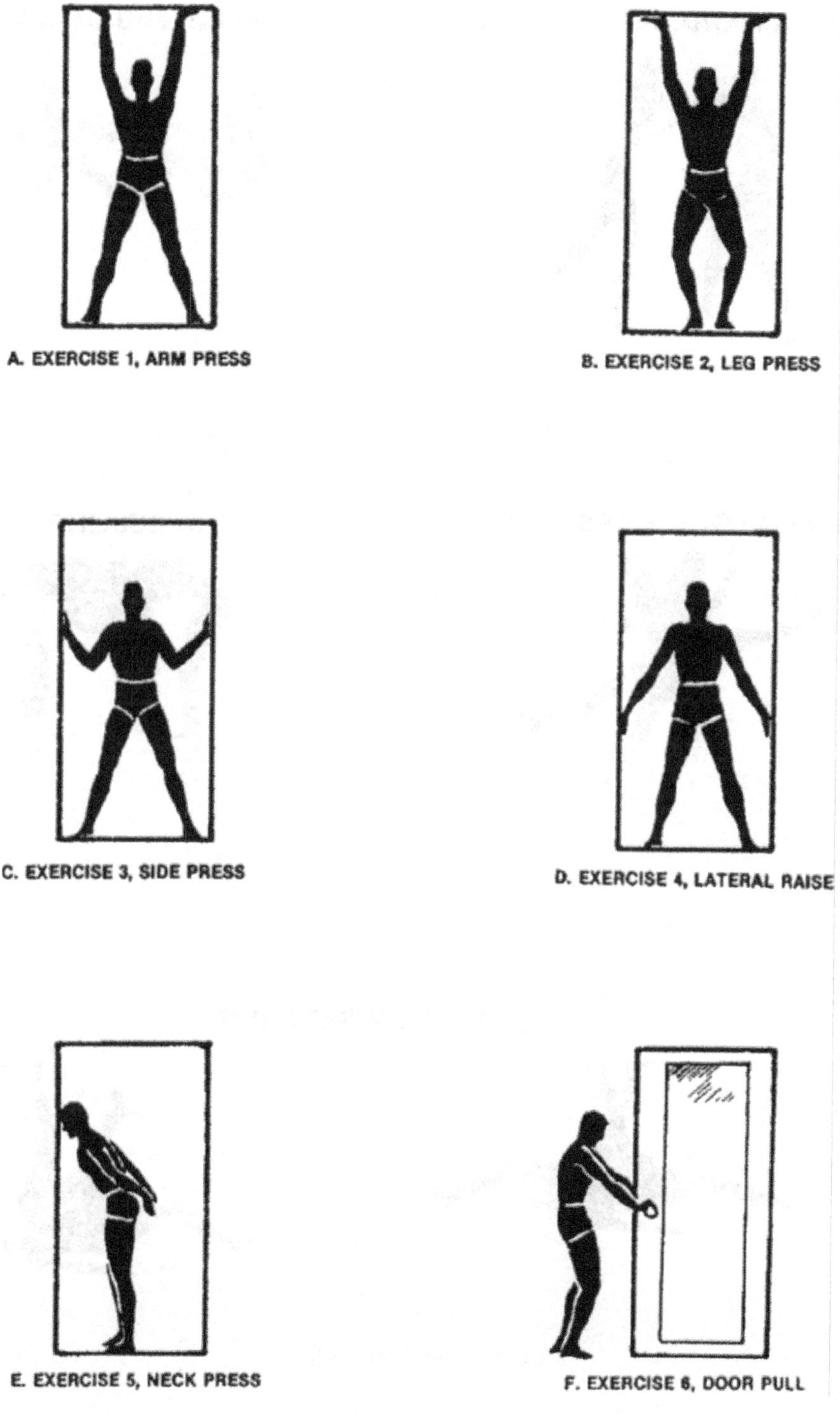

Figure 3-43. Isometric Exercises.

Figure 5-1.Combatives Table I.

Figure 5-2.Combatives Table II.

Figure. 5-3.Combatives Table III.

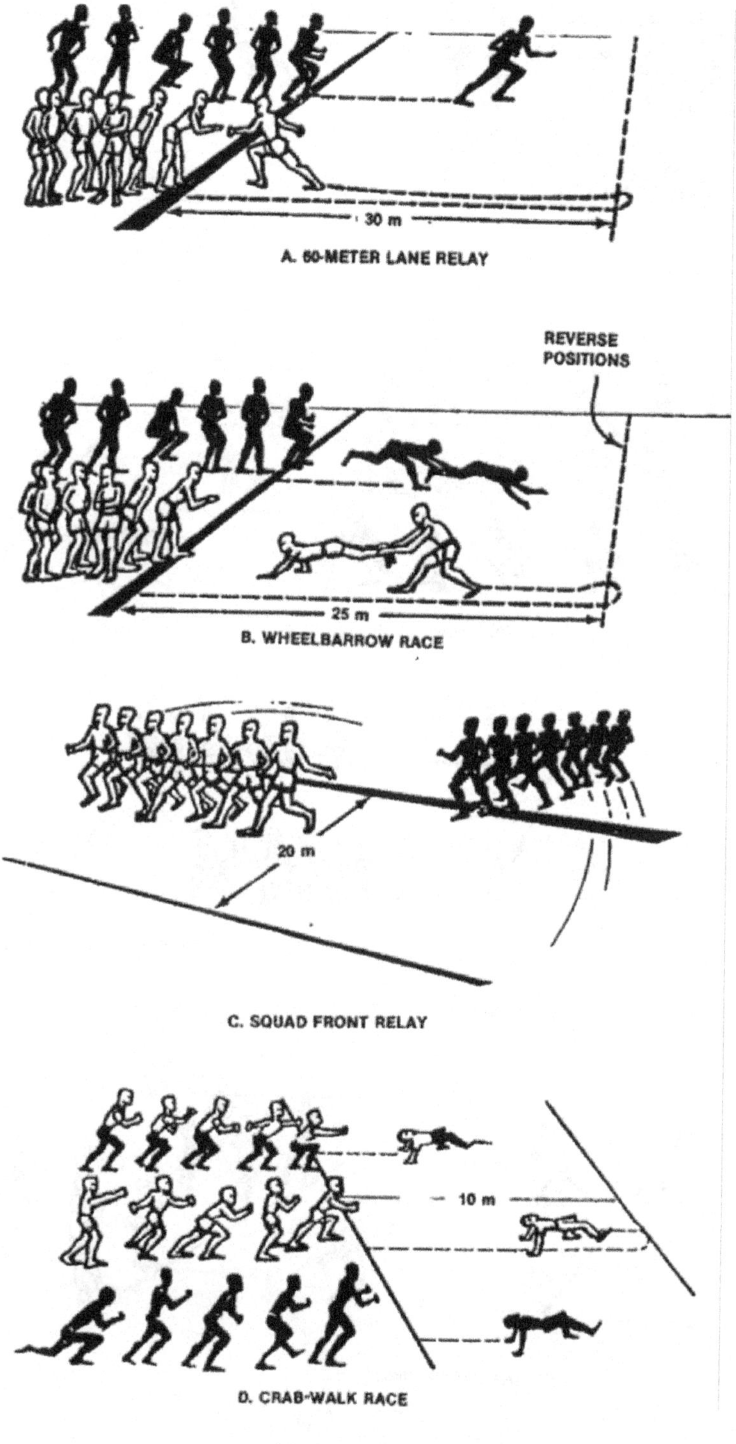

Figure 5-4.Relay Table I.

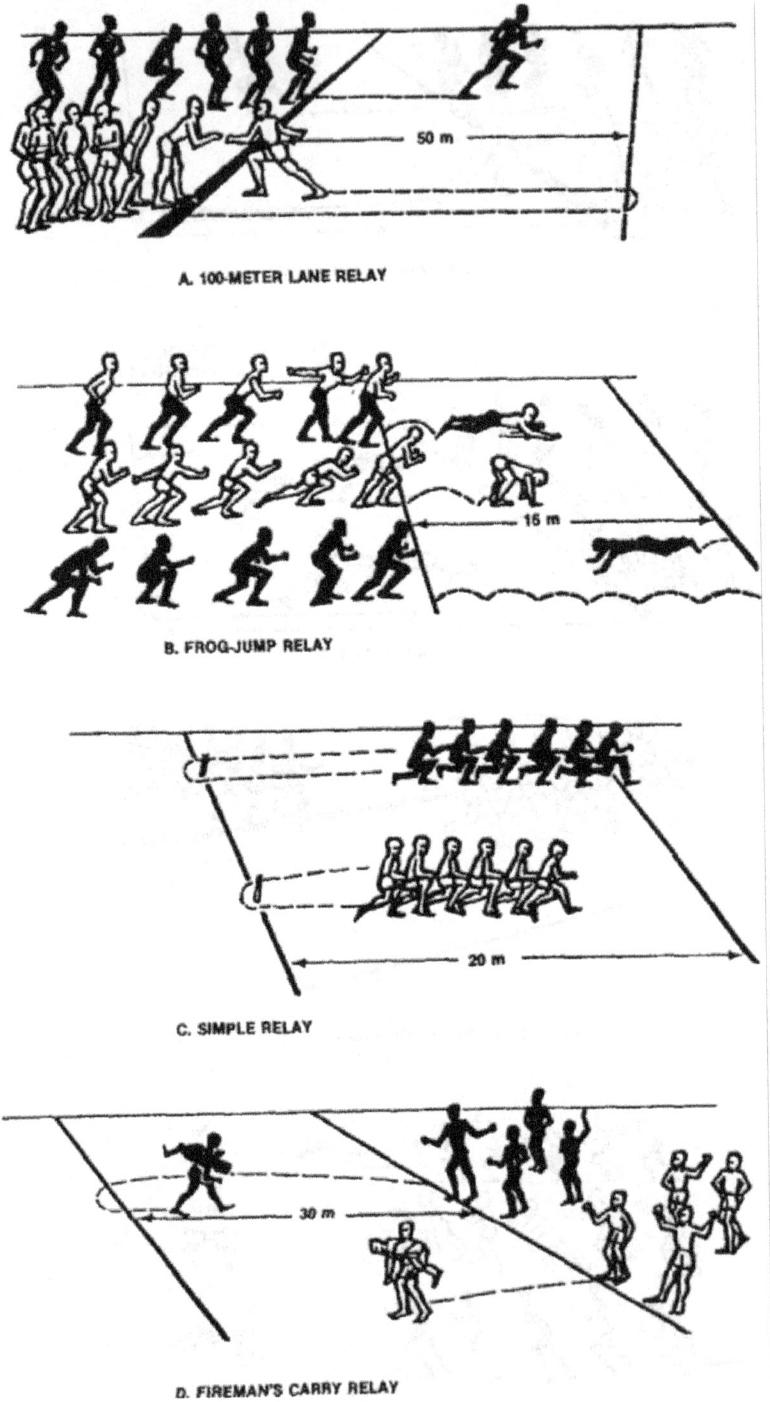

Figure 5-5.Relay Table II.

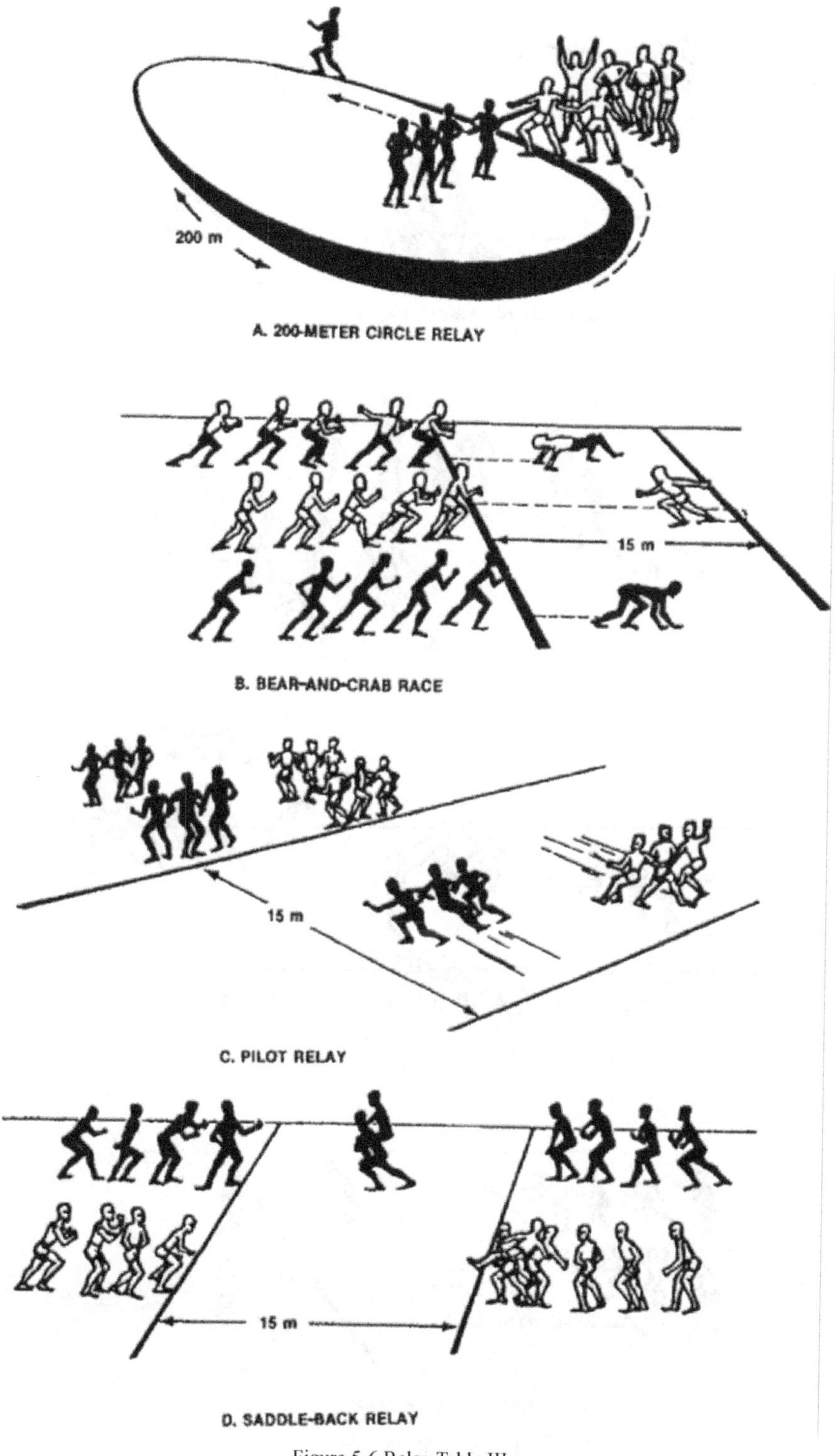

Figure 5-6.Relay Table III.

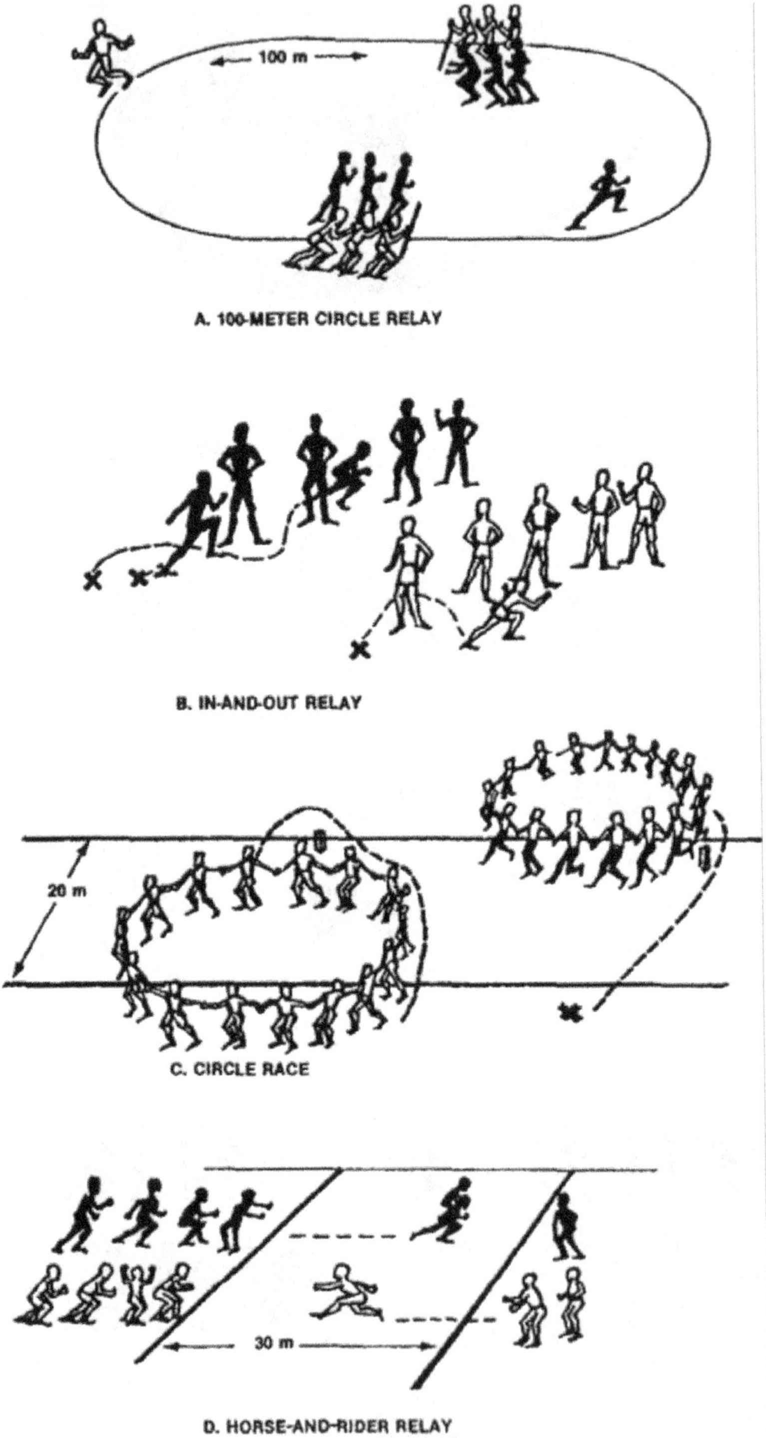

Figure 5-7.Relay Table IV.

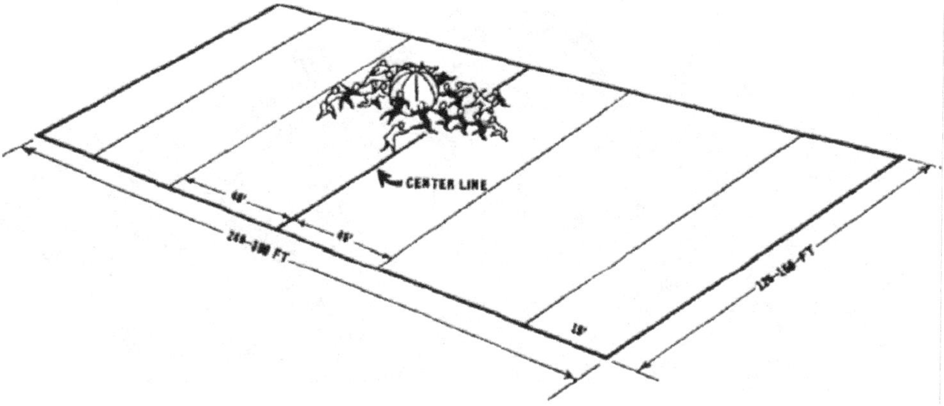

Figure 5-8.Pushball.

Figure 5-9.Line Rush.

Figure. 5-10. Human Tug-of-War.

Figure 5-11. Master of the Ring.

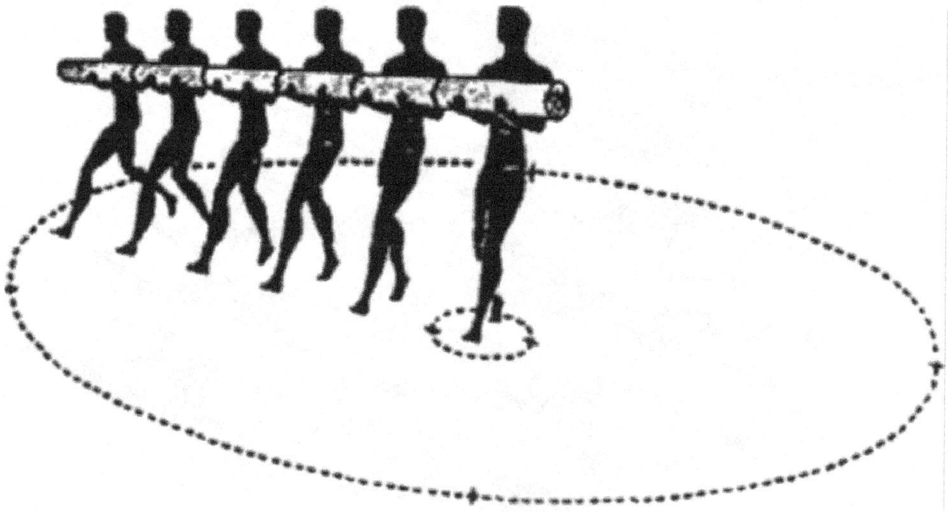

Figure 5-12.Log Pivot Circle.

Figure 5-13.Log-Rolling Race.

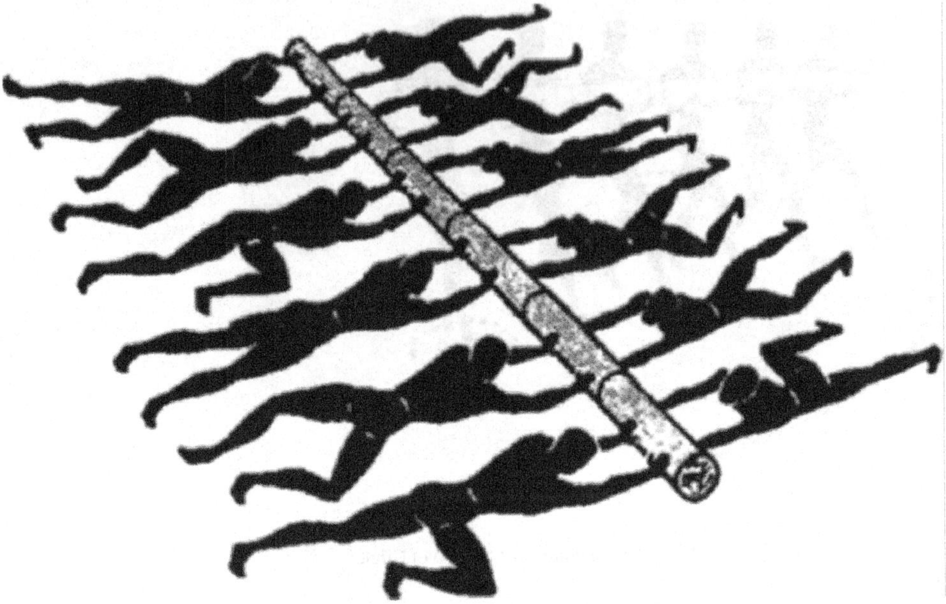

Figure 5-14. Prone Pushing Contest.

Figure 5-15. Shuttle Relay Race.

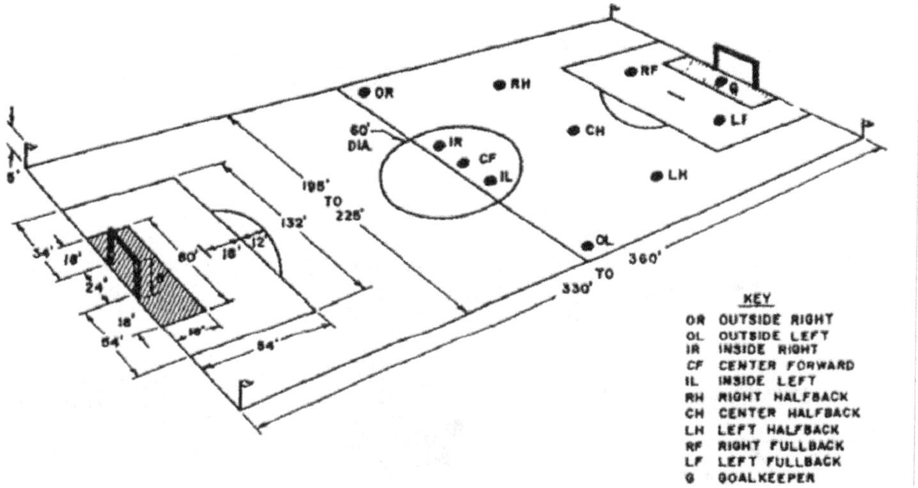

Figure 5-16. Soccer.

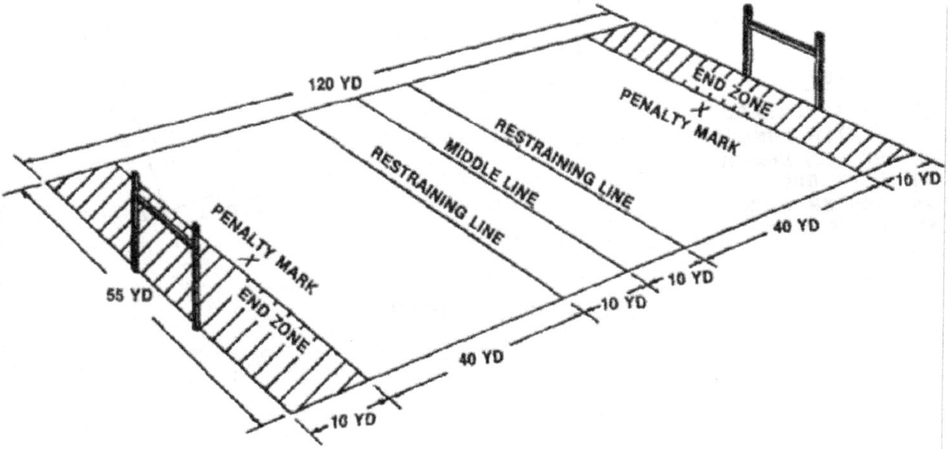

Figure 5-17.Speedball Field.

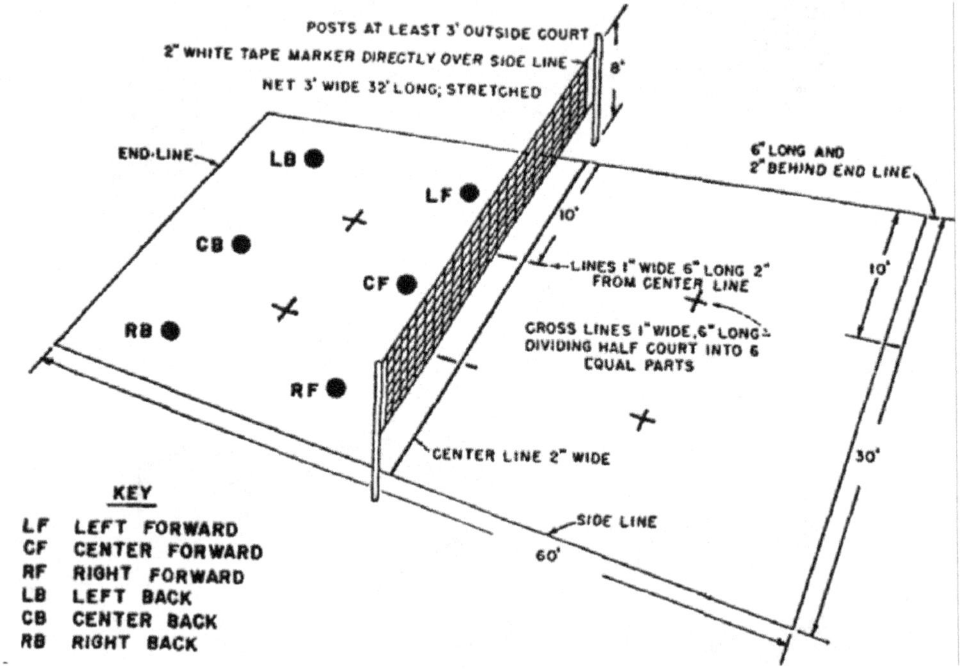

Figure 5-18. Volleyball.

PTS	UPHILL CLIMB	ROPE CLIMB	EVACUATION	FIRE AND MANEUVER	FORCED MARCH
100	1:20-BELOW	0:10-BELOW	0:25-BELOW	1:20-BELOW	26:00-BELOW
99	1:21-1:22	0:11		1 21-1:22	26:01-26:12
98	1:23-1:24	0:12	0:26	1:23-1:24	26:13-26:24
97	1:25-1:26	0:13		1:25-1:26	26:25-26:36
96	1:27-1:28	0:14	0:27	1:27-1:28	26:37-26:48
95	1:29-1:30	0:15		1:29-1:30	26:49-27:00
94	1:31-1:32	0:16	0:28	1:31-1:32	27:01-27:12
93	1:33-1:34	0:17		1 33-1.34	27:13-27:24
92	1:35-1:36	0:18	0:29	1:35-1.36	27:25-27:36
91	1:37-1:38	0:19		1:37-1:38	27:37-27:48
90	1:39-1:40	0:20	0:30	1:39-1:40	27:49-28:00
89	1:41-1:42	0:21	0:31	1:41-1:42	28:01-28:12
83	1:43-1:44	0:22	0:32	1:43-1:44	28:13-28:24
87	1:45-1:46	0:23	0:33	1:45-1:46	28:25-28:36
86	1:47-1:48	0:24	0:34	1.47-1:48	28:37-28:48
85	1:49-1:50	0:25	0:35	1.49-1:50	28:49-29:00
84	1:51-1:52	0:26	0:36	1:51-1:52	29:01-29:12
83	1:53-1:54	0:27	0.37	1:53-1:54	29:13-29:24
82	1:55-1:56	0:28	0:38	1:55-1:56	29:25-29:36
81	1:57-1:58	0:29	0:39	1:58-1:58	29:37-29:48
80	1:59-2.00	0:30	0:40	1:59-2:00	29:49-30:00
79	2.01-2:02	0:31	0:41	2:01-2:02	30:01-30:12
78	2:03-2:04	0:32	0:42	2:03-2.04	30:13-30:24
77	2:05-2:06	0.33	0:43	2:05-2:06	30:25-30:36
76	2:07-2:08	0.34	0.44	2:07-2:08	30:37-30:48
75	2:09-2:10	0:35	0:45	2:09-2:10	30:49-31:00
74	2:11-2:12	0:36	0:46	2:11-2 12	31.01-31:12
73	2:13-2:14	0:37	0.47	2.13-2:14	31:13-31:24
72	2:15-2:16	0:38	0:48	2.15-2:16	31:25-31:36
71	2:17-2:18	0:39	0:49	2:17-2:18	31:37-31:48
70	2:19-2:20	0:40	0:50	2:19-2.20	31:49-32.00
69	2.21-2:22	0:41	0:51	2:21-2:22	32:01-32:12
68	2:23-2:24	0:42	0:52	2:23-2:24	32:13-32:24
67	2:25-2.26	0:43	0:53	2:25-2:26	32:25-32:36
66	2:27-2:28	0:44	0:54	2.27-2:28	32.37-32:48
65	2:29-2:30	0.45	0.55	2.20-2 30	32.49-33:00

Figure 6-1.Physical Readiness Test Scoring Table for Male.

	EXAMPLE 1	EXAMPLE 2	EXAMPLE 3
Event #1	100 pts	100 pts	99 pts
Event #2	100 pts	99 pts	86 pts
Event #3	85 pts	75 pts (F)	80 pts
Event #4	90 pts	79 pts (F)	100 pts
Event #5	79 pts	85 pts	90 pts
Pts	454 pts	438 pts	455 pts
Final Score	79.0% (F)	69.0% (F)	91.0%
(F) = Failure			

Figure 6-2.Scoring Examples.

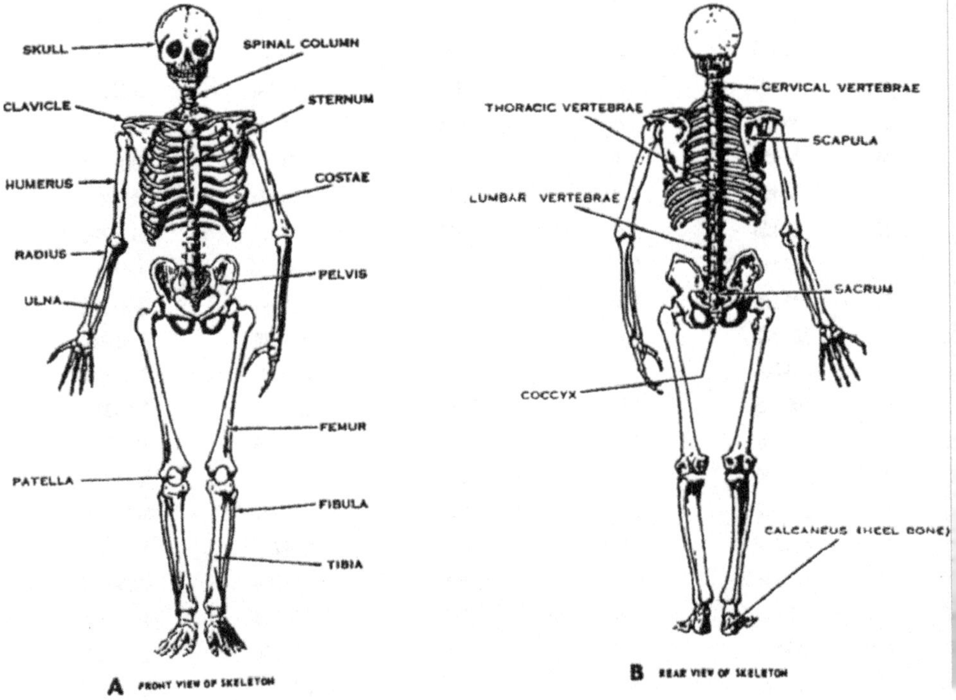

Figure 7-1.Front and Rear Views of Skeleton.

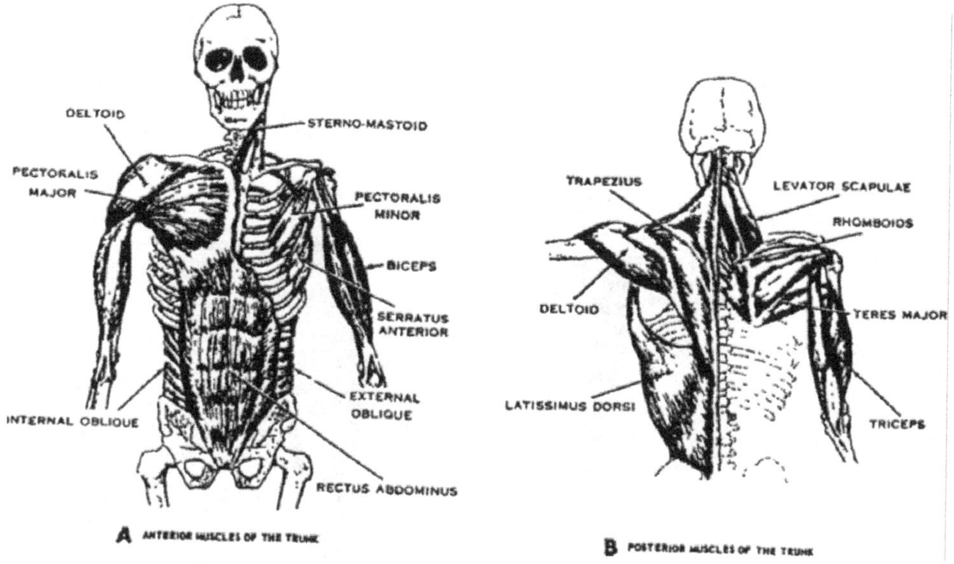

Figure 7-2.Muscles of the Trunk.

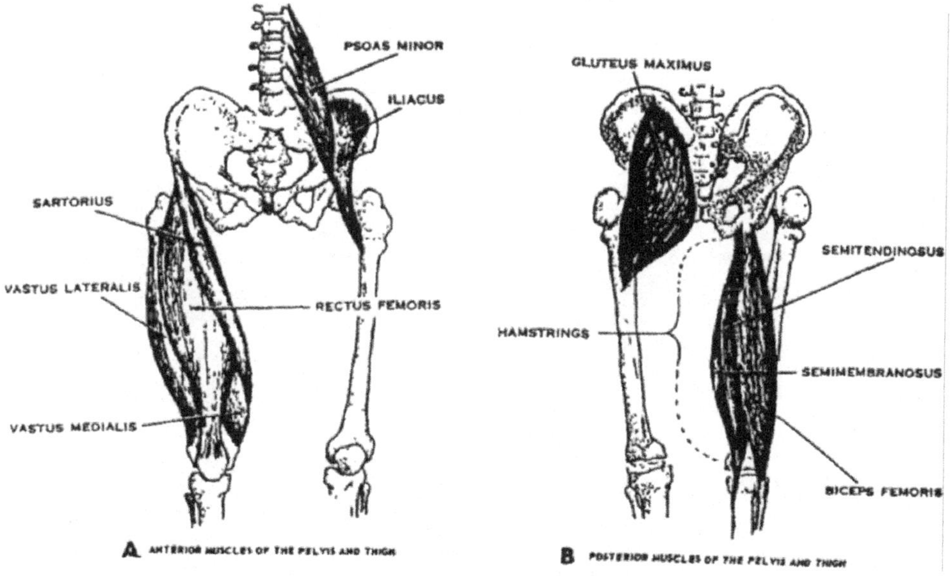

Figure 7-3.Anterior and Posterior Muscles of the Pelvisand Thigh.

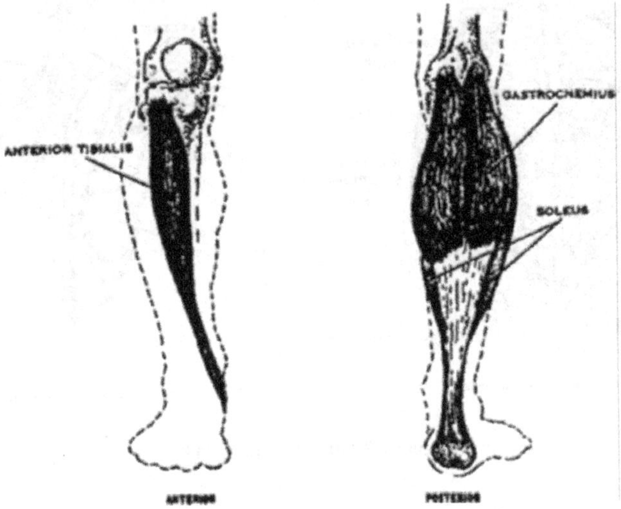

Figure 7-4. Muscles of Lower Leg.

Figure 7-5.Circulation of Blood from Heart to the Body.

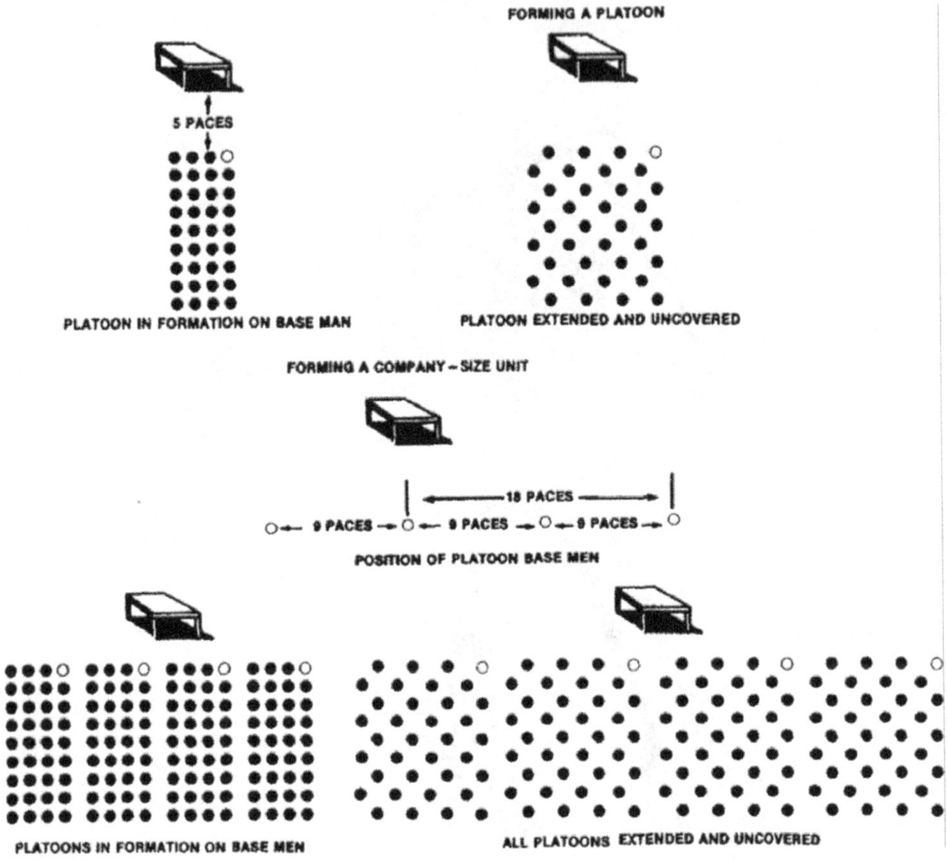

Figure A-l. Forming the Extended Rectangular Formation.

Figure A-2.Circle Formation.

Desert Operations

Department of Defense

Pentagon
Publishing

* FM 90-3
FMFM 7-27

Headquarters
Department of the Army
The Marine Corps
Washington, DC
24 August 1993

FM 90-3
DESERT OPERATIONS

Table of Contents

PREFACE

INTRODUCTION

CHAPTER 1 - THE ENVIRONMENT AND ITS EFFECTS ON PERSONNEL AND EQUIPMENT

- SECTION I. THE ENVIRONMENT
- SECTION II. ENVIRONMENTAL EFFECTS ON PERSONNEL
- SECTION III. ENVIRONMENTAL EFFECTS ON EQUIPMENT

CHAPTER 2 - PREPARATIONS FOR DESERT OPERATIONS

SECTION I. FACTORS TO BE CONSIDERED FOR DESERT OPERATIONS

SECTION II. INDIVIDUAL TRAINING

SECTION III. UNIT TRAINING

CHAPTER 3 - OPERATIONS IN DESERT CONDITIONS

SECTION I. HOW THE DESERT ENVIRONMENT AFFECTS TACTICAL OPERATIONS

SECTION II. OFFENSIVE OPERATIONS

SECTION III. DEFENSIVE OPERATIONS

CHAPTER 4 - COMBAT SERVICE SUPPORT

SECTION I. BASE DEVELOPMENT PLAN

SECTION II. THEATER SUPPORT

SECTION III. CORPS SUPPORT COMMAND

SECTION IV. DIVISION SUPPORT COMMAND

SECTION V. COMBAT SERVICE SUPPORT ELEMENT

SECTION VI. SUPPORT OPERATIONS

SECTION VII. SECURITY OF SUPPLY ROUTES

SECTION VIII. SUPPLY

SECTION IX. MAINTENANCE

SECTION X. PERSONNEL SUPPORT

SECTION XI. HEALTH SERVICES

SECTION XII. NAVAL AND AIR FORCE ASSISTANCE

SECTION XIII. OTHER COMBAT SERVICE SUPPORT ISSUES

APPENDIX A - DESERT COUNTRIES OF THE WORLD

APPENDIX B - EMPLOYMENT OF AIRCRAFT IN DESERT OPERATIONS

APPENDIX C - TECHNIQUES FOR OPERATING EQUIPMENT IN THE DESERT

APPENDIX D - EFFECTS OF THE ENVIRONMENT ON NUCLEAR , BIOLOGICAL, AND CHEMICAL WEAPONS

APPENDIX E - DESERT CONCEALMENT AND CAMOUFLAGE

APPENDIX F - OPERATIONS IN MOUNTAINS

APPENDIX G - WATER USAGE IN DESERT OPERATIONS

APPENDIX H - FRATRICIDE REDUCTION

GLOSSARY

REFERENCES

AUTHORIZATION LETTER

DISTRIBUTION RESTRICTION: Approved for public release; distribution is unlimited.

* This publication supersedes FM 90-3, 19 August 1977.
U.S. Marine Corps: PCN 139 000670 00

INTRODUCTION

Arid regions make up about one-third of the earth's land surface, a higher percentage then that of any other type of climate. As we have seen in the recent past, some of these regions—because of diverse and conflicting culturs, strategic importance, and natural resources—have become centers of conflict.

Military leaders have long recognized the potential for US involvement in conflict In these regions. Exercises at the Array's National Training Center, Fort Irwin and the marine Corps' Marine Air Ground Combat Center, Twentynine Palms. California, have provided an opportunity for virtually all our ground forces to experience desert conditions, The success of Operation Desert Storm can be directly attributed to this realistic training.

Desert operations demand adaptation to the environment and to the limitations imposed by terrain and climate, Success depends on an appreciation of the effects of arid conditions on soldiers (both physically and psychologically), on equipment and facilities, and on eombot and support operations. Leaders and soldiers must continually evaluate the situation and be ready to react to changing conditions. Equipment and tactics must be modified and adapted to a dusty, rugged landscape where temperatures vary from extreme highs to freezing lows and where visibility can clange from 30 miles to 30 feet in a matter of minutes.

The key to success in desert operations is mobility. This was clearly evident in the ground operations of Desert Storm, The tactics employed to achieve victory over Iraq were wide, rapid flanking movements similar to those executed by Montgomery and Rommel during World War II, During Desert Storm, however, new technologies increased higher-echelon headquarter' ability to target, attack, and fight deep operations simultaneously. Modem weapon systems like the

M1 A1 Abrams tank, Bradley fighting vehicle, light armored vehicle, and assault amphibious vehicle, coupled with newly developed navigation and targeting devices, contributed immeasurably. Taclictl units were able to fight battles with minimal direction; leaders were able to exercise initiative based on a clear understanding of their commanders' intent. Current doctrine—Focused on improving mobility and implemented through, the planning, preparation, and execution processes, battle drills, and tactical SOPs, paved the way for the overwhelming triumph.

Arid regions create ***bold*** opportunities and resurants **for** soldiers and marines at all levels. The US military's performance in Desert Storm shows it understands these factors and has successfully addressed the effects of desert warfare TO troops, equipment, and operations. As they prepare for the future, leaders, soldiers, and marines must study past carnpaigns and use the lessons they learn **to** reduce casualties, use the environment to their advantage, and ensure victory on the desert battlefield.

CHAPTER 1

THE ENVIRONMENT AND ITS EFFECTS ON PERSONNEL AND EQUIPMENT

This chapter describes the desert environment and how it affects personnel and equipment.

Section I. The Environment

Successful desert operations require adaptation to the environment and to the limitations its terrain and climate impose. Equipment and tactics must be modified and adapted to a dusty and rugged landscape where temperatures vary from extreme highs down to freezing and where visibility may change from 30 miles to 30 feet in a matter of minutes. Deserts are arid, barren regions of the earth incapable of supporting normal life due to lack of water. See Figure 1-1 for arid regions of the world. Temperatures vary according to latitude and season, from over 136 degrees Fahrenheit in the deserts of Mexico and Libya to the bitter cold of winter in the Gobi (East Asia). In some deserts, day-to-night temperature fluctuation exceeds 70 degrees Fahrenheit. Some species of animal and plant life have adapted successfully to desert conditions where annual rainfall may vary from 0 to 10 inches.

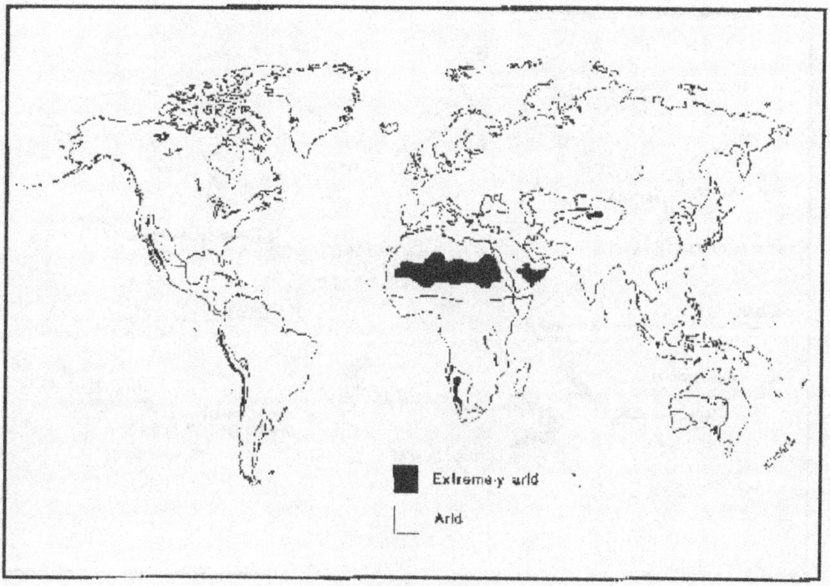

Figure 1-1. Desert of the world.

Desert terrain also varies considerably from place to place, the sole common denominator being lack of water with its consequent environmental effects, such as sparse, if any, vegetation. The basic land forms are similar to those in other parts of the world, but the topsoil has been eroded due to a combination of lack of water, heat, and wind to give deserts their characteristic barren appearance. The bedrock may be covered by a flat layer of sand, or gravel, or may have been exposed by erosion. Other common features are sand dunes, escarpments, wadis, and depressions. This environment can profoundly affect military operations. See Figure 1-2 for locations of major deserts of the world, and Appendix A for additional information on desert countries of the world.

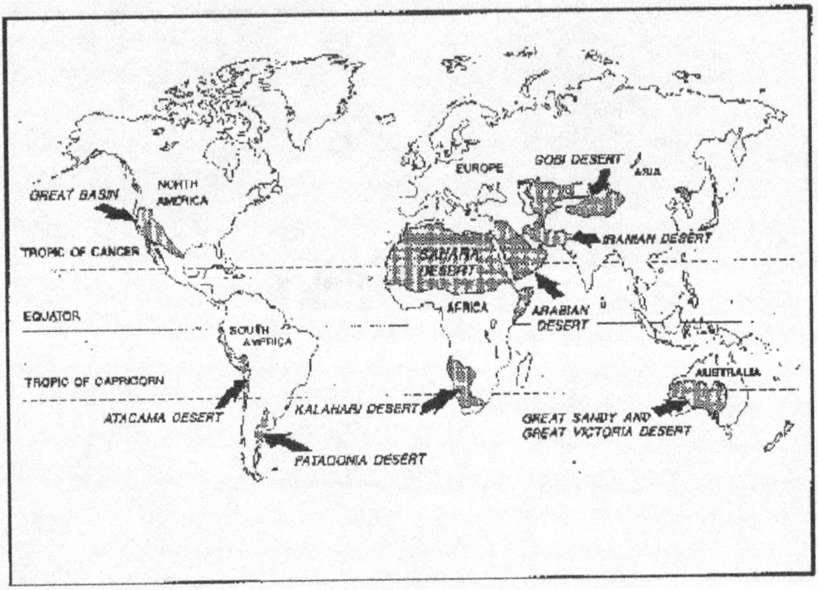

It is important to realize that deserts are affected by seasons. Those in the Southern Hemisphere have summer between 21 December and 21 March. This 6-month difference from the United States is important when considering equipping and training nonacclimatized soldiers/marines for desert operations south of the equator.

TERRAIN

Key terrain in the desert is largely dependent on the restrictions to movement that are present. If the desert floor will not support wheeled vehicle traffic, the few roads and desert tracks become key terrain. Crossroads are vital as they control military operations in a large area. Desert warfare is often a battle for control of the lines of communication (LOC). The side that can protect its own LOC while interdicting those of the enemy will prevail. Water sources are vital, especially if a force is incapable of long distance resupply of its water requirements. Defiles play an important role, where they exist. In the Western Desert of Libya, an escarpment that paralleled the coast was a barrier to movement except through a few passes. Control of these passes was vital. Similar escarpments are found in Saudi Arabia and Kuwait.

Types of Desert Terrain

There are three types of desert terrain: mountain, rocky plateau, and sandy or dune terrain. The following paragraphs discuss these types of terrain.

Mountain Deserts

Mountain deserts are characterized by scattered ranges or areas of barren hills or mountains, separated by dry, flat basins. See Figure 1-3 for an example of mountain desert terrain. High ground may rise gradually or abruptly from flat areas, to a height of several thousand feet above sea level. Most of the infrequent rainfall occurs on high ground and runs off in the form of flash floods, eroding deep gullies and ravines and depositing sand and gravel around the edges of the basins. Water evaporates rapidly, leaving the land as barren as before, although there may be short-lived vegetation. If sufficient water enters the basin to compensate for the rate of evaporation, shallow lakes may develop, such as the Great Salt Lake in Utah or the Dead Sea; most of these have a high salt content.

Figure 1-3, Example of mountain desert terrian,

Rocky Plateau Deserts

Rocky plateau deserts are extensive flat areas with quantities of solid or broken rock at or near the surface. See Figure 1-4 for an example of a rocky plateau desert. They may be wet or dry, steep-walled eroded valleys, known as wadis, gulches, or canyons. Narrow valleys can be extremely dangerous to men and materiel due to flash flooding after rains; although their flat bottoms may be superficially attractive as assembly areas. The National Training Center and the Golan Heights are examples of rocky plateau deserts.

figure1-4, Example of rocky plateau desert terrain.

Sandy or Dune Deserts

Sandy or dune deserts are extensive flat areas covered with sand or gravel, the product of ancient deposits or modern wind erosion. "Flat" is relative in this case, as some areas may contain sand dunes that are over 1,000 feet high and 10-15 miles long; trafficability on this type of terrain will depend on windward/leeward gradients of the dunes and the texture of the sand. See Figure 1-5 for an example of a sandy desert. Other areas, however, may be totally flat for distances of 3,000 meters and beyond. Plant life may vary from none to scrub, reaching over 6 feet high. Examples of this type of desert include the ergs of the Sahara, the Empty Quarter of the Arabian desert, areas of California and New Mexico, and the Kalahari in South Africa. See Figure 1-6 for an example of a dune desert.

figure 1 -3. Example of sandy desert terrain,

Figure 1-6. Example of dune desert terrain.

Trafficability

Roads and trails are rare in the open desert. Complex road systems beyond simple commercial links are not needed. Road systems have been used for centuries to connect centers of commerce, or important religious shrines such as Mecca and Medina in Saudi Arabia. These road systems are supplemented by routes joining oil or other mineral deposits to collection outlet points. Some surfaces, such as lava beds or salt marshes, preclude any form of routine vehicular movement, but generally ground movement is possible in all directions. Speed of movement varies depending on surface texture. Rudimentary trails are used by minor caravans and nomadic tribesmen, with wells or oases approximately every 20 to 40 miles; although there are some waterless stretches which extend over 100 miles. Trails vary in width from a few meters to over 800 meters.

Vehicle travel in mountainous desert country may be severely restricted. Available mutes can be easily blinked by the enemy or by climatic conditions. Hairpin turns are common on the edges of precipitous mountain gorges, and the higher passes may be blocked by snow in the winter.

Natural Factors

The following terrain features require special considerations regarding trafficability.

Wadis or dried water courses, vary from wide, but barely perceptible depressions of soft sand, dotted with bushes, to deep, steep-sided ravines. There frequently is a passable route through the bottom of a dried wadi. Wadis can provide cover from ground observation and camouflage from visual air reconnaissance. The threat of flash floods after heavy rains poses a significant danger to troops and equipment downstream. Flooding may occur in these areas even if it is not raining in the immediate area. See Figure 1-7 for an example of a wadi.

Figure 1-7. Example of a wadi

Salt marsh (sebkha) terrain is impassable to tracks and wheels when wet. When dry it has a brittle, crusty surface, negotiable by light wheel vehicles only. Salt marshes develop at points where the water in the subsoil of the desert rose to the surface. Because of the constant evaporation in the desert, the salts carried by the water are deposited, and results in a hard, brittle crust.

Salt marshes are normally impassable, the worst type being those with a dry crust of silt on top. Marsh mud used on desert sand will, however, produce an excellent temporary road. Many desert areas have salt marshes either in the center of a drainage basin or near the sea coast. Old trails or paths may cross the marsh, which are visible during the dry season but not in the wet season. In the wet season trails are indicated by standing water due to the crust being too hard or too thick for it to penetrate. However, such routes should not be tried by load-carrying vehicles without prior reconnaissance and marking. Vehicles may become mired so severely as to render equipment and units combat ineffective. Heavier track-laying vehicles, like tanks, are especially susceptible to these areas, therefore reconnaissance is critical.

Man-made Factors

The ruins of earlier civilizations, scattered across the deserts of the world, often are sited along important avenues of approach and frequently dominate the only available passes in difficult terrain. Control of these positions maybe imperative for any force intending to dominate the immediate area. Currently occupied dwellings have little impact on trafficability except that they are normally located near roads and trails. Apart from nomadic tribesmen who live in tents (see Figure 1-8 for an example of desert nomads), the population lives in thick-walled structures with small windows, usually built of masonry or a mud and straw (adobe) mixture. Figure 1-9 shows common man-made desert structures.

Figure 1-8. Example of desert nomads

Figure 1-9. Common man_made desert.

Because of exploration for and production of oil and other resources, wells, pipelines, refineries, quarries, and crushing plants may be of strategic importance in the desert. Pipelines are often raised 1 meter off the ground-where this is the case, pipelines will inhibit movement. Subsurface pipelines can also be an obstacle. In Southwest Asia, the subsurface pipelines were indicated on maps. Often they were buried at such a shallow depth that they could be damaged by heavy vehicles traversing them. Furthermore, if a pipeline is ruptured, not only is the spill of oil a consideration, but the fumes maybe hazardous as well.

Agriculture in desert areas has little effect on trafficability except that canals limit surface mobility. Destruction of an irrigation system, which may be a result of military operations, could have a devastating effect on the local population and should be an important consideration in operational estimates. Figure 1-10 shows an irrigation ditch.

Figure 1-10. Irrigation ditch

TEMPERATURE

The highest known ambient temperature recorded in a desert was 136 degrees Fahrenheit (58 degrees Celsius). Lower temperatures than this produced internal tank temperatures approaching 160 degrees Fahrenheit (71 degrees Celsius) in the Sahara Desert during the Second World War. Winter temperatures in Siberian deserts and in the Gobi reach minus 50 degrees Fahrenheit (minus 45 degrees Celsius). Low temperatures are aggravated by very strong winds producing high windchill factors. The cloudless sky of the desert permits the earth to heat during sunlit hours, yet cool to near freezing at night. In the inland Sinai, for example, day-to-night temperature fluctuations are as much as 72 degrees Fahrenheit.

WINDS

Desert winds can achieve velocities of near hurricane force; dust and sand suspended within them make life intolerable, maintenance very difficult, and restrict visibility to a few meters. The Sahara "Khamseen", for example, lasts for days at a time; although it normally only occurs in the spring and summer. The deserts of Iran are equally well known for the "wind of 120 days," with sand blowing almost constantly from the north at wind velocities of up to 75 miles per hour.

Although there is no danger of a man being buried alive by a sandstorm, individuals can become separated from their units. In all deserts, rapid temperature changes invariably follow strong winds. Even without wind, the telltale clouds raised by wheels, tracks, and marching troops give away movement. Wind aggravates the problem. As the day gets warmer the wind increases and the dust signatures of vehicles may drift downwind for several hundred meters.

In the evening the wind normally settles down. In many deserts a prevailing wind blows steadily from one cardinal direction for most of the year, and eventually switches to another direction for the remaining months. The equinoctial gales raise huge sandstorms that rise to several thousand feet and may last for several days. Gales and sandstorms in the winter months can be bitterly cold. See Figure 1-11 for an example of wind erosion.

Figure 1-11. Example of wind crosion

Sandstorms are likely to form suddenly and stop just as suddenly. In a severe sandstorm, sand permeates everything making movement nearly impossible, not only because of limited visibility, but also because blowing sand damages moving parts of machinery.

WATER

The lack of water is the most important single characteristic of the desert. The population, if any, varies directly with local water supply. A Sahara oasis may, for its size, be one of the most densely occupied places on earth (see Figure 1-12 for a typical oasis).

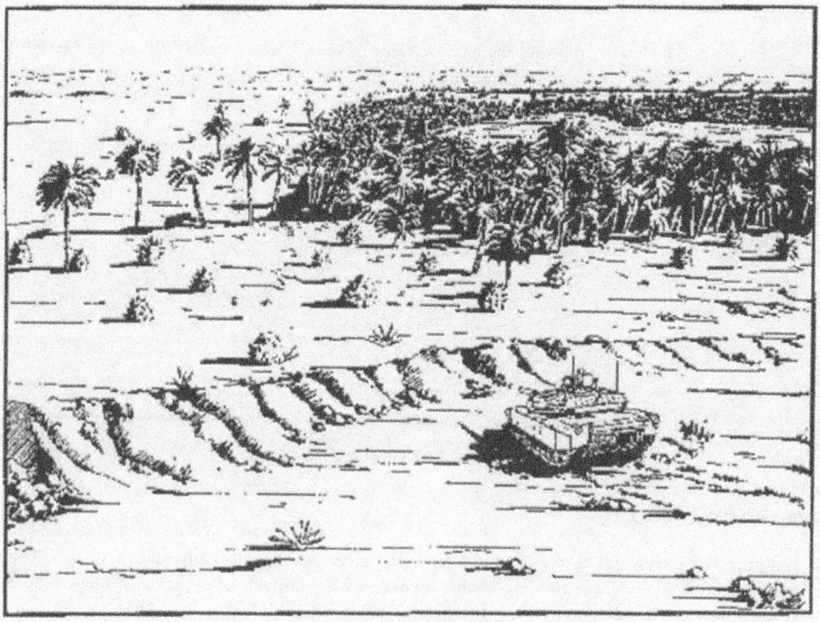

Figure 1-12. Typical oasis

Desert rainfall varies from one day in the year to intermittent showers throughout the winter. Severe thunderstorms bring heavy rain, and usually far too much rain falls far too quickly to organize collection on a systematic basis. The water soon soaks into the ground and may result in flash floods. In some cases the rain binds the sand much like a beach after the tide ebbs allowing easy maneuver however, it also turns loam into an impassable quagmire obstacle. Rainstorms tend to be localized, affecting only a few square kilometers at a time. Whenever possible, as storms approach, vehicles should move to rocky areas or high ground to avoid flash floods and becoming mired.

Permanent rivers such as the Nile, the Colorado, or the Kuiseb in the Namib Desert of Southwest Africa are fed by heavy precipitation outside the desert so the river survives despite a high evaporation rate.

Subsurface water may be so far below the surface, or so limited, that wells are normally inadequate to support any great number of people. Because potable water is absolutely vital, a large natural supply may be both tactically and strategically important. Destruction of a water supply system may become a political rather than military decision, because of its lasting effects on the resident civilian population.

Finding Water

When there is no surface water, tap into the earth's water table for ground water. Access to this table and its supply of generally pure water depends on the contour of the land and the type of soil. See Figure 1-13 for water tables.

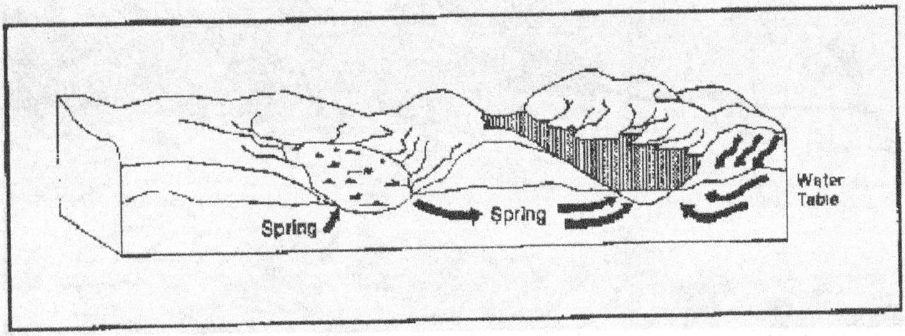

Figure 1-13. Water tables.

From Rocky Soil

Look for springs and seepages. Limestone has more and larger springs than any other type rock. Because limestone is easily dissolved, caverns are readily etched in it by ground water. Look in these caverns for springs. Lava rock is a good source of seeping ground water because it is porous. Look for springs along the walls of valleys that cross the lava flow. Look for seepage where a dry canyon cuts through a layer of porous sandstone.

Watch for water indicators in desert environments. Some signs to look for are the direction in which certain birds fly, the location of plants, and the convergence of game trails. Asian sand grouse, crested larks, and zebra birds visit water holes at least once a day. Parrots and pigeons must live within reach of water. Cattails, greasewoods, willows, elderberry, rushes, and salt grass grow only where ground water is near the surface. Look for these signs and dig. If you do not have a bayonet or entrenching tool, dig with a flat rock or sharp stick.

Desert natives often know of lingering surface pools in low places. They cover their surface pools, so look under brush heaps or in sheltered nooks, especially in semiarid and brush country.

Places that are visibly damp, where animals have scratched, or where flies hover, indicate recent surface water. Dig in such places for water. Collect dew on clear nights by sponging it up with a handkerchief. During a heavy dew you should be able to collect about a pint an hour.

Dig in dry stream beds because water may be found under the gravel. When in snow fields, put in a water container and place it in the sun out of the wind.

From Plants

If unsuccessful in your search for ground or runoff water, or if you do not have time to purify the questionable water, a water-yielding plant may be the best source. Clear sap from many plants is easily obtained. This sap is pure and is mostly water.

Plant tissues. Many plants with fleshy leaves or stems store drinkable water. Try them wherever you find them. The barrel cactus of the southwestern United States is a possible source of water (see Figure 1-14). Use it only as a last resort and only if you have the energy to cut through the tough, spine-studded outer rind. Cut off the top of the cactus and smash the pulp within the plant. Catch the liquid in a container. Chunks may be carried as an emergency water source. A barrel cactus 3-1/2 feet high will yield about a quart of milky juice and is an exception to the rule that milky or colored sap-bearing plants should not be eaten.

figure 1-14. Barrel cactus of possible source of water.

Roots of desert plants. Desert plants often have their roots near the surface. The Australian water tree, desert oak, and bloodwood are some examples. Pry these roots out of the ground, cut them into 24-36 inch lengths, remove the bark, and suck the water.

Vines. Not all vines yield palatable water, but try any vine found. Use the following method for tapping a vine--it will work on any species:

Step 1. Cut a deep notch in the vine as high up as you can reach.

Step 2. Cut the vine off close to the ground and let the water drip into your mouth or into a container.

Step 3. When the water ceases to drip, cut another section off the top. Repeat this until the supply of fluid is exhausted.

Palms. Burl, coconut, sugar and nipa palms contain a drinkable sugary fluid. To start the flow in coconut palm, bend the flower stalk downward and cut off the top. If a thin slice is cut off the stalk every 12 hours, you can renew the flow and collect up to a quart a day.

Coconut. Select green coconuts. They can be opened easily with a knife and they have more milk than ripe coconuts. The juice of a ripe coconut is extremely laxative; therefore, do not drink more than three or four cups a day.

The milk of a coconut can be obtained by piercing two eyes of the coconut with a sharp object such as a stick or a nail. To break off the outer fibrous covering of the coconut without a knife, slam the coconut forcefully on the point of a rock or protruding stump.

Survival Water Still

You can build a cheap and simple survival still that will produce drinking water in a dry desert. Basic materials for setting up this still are-

- 6-foot square sheet of clean plastic.
- A 2- to 4-quart capacity container.
- A 5-foot piece of flexible plastic tubing.

Pick an unshaded spot for the still, and dig a hole. If no shovel is available, use a stick or even your hands. The hole should be about 3 feet across for a few inches down, then slope the hole toward the bottom as shown in Figure 1-15 which depicts a cross section of a survival still. The hole should be deep enough so the point of the plastic cone will be about 18 inches below ground and will still clear the top of the container. Once the hole is properly dug, tape one end of the plastic drinking tube inside the container and center the container in the bottom of the hole. Leave the top end of the drinking tube free, lay the plastic sheet over the hole, and pile enough dirt around the edge of the plastic to hold it securely. Use a fist-size rock to weight down the center of the plastic; adjust the plastic as necessary to bring it within a couple of inches of the top of the container. Heat from the sun vaporizes the ground water. This vapor condenses under the plastic, trickles down, and drops into the container.

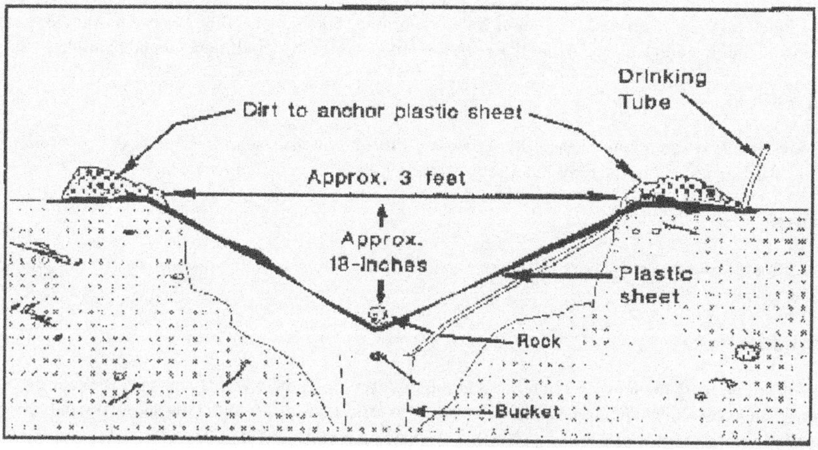

Figure 1-15. Cross-section of a survival still.

VEGETATION

The indigenous vegetation and wildlife of a desert have physiologically adapted to the conditions of the desert environment. For example, the cacti of the American desert store moisture in enlarged stems. Some plants have drought-resistant seeds that may lie dormant for years, followed by a brief, but colorful display of growth after a rainstorm. The available vegetation is usually inadequate to provide much shade, shelter, or concealment, especially from the air. Some plants, like the desert gourd, have vines which grow to 4.5 meters (15 feet). Others have wide lateral roots just below the surface to take advantage of rain and dew, while still others grow deep roots to tap subsurface water. Presence of palm trees usually indicates water within a meter of the surface, salt grass within 2 meters, Cottonwood and willows up to 4 meters. In addition to indicating the presence of water, some plants are edible.

WILDLIFE

Invertebrates such as ground-dwelling spiders, scorpions, and centipedes, together with insects of almost every type, are in the desert. Drawn to man as a source of moisture or food, lice, mites, and flies can be extremely unpleasant and carry diseases such as scrub typhus and dysentery. The stings of scorpions and the bites of centipedes and spiders are extremely painful, though seldom fatal. Some species of scorpion, as well as black widow and recluse spiders, can cause death. The following paragraphs describe some of the wildlife that are encountered in desert areas and the hazards they may pose to man.

Scorpions

Scorpions are prevalent in desert regions, particularly active at night. Scorpions are They prefer damp locations and are easily recognizable by their crab-like appearance, and by their long tail which ends in a sharp stinger. Adult scorpions vary from less than an inch to almost 8 inches in length. Colors range from nearly black to straw to striped. Scorpions hide in clothing, boots, or bedding, so troops should routinely shake these items before using. Although scorpion stings are rarely fatal, they can be painful.

Flies

Flies are abundant throughout desert environments. Filth-borne disease is a major health problem posed by flies. Dirt or insects in the desert can cause infection in minor cuts and scratches.

Fleas

Avoid all dogs and rats which are the major carriers of fleas. Fleas are the primary carriers of plague and murine typhus.

Reptiles

Reptiles are perhaps the most characteristic group of desert animals. Lizards and snakes occur in quantity, and crocodiles are common in some desert rivers. Lizards are normally harmless and can be ignored; although exceptions occur in North America and Saudi Arabia.

Snakes, ranging from the totally harmless to the lethal, abound in the desert. A bite from a poisonous snake under two feet long can easily become infected. Snakes seek shade (cool areas) under bushes, rocks, trees, and shrubs. These areas should be checked before sitting or resting. Troops should always check clothing and boots before putting them on. Vehicle operators should look for snakes when initially conducting before-operations maintenance. Look for snakes in and around suspension components and engine compartments as snakes may seek the warm areas on recently parked vehicles to avoid the cool night temperatures.

Sand vipers have two long and distinctive fangs that may be covered with a curtain of flesh or folded back into the mouth. Sand vipers usually are aggressive and dangerous in spite of their size. A sand viper usually buries itself in the sand and may strike at a passing man; its presence is alerted by a characteristic coiling pattern left on the sand.

The Egyptian cobra can be identified by its characteristic cobra combative posture. In this posture, the upper portion of the body is raised vertically and the head tilted sharply forward. The neck is usually flattened to form a hood. The Egyptian cobra is often found around rocky places and ruins and is fairly common. The distance the cobra can strike in a forward direction is equal to the distance the head is raised above the ground. Poking around in holes and rock piles is particularly dangerous because of the likelihood of encountering a cobra. See Figure 1-16 for an example of a viper and cobra.

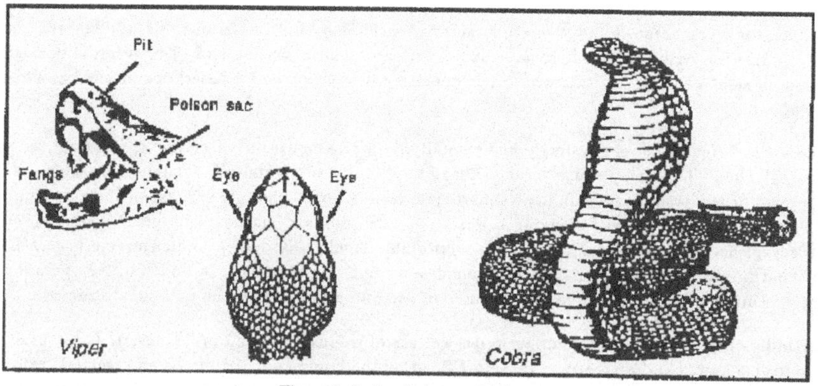

Figure 1-6. Sand viper and cobra

Mammals

The camel is the best known desert mammal. The urine of the camel is very concentrated to reduce water loss, allowing it to lose 30 percent of its body weight without undue distress. A proportionate loss would be fatal to man. The camel regains this weight by drinking up to 27 gallons (120 liters) of water at a time. It cannot, however, live indefinitely without water and will die of dehydration as readily as man in equivalent circumstances. Other mammals, such as gazelles, obtain most of their required water supply from the vegetation they eat and live in areas where there is no open water. Smaller animals, including rodents, conserve their moisture by burrowing underground away from the direct heat of the sun, only emerging for foraging at night. All these living things have adapted to the environment over a period of thousands of years; however, man has not made this adaptation and must carry his food and water with him and must also adapt to this severe environment.

Dogs are often found near mess facilities and tend to be in packs of 8 or 10. Dogs are carriers of rabies and should be avoided. Commanders must decide how to deal with packs of dogs; extermination and avoidance are two options. Dogs also carry fleas which may be transferred upon bodily contact. Rabies is present in most desert mammal populations. Do not take any chances of contracting fleas or rabies from any animal by adopting pets.

Rats are carriers of various parasites and gastrointestinal diseases due to their presence in unsanitary locations.

Section II. Environmental Effects on Personnel

There is no reason to fear the desert environment, and it should not adversely affect the morale of a soldier/marine who is prepared for it. Lack of natural concealment has been known to induce temporary agoraphobia (fear of open spaces) in some troops new to desert conditions, but this fear normally disappears with acclimatization. Remember that there is nothing unique about either living or fighting in deserts; native tribesmen have lived in the Sahara for thousands of years. The British maintained a field army and won a campaign in the Western Desert in World War II at the far end of a 12,000-mile sea line of communication with equipment considerably inferior to that in service now. The desert is neutral, and affects both sides equally; the side whose personnel are best prepared for desert operations has a distinct advantage.

The desert is fatiguing, both physically and mentally. A high standard of discipline is essential, as a single individual's lapse may cause serious damage to his unit or to himself. Commanders must exercise a high level of leadership and train their subordinate leaders to assume greater responsibilities required by the wide dispersion of units common in desert warfare. Soldiers/marines with good leaders are more apt to accept heavy physical exertion and uncomfortable conditions. Every soldier/marine must clearly understand why he is fighting in such harsh conditions and should be kept informed of the operational situation. Ultimately, however, the maintenance of discipline will depend on individual training.

Commanders must pay special attention to the welfare of troops operating in the desert, as troops are unable to find any "comforts" except those provided by the command. Welfare is an essential factor in the maintenance of morale in a harsh environment, especially to the inexperienced. There is more to welfare than the provision of mail and clean clothing. Troops must be kept healthy and physically fit; they must have adequate, palatable, regular food, and be allowed periods of rest and sleep. These things will not always be possible and discomfort is inevitable, but if troops know that their commanders are doing everything they can to make life tolerable, they will more readily accept the extremes brought on by the environment.

HEAT

The extreme heat of the desert can cause heat exhaustion and heatstroke and puts troops at risk of degraded performance. For optimum mental and physical performance, body temperatures must be maintained within narrow limits. Thus, it is important that the body lose the heat it gains during work. The amount of heat accumulation in the human body depends upon the amount of physical activity, level of hydration, and the state of personal heat acclimatization. Unit leaders must monitor their troops carefully for signs of heat distress and adjust schedules, work rates, rest, and water consumption according to conditions.

Normally, several physical and physiological mechanisms (e.g., convection and evaporation) assure transfer of excess body heat to the air. But when air temperature is above skin temperature (around 92 degrees Fahrenheit) the evaporation of sweat is the only operative mechanism. Following the loss of sweat, water must be consumed to replace the body's lost fluids. If the body fluid lost through sweating is not replaced, dehydration will follow. This will hamper heat dissipation and can lead to heat illness. When humidity is high, evaporation of sweat is inhibited and there is a greater risk of dehydration or heat stress. Consider the following to help prevent dehydration:

- Heat, wind, and dry air combine to produce a higher individual water requirement, primarily through loss of body water as sweat. Sweat rates can be high even when the skin looks and feels dry.
- Dehydration nullifies the benefits of heat acclimatization and physical fitness, it increases the susceptibility to heat injury, reduces the capacity to work, and decreases appetite and alertness. A lack of alertness can indicate early stages of dehydration.

- Thirst is not an adequate indicator of dehydration. The soldier/marine will not sense when he is dehydrated and will fail to replace body water losses, even when drinking water is available. The universal experience in the desert is that troops exhibit "voluntary dehydration" that is, they maintain their hydration status at about 2 percent of body weight (1.5 quarts) below their ideal hydration status without any sense of thirst.

Chronic dehydration increases the incidence of several medical problems: constipation (already an issue in any field situation), piles (hemorrhoids), kidney stones, and urinary infections. The likelihood of these problems occurring can be reduced by enforcing mandatory drinking schedules.

Resting on hot sand will increase heat stress-the more a body surface is in contact with the sand, the greater the heat stress. Ground or sand in full sun is hot, usually 30-45 degrees hotter than the air, and may reach 150 degrees Fahrenheit when the air temperature is 120 degrees Fahrenheit. Cooler sand is just inches below the surface; a shaded trench will provide a cool resting spot.
At the first evidence of heat illness, have the troops stop work, get into shade, and rehydrate. Early intervention is important. Soldiers/ marines who are not taken care of can become more serious casualties.

WARNING

One heat casually Is usually followed by Others and Is a warning that the entire unit may be at risk, This is the "Week Link Rule." The slalus Of the whole unit Is assessed at this point.

ACCLIMATIZATION

Acclimatization to heat is necessary to permit the body to reach and maintain efficiency in its cooling process. A period of approximately 2 weeks should be allowed for acclimatization, with progressive increases in heat exposure and physical exertion. Significant acclimatization can be attained in 4-5 days, but full acclimatization takes 7-14 days, with 2-3 hours per day of exercise in the heat. Gradually increase physical activity until full acclimatization is achieved.

Acclimatization does not reduce, and may increase, water requirements. Although this strengthens heat resistance, there is no such thing as total protection against the debilitating effects of heat. Situations may arise where it is not possible for men to become fully acclimatized before being required to do heavy labor. When this happens heavy activity should be limited to cooler hours and troops should be allowed to rest frequently. Check the weather daily. Day-to-day and region-to-region variations in temperatures, wind, and humidity can be substantial.

CLIMATIC STRESS

Climatic stress on the human body in hot deserts can be caused by any combination of air temperature, humidity, air movement, and radiant heat. The body is also adversely affected by such factors as lack of acclimatization, being overweight, dehydration, alcohol consumption, lack of sleep, old age, and poor health.

The body maintains its optimum temperature of 98.6 degrees Fahrenheit by conduction/convection, radiation, and evaporation (sweat). The most important of these in the daytime desert is evaporation, as air temperature alone is probably already above skin temperature. If, however, relative humidity is high, air will not easily evaporate sweat and the cooling effect is reduced. The following paragraphs describe the effects of radiant light, wind, and sand on personnel in desert areas.

Radiant Light

Radiant light comes from all directions. The sun's rays, either direct or reflected off the ground, affect the skin and can also produce eyestrain and temporarily impaired vision. Not only does glare damage the eyes but it is very tiring; therefore, dark glasses or goggles should be worn.

Overexposure to the sun can cause sunburn. Persons with fair skin, freckled skin, ruddy complexions, or red hair are more susceptible to sunburn than others, but all personnel are susceptible to some degree. Personnel with darker complexions can also sunburn. This is difficult to monitor due to skin pigmentation, so leaders must be ever vigilant to watch for possible sunburn victims. Sunburn is characterized by painful reddened skin, and can result in blistering and lead to other forms of heat illness.

Soldier/marines should acquire a suntan in gradual stages (preferably in the early morning or late afternoon) to gain some protection against sunburn. They should not be permitted to expose bare skin to the sun for longer than five minutes on the first day, increasing exposure gradually at the rate of five minutes per day. They should be fully clothed in loose garments in all operational situations. This will also reduce sweat loss. It is important to remember that—

- The sun is as dangerous on cloudy days as it is on sunny days.
- Sunburn ointment is not designed to give complete protection against excessive exposure.
- Sunbathing or dozing in the desert sun can be fatal.

Wind

The wind can be as physically demanding as the heat, burning the face, arms, and any exposed skin with blown sand. Sand gets into eyes, nose, mouth, throat, lungs, ears, and hair, and reaches every part of the body. Even speaking and listening can be difficult. Continual exposure to blown sand is exhausting and demoralizing. Technical work spaces that are protected from dust and sand are likely to be very hot. Work/rest cycles and enforced water consumption will be required.

The combination of wind and dust or sand can cause extreme irritation to mucous membranes, chap the lips and other exposed skin surfaces, and can cause nosebleed. Cracked, chapped lips make eating difficult and cause communication problems. Irritative conjunctivitis, caused when fine particles enter the eyes, is a frequent complaint of vehicle crews, even those wearing goggles. Lip balm and skin and eye ointments must be used by all personnel. Constant wind noise is tiresome and increases soldier/marine fatigue, thus affecting alertness.

When visibility is reduced by sandstorms to the extent that military operations are impossible, soldiers/marines should not be allowed to leave their group for any purpose unless secured by lines for recovery.

The following are special considerations when performing operations in dust or sand:

- Contact lenses are very difficult to maintain in the dry dusty environment of the desert and should not be worn except by military personnel operating in air conditioned environments, under command guidance.
- Mucous membranes can be protected by breathing through a wet face cloth, snuffing small amounts of water into nostrils (native water is not safe for this purpose) or coating the nostrils with

a small amount of petroleum jelly. Lips should be protected by lip balm.

- Moving vehicles create their own sandstorms and troops traveling in open vehicles should be protected.
- Scarves and bandannas can be used to protect the head and face.
- The face should be washed as often as possible. The eyelids should be cleaned daily.

BASIC HEAT INJURY PREVENTION

The temperature of the body is regulated within very narrow limits. Too little salt causes heat cramps; too little salt and insufficient water causes heat exhaustion. Heat exhaustion will cause a general collapse of the body's cooling mechanism. This condition is heatstroke, and is potentially fatal. To avoid these illnesses, troops should maintain their physical fitness by eating adequately, drinking sufficient water, and consuming adequate salt. If soldiers/marines expend more calories than they take in, they will be more prone to heat illnesses. Since troops may lose their desire for food in hot climates, they must be encouraged to eat, with the heavier meal of the day scheduled during the cooler hours.

It is necessary to recognize heat stress symptoms quickly. When suffering from heatstroke, the most dangerous condition, there is a tendency for a soldier/marine to creep away from his comrades and attempt to hide in a shady and secluded spot; if not found and treated, he will die. When shade is required during the day, it can best be provided by tarpaulins or camouflage nets, preferably doubled to allow air circulation between layers and dampened with any surplus water.

WARNING

Drink water even if not thirsty. Thirst Is not an adequate warning of dehydration.

Approximately 75 percent of the human body is fluid. All chemical activities in the body occur in a water solution, which assists in the removal of toxic body wastes and plays a vital part in the maintenance of an even body temperature. A loss of 2 quarts of body fluid (2.5 percent of body weight) decreases efficiency by 25 percent and a loss of fluid equal to 15 percent of body weight is usually fatal. The following are some considerations when operating in a desert environment:

- Consider water a tactical weapon. Reduce heat injury by forcing water consumption. Soldiers/marines in armored vehicles, MOPP gear, and in body armor need to increase their water intake.
- When possible, drink Cool (50-55 degrees Fahrenheit) water.
- Drink one quart of water in the morning, at each meal, and before strenuous work. In hot climates drink at least one quart of water each hour. At higher temperatures hourly water requirements increase to over two quarts.
- Take frequent drinks since they are more often effective than drinking the same amount all at once. Larger soldiers/marines need more water.
- Replace salt loss through eating meals.
- When possible, work loads and/or duration of physical activity should be less during the first days of exposure to heat, and then should gradually be increased to follow acclimatization.

- Modify activities when conditions that increase the risk of heat injury (fatigue/loss of sleep, previous heat exhaustion, taking medication) are present.
- Take frequent rest periods in the shade, if possible. Lower the work rate and work loads as the heat condition increases.
- Perform heavy work in the cooler hours of the day such as early morning or late evening, if possible.

A description of the symptoms and treatment for heat illnesses follows:

- Heat cramps.

  - Symptoms: Muscle cramps of arms, legs, and/or stomach. Heavy sweating (wet skin) and extreme thirst.

  - First aid: Move soldier/marine to a shady area and loosen clothing. Slowly give large amounts of cool water. Watch the soldier/marine and continue to give him water, if he accepts it. Get medical help if cramps continue.

- Heat exhaustion.

  - Symptoms: Heavy sweating with pale, moist, cool skin; headache, weakness, dizziness, and/or loss of appetite; heat cramps, nausea (with or without vomiting), rapid breathing, confusion, and tingling of the hands and/or feet.

  - First aid: Move the soldier/marine to a cool, shady area and loosen/remove clothing. Pour water on the soldier/marine and fan him to increase the cooling effect. Have the soldier/ marine slowly drink at least one full canteen of water. Elevate the soldier's/marine's legs. Get medical help if symptoms continue; watch the soldier/marine until the symptoms are gone or medical aid arrives.

- Heatstroke.

  - Symptoms: Sweating stops (red, flushed, hot dry skin).

  - First aid: Evacuate to a medical facility immediately. Move the soldier/marine to a cool, shady area and loosen or remove clothing if the situation permits. Start cooling him immediately. Immerse him in water and fan him. Massage his extremities and skin and elevate his legs. If conscious, have the soldier/marine slowly drink one full canteen of water.

DANGER

Heatstroke is a medical emergency. Seek medical attention immediately.

WATER SUPPLY

Maintaining safe, clean, water supplies is critical. The best containers for small quantities of water (5 gallons) are plastic water cans or coolers. Water in plastic cans will be good for up to 72 hours; storage in metal containers is safe only for 24 hours. Water trailers, if kept cool, will keep water fresh up to five days. If the air temperature exceeds 100 degrees Fahrenheit, the water temperature must be monitored. When the temperature exceeds 92 degrees Fahrenheit, the water should be changed, as bacteria will multiply. If the water is not changed the water can become a source of sickness, such as diarrhea. Ice in containers keeps water cool. If ice is put in water trailers, the ice must be removed prior to moving the trailer to prevent damage to the inner lining of the trailer.

Potable drinking water is the single most important need in the desert. Ensure nonpotable water is never mistaken for drinking water. Water that is not fit to drink but is not otherwise dangerous (it may be merely oversalinated) may be used to aid cooling. It can be used to wet clothing, for example, so the body does not use too much of its internal store of water.

Use only government-issued water containers for drinking water. Carry enough water on a vehicle to last the crew until the next planned resupply. It is wise to provide a small reserve. Carry water containers in positions that-

- Prevent vibration by clamping them firmly to the vehicle body.
- Are in the shade and benefit from an air draft.
- Are protected from puncture by shell splinters.
- Are easily dismounted in case of vehicle evacuation.

Troops must be trained not to waste water. Water that has been used for washing socks, for example, is perfectly adequate for a vehicle cooling system.

Obtain drinking water only from approved sources to avoid disease or water that may have been deliberately polluted. Be careful to guard against pollution of water sources. If rationing is in effect, water should be issued under the close supervision of officers and noncommissioned officers.

Humans cannot perform to maximum efficiency on a decreased water intake. An acclimatized soldier/marine will need as much (if not more) water as the nonacclimatized soldier/marine, as he sweats more readily. If the ration water is not sufficient, there is no alternative but to reduce physical activity or restrict it to the cooler parts of the day.

In very hot conditions it is better to drink smaller quantities of water often rather than large quantities occasionally. Drinking large quantities causes excessive sweating and may induce heat cramps. Use of alcohol lessens resistance to heat due to its dehydrating effect. As activities increase or conditions become more severe, increase water intake accordingly.

The optimum water drinking temperature is between 10 degrees Celsius and 15.5 degrees Celsius (50-60 degrees Fahrenheit). Use lister bags or even wet cloth around metal containers to help cool water.

Units performing heavy activities on a sustained basis, such as a forced march or digging in, at 80 degrees wet bulb globe temperature index, may require more than 3 gallons of drinking water per man. Any increase in the heat stress will increase this need. In high temperatures, the average soldier/marine will require 9 quarts of water per day to survive, but 5 gallons are recommended. Details on water consumption and planning factors are contained in Appendix G.

HINT: Taking a military issue wool sock, soaking it in water, and wrapping it around a canteen or placing it on a water bottle, will cool the water through evaporation.

While working in high desert temperatures, a man at rest may lose as much as a pint of water per hour from sweating. In very high temperatures and low humidity, sweating is not noticeable as it evaporates so fast the skin will appear dry. Whenever possible, sweat should be retained on the skin to improve the cooling process; however, the only way to do this is to avoid direct sun on the skin. This is the most important reason why desert troops must remain fully clothed. If a soldier/marine is working, his water loss through sweating (and subsequent requirement for replenishment) increases in proportion to the amount of work done (movement). Troops will not always drink their required amount of liquid readily and will need to be encouraged or ordered to drink more than they think is necessary as the sensation of thirst is not felt until there is a body deficit of 1 to 2 quarts of water. This is particularly true during the period of acclimatization. Packets of artificial fruit flavoring encourages consumption due to the variety of pleasant tastes.

All unit leaders must understand the critical importance of maintaining the proper hydration status. Almost any contingency of military operations will act to interfere with the maintenance of hydration. Urine provides the best indicator of proper hydration. The following are considerations for proper hydration during desert operations:

- Water is the key to your health and survival. Drink before you become thirsty and drink often, When you become thirsty you will be about a "quart and a half low".
- Carry as much water as possible when away from approved sources of drinking water. Man can live longer without food than without water.
- Drink before you work; carry water in your belly, do not "save" it in your canteen. Learn to drink a quart or more of water at one time and drink frequently to replace sweat losses.
- Ensure troops have at least one canteen of water in reserve, and know where and when water resupply will be available.
- Carbohydrate/electrolyte beverages (e.g., Gatorade) are not required, and if used, should not be the only source of water. They are too concentrated to be used alone. Many athletes prefer to dilute these 1:1 with water. Gaseous drinks, sodas, beer, and milk are not good substitutes for water because of their dehydrating effects.
- If urine is more colored than diluted lemonade, or the last urination cannot be remembered, there is probably insufficient water intake. Collect urine samples in field expedient containers and spot check the color as a guide to ensuring proper hydration. Very dark urine warns of dehydration. Soldiers/marines should observe their own urine, and use the buddy system to watch for signs of dehydration in others.
- Diseases, especially diarrheal diseases, will complicate and often prevent maintenance of proper hydration.

Salt, in correct proportions, is vital to the human body; however, the more a man sweats, the more salt he loses. The issue ration has enough salt for a soldier/marine drinking up to 4 quarts of water per day. Unacclimatized troops need additional salt during their first few days of exposure and all soldiers/marines need additional salt when sweating heavily. If the water demand to balance sweat loss rises, extra salt must be taken under medical direction. Salt, in excess of body requirements, may cause increased thirst and a feeling of sickness, and can be dangerous. Water must be tested before adding salt as some sources are already saline, especially those close to the sea.

COLD

The desert can be combine to produce dangerously cold. The dry air, wind, and clear sky can bone-chilling discomfort and even injury. The ability of the body to maintain body temperature within a narrow range is as important in the cold as in the heat. Loss of body heat to the environment can lead to cold injury; a genera] lowering of the body temperature can result in hypothermia, and local freezing of body tissues can lead to frostbite. Hypothermia is the major threat from the cold in the desert, but frostbite also occurs.

Troops must have enough clothing and shelter to keep warm. Remember, wood is difficult to find; any that is available is probably already in use. Troops maybe tempted to leave clothing and equipment behind that seems unnecessary (and burdensome) during the heat of the day. Cold-wet injuries (immersion foot or trench foot) may be a problem for dismounted troops operating in the coastal marshes of the Persian Gulf during the winter. Some guidelines to follow when operating in the cold are—

- Anticipate an increased risk of cold-wet injuries if a proposed operation includes lowland or marshes. Prolonged exposure of the feet in cold water causes immersion foot injury, which is completely disabling.
- Check the weather-know what conditions you will be confronting. The daytime temperature is no guide to the nighttime temperature; 90-degree-Fahrenheit days can turn into 30-degree-Fahrenheit nights.
- The effects of the wind on the perception of cold is well known. Windchill charts contained in FM 21-10 allow estimation of the combined cooling power of air temperature and wind speed compared to the effects of an equally cooling still-air temperature.

CLOTHING

Uniforms should be worn to protect against sunlight and wind. Wear the uniform loosely. Use hats, goggles, and sunscreen. Standard lightweight clothing is suitable for desert operations but should be camouflaged in desert colors, not green. Wear nonstarched long-sleeved shirts, and full-length trousers tucked into combat boots. Wear a scarf or triangular bandanna loosely around the neck (as a sweat rag) to protect the face and neck during sandstorms against the sand and the sun. In extremely hot and dry conditions a wet sweat rag worn loosely around the neck will assist in body cooling.

Combat boots wear out quickly in desert terrain, especially if the terrain is rocky. The leather drys out and cracks unless a nongreasy mixture such as saddle soap is applied. Covering the ventilation holes on jungle boots with glue or epoxies prevents excessive sand from entering the boots. Although difficult to do, keep clothing relatively clean by washing in any surplus water that is available. When water is not available, air and sun clothing to help kill bacteria.

Change socks when they become wet. Prolonged wear of wet socks can lead to foot injury. Although dry desert air promotes evaporation of water from exposed clothing and may actually promote cooling, sweat tends to accumulate in boots.

Soldier/marines may tend to stay in thin clothing until too late in the desert day and become susceptible to chills--so respiratory infections may be common. Personnel should gradually add layers of clothing at night (such as sweaters), and gradually remove them in the morning. Where the danger of cold weather injury exists in the desert, commanders must guard against attempts by inexperienced troops to discard cold weather clothing during the heat of the day.

Compared to the desert battle dress uniform (DBDU) the relative impermeability of the battle dress overgarment (BDO) reduces evaporative cooling capacity. Wearing underwear and the complete DBDU, with sleeves rolled down and under the chemical protective garment, provides additional protection against chemical poisoning. However, this also increases the likelihood of heat stress casualties.

HYGIENE AND SANITATION

Personal hygiene is absolutely critical to sustaining physical fitness. Take every opportunity to wash. Poor personal hygiene and lack of attention to siting of latrines cause more casualties than actual combat. Field Marshal Rommel lost over 28,400 soldiers of his Afrika Corps to disease in 1942. During the desert campaigns of 1942, for every one combat injury, there were three hospitalized for disease. Hygiene and sanitation are covered in detail in FM 21-10. This section highlights some of the points that are of special importance to the commander in the desert.

Proper standards of personal hygiene must be maintained not only as a deterrent to disease but as a reinforcement to discipline and morale. If water is available, shave and bathe daily. Cleaning the areas of the body that sweat heavily is especially important; change underwear and socks frequently, and use foot powder often. Units deployed in remote desert areas must have a means of cutting hair therefore, barber kits should be maintained and inventoried prior to any deployment. If sufficient water is not available, troops should clean themselves with sponge baths, solution-impregnated pads, a damp rag, or even a dry, clean cloth. Ensure that waste water is disposed of in an approved area to prevent insect infestation. If sufficient water is not available for washing, a field expedient alternative is powder baths, that is, using talcum or baby powder to dry bathe.

Check troops for any sign of injury, no matter how slight, as dirt or insects can cause infection in minor cuts and scratches. Small quantities of disinfectant in washing water reduces the chance of infection. Minor sickness can have serious effects in the desert. Prickly heat for example, upsets the sweating mechanism and diarrhea increases water loss, making the soldier/marine more prone to heat illnesses. The buddy system helps to ensure that prompt attention is given to these problems before they incapacitate individuals.

Intestinal diseases can easily increase in the desert. Proper mess sanitation is essential. Site latrines well away and downwind of troop areas and lagers. Trench-type latrines should be used where the soil is suitable but must be dug deeply, as shallow latrines become exposed in areas of shifting sand. Funnels dug into a sump work well as urinals. Layer the bottom of slit trenches with lime and cover the top prior to being filled in. Ensure lime is available after each use of the latrine. Flies area perpetual source of irritation and carry infections. Only good sanitation can keep the fly problem to a minimum. Avoid all local tribe camps since they are frequently a source of disease and vermin.

DESERT SICKNESS

Diseases common to the desert include plague, typhus, malaria, dengue fever, dysentery, cholera, and typhoid. Diseases which adversely impact hydration, such as those which include nausea, vomiting, and diarrhea among their symptoms, can act to dramatically increase the risk of heat (and cold) illness or injury. Infectious diseases can result in a fever; this may make it difficult to diagnose heat illness. Occurrences of heat illness in troops suffering from other diseases complicate recovery from both ailments.

Many native desert animals and plants are hazardous. In addition to injuries as a result of bites, these natural inhabitants of the desert can be a source of infectious diseases.

Many desert plants and shrubs have a toxic resin that can cause blisters, or spines that can cause infection. Consider milky sap, all red beans, and smoke from burning oleander shrubs, poisonous. Poisonous snakes, scorpions, and spiders are common in all deserts. Coastal waters of the Persian Gulf contain hazardous marine animals including sea snakes, poisonous jellyfish, and sea urchins.

Skin diseases can result from polluted water so untreated water should not be used for washing clothes; although it can be used for vehicle cooling systems or vehicle decontamination.

The excessive sweating common in hot climates brings on prickly heat and some forms of fungus infections of the skin. The higher the humidity, the greater the possibility of their occurrence. Although many deserts are not humid, there are exceptions, and these ailments are common to humid conditions.

The following are additional health-related considerations when operating in a desert environment:

- The most common and significant diseases in deserts include diarrheal and insectborne febrile (i.e., fever causing) illnesses-both types of these diseases are preventable.
- Most diarrheal diseases result from ingestion of water or food contaminated with feces. Flies, mosquitoes, and other insects carry fever-causing illnesses such as malaria, sand fly fever, dengue (fever with severe pain in the joints), typhus, and tick fevers.
- There are no safe natural water sources in the desert. Standing water is usually infectious or too brackish to be safe for consumption. Units and troops must always know where and how to get safe drinking water.
- Avoid brackish water (i.e., salty). It, like sea water, increases thirst; it also dehydrates the soldier/marine faster than were no water consumed. Brackish water is common even in public water supplies, Iodine tablets only kill germs, they do not reduce brackishness.
- Water supplies with insufficient chlorine residuals, native food and drink, and ice from all sources are common sources of infective organisms.

PREVENTIVE MEASURES

Both diarrheal and insectborne diseases can be prevented through a strategy which breaks the chain of transmission from infected sources to susceptible soldiers/marines by effectively applying the preventive measures contained in FM 21-10. Additional preventive measures are described below:

- Careful storage, handling and purification/preparation of water and food are the keys to prevention of diarrheal disease. Procure all food, water, ice, and beverages from US military approved sources and inspect them routinely.
- Well-cooked foods that are "steaming hot" when eaten are generally safe, as are peeled fruits and vegetables.
- Local dairy products and raw leafy vegetables are generally unsafe.
- Consider the food in native markets hazardous. Avoid local food unless approved by medical personnel officials.

- Assume raw ice and native water to be contaminated-raw ice cannot be properly disinfected. Ice has been a major source of illness in all prior conflicts; therefore, use ice only from approved sources.
- If any uncertainty exists concerning the quality of drinking water, troops should disinfect their supplies using approved field-expedient methods (e.g., hypochlorite for lister bags, iodine tablets for canteens, boiling).
- Untreated water used for washing or bathing risks infection.
- Hand washing facilities should be established at both latrines and mess facilities. Particular attention should be given to the cleanliness of hands and fingernails. Dirty hands are the primary means of transmitting disease.
- Dispose of human waste and garbage as specified in FM 21-10. Additional considerations regarding human waste and garbage are—

    - Sanitary disposal is important in preventing the spread of disease from insects, animals, and infected individuals, to healthy soldiers/marines.

    - Construction and maintenance of sanitary latrines are essential.

    - Burning is the best solution for waste.

    - Trench latrines can be used if the ground is suitable, but they must be dug deeply enough so that they are not exposed to shifting sand, and they must have protection against flies and other insects that can use them as breeding places.

    - Food and garbage attract animals-do not sleep where you eat and keep refuse areas away from living areas.
- Survey the unit area for potential animal hazards.
- Shake out boots, clothing, and bedding before using them.

Section III. Environmental Effects on Equipment

Conditions in an arid environment can damage military equipment and facilities. Temperatures and dryness are major causes of equipment failure, and wind action lifts and spreads sand and dust, clogging andjamming anything that has moving parts. Vehicles, aircraft, sensors, and weapons are all affected. Rubber components such as gaskets and seals become brittle, and oil leaks are more frequent. Ten characteristics of the desert environment may adversely affect equipment used in the desert:

- Terrain.
- Heat.
- Winds.
- Dust and sand.
- Humidity.
- Temperature variations.
- Thermal bending.
- Optical path bending.

- Static electricity.
- Radiant light.

The relative importance of each characteristic varies from desert to desert. Humidity, for example, can be discounted in most deserts but is important in the Persian Gulf.

TERRAIN

Terrain varies from nearly flat, with high trafficability, to lava beds and salt marshes with little or no trafficability. Drivers must be well trained in judging terrain over which they are driving so they can select the best method of overcoming the varying conditions they will encounter. Techniques for driving and operating equipment in desert conditions are contained in Appendix C.

Track vehicles are well suited for desert operations. Wheel vehicles may be acceptable as they can go many places that track vehicles can go; however, their lower speed average in poor terrain maybe unacceptable during some operations. Vehicles should be equipped with extra fan belts, tires, (and other items apt to malfunction), tow ropes (if not equipped with a winch), extra water cans, and desert camouflage nets. Air-recognition panels, signal mirrors, and a tarpaulin for crew sun protection are very useful. Wheel vehicles should also carry mats, or channels as appropriate, to assist in freeing mired vehicles.

The harsh environment requires a very high standard of maintenance, which may have to be performed well away from specialized support personnel. Operators must be fully trained in operating and maintaining their equipment. Some types of terrain can have a severe effect on suspension and transmission systems, especially those of wheel vehicles. Tanks are prone to throw tracks when traveling over rocks.

Track components require special care in the desert. Grit, heat, and bad track tension accelerate track failure. Sprockets wear out quickly in sandy conditions. Track pins break more easily in high temperatures and high temperatures also increase rubber/metal separation on road wheels. Proper track tension is critical, as loose track is easily thrown and excessive tension causes undue stress on track components.

Increase the unit PLL of tires and tracks as sand temperatures of 165 degrees Fahrenheit are extremely detrimental to rubber, and weaken resistance to sharp rocks and plant spines, Items affected by mileage such as wheels, steering, track wedge bolts and sprocket nuts, and transmission shafts, must be checked for undue wear when conducting before-, during-, and after-operations maintenance.

HEAT

Vehicle coding and lubrication systems are interdependent. A malfunction by one will rapidly place the other system under severe strain. In temperature extremes, all types of engines are apt to operate above optimum temperatures, leading to excessive wear, or leaking oil seals in the power packs, and ultimately, engine failure. Commanders should be aware which types of vehicles are prone to excessive overheating, and ensure extra care is applied to their maintenance. The following are considerations for ensuring engines do not overheat:

- Check oil levels frequently to ensure proper levels are maintained (too high may be as bad as too low), that seals are not leaking, and oil consumption is not higher than normal.

- Keep radiators and air flow areas around engines clean and free of debris and other obstructions.
- Fit water-cooled engines with condensers to avoid steam escaping through the overflow pipe.
- Cooling hoses must be kept tight (a drip a second loses 7 gallons of fluid in 24 hours).
- Operators should not remove hood side panels from engine compartments while the engine is running as this causes air turbulence and leads to ineffective cooling.

Batteries do not hold their charge efficiently in intense heat. Check them twice daily. The following are additional considerations for maintaining batteries in intense heat:

- Change battery specific gravity to adjust to the desert environment (see vehicle TMs for details).
- Keep batteries full, but not overfilled, and carry a reserve of distilled water.
- Keep air vents clean, or vapors may build up pressure and cause the battery to explode.
- Set voltage regulators as low as practical.
- Increase dry battery supplies to offset high attrition rate caused by heat exposure.

Severe heat increases pressure in closed pressurized systems such as the M2 fire burner unit, and increases the volume of liquids. Ensure that the working pressure of all equipment is within safety limits and be careful when removing items such as filler caps.

Treat Halon fire extinguishers with care. High temperatures may cause them to discharge spontaneously. Put wet rags on them during the hottest part of the day to keep them cooler.

Some items of equipment are fitted with thermal cutouts that open circuit breakers whenever equipment begins to overheat. Overheating is often caused by high ambient temperatures, and can be partly avoided by keeping the item in the shade or wrapping it in a wet cloth to maintain a lower temperature by evaporation.

Flying time and performance of helicopters are degraded as the altitude and heat increases. Helicopter performance is also affected by humidity. Aircraft canopies have been known to bubble under direct heat and should be covered when not in use.

Ammunition must be out of direct heat and sunlight. Use camouflage nets and tarpaulins to provide cover. Ammunition cool enough to be held by bare hands is safe to fire.

Wood shrinks in a high-temperature, low-humidity environment. Equipment, such as axes carried on track vehicles, can become safety hazards as heads are likely to fly off shrunken handles.

Radiators require special attention. Proper cooling-system operation is critical in high-temperature environments. Check cooling systems for serviceability prior to deployment. Local water maybe high in mineral content which will calcify in cooling systems. Distilled water is better since tap water contains chemicals that will form a crusty coating inside the radiator and will ultimately clog it. A mixture of 40 percent antifreeze and 60 percent water is usually acceptable-check your appropriate technical manual to be certain.

During movement, and at operation sites where extremely hot temperatures exist, continuous protection from the heat is necessary for medical items and supplies, which deteriorate rapidly.

Air and all fluids expand and contract according to temperature. If tires are inflated to correct pressure during the cool of night, they may burst during the heat of day. If fuel tanks are filled to the brim at night, they will overflow at midday. Servicing these items during the heat of day can result in low tire pressure, overheating of tires, and a lack of endurance if the fuel tanks were not filled to their correct levels. Air pressure in tires must be checked when the equipment is operating at efficient working temperatures, and fuel tanks must be filled to their correct capacity as defined in the appropriate technical manual. These items should be checked several times a day and again at night.

The major problem with radios in a desert environment is overheating. The following steps can help prevent overheating of radios:

- Keep radios out of direct sunlight.
- Place a piece of wood on top of the radio. Leaving space between the wood and the top of the radio will help cool the equipment. Operating on low power whenever possible will also help.
- Place wet rags on top of radios to help keep them cool and operational. Do not cover the vents.

Any oil or fuel blown onto a cooler (heat exchanger) will gather and quickly degrade cooling. Fix even slight leaks promptly. Do not remove cooling ducts or shrouds. Check them for complete coverage-use tape to seal cracks. Do not remove serviceable thermostats if overheating occurs.

WINDS

Desert winds, by their velocity alone, can be very destructive to large and relatively light materiel such as aircraft, tentage, and antenna systems. To minimize the possibility of wind damage, materiel should be sited to benefit from wind protection and should be firmly picketed to the ground.

DUST AND SAND

Dust and sand are probably the greatest danger to the efficient functioning of equipment in the desert. It is almost impossible to avoid particles settling on moving parts and acting as an abrasive. Sand mixed with oil forms an abrasive paste.

Lubricants must be of the correct viscosity for the temperature and kept to the recommended absolute minimum in the case of exposed or semiexposed moving parts. Lubrication fittings are critical items and should be checked frequently. If they are missing, sand will enter the housing causing bearing failure. Teflon bearings require constant inspection to ensure that the coating is not being eroded.

Proper lubrication is crucial for success. Wipe off all grease fittings before you attach the grease gun and after use. Keep cans of grease covered to prevent sand contamination. Preserve opened grease containers by covering and sealing with plastic bags. Use of grease cartridges in lieu of bulk grease is preferred. All POL dispensing tools must be stored in protected areas to prevent contamination. Place a tarpaulin, or other material, under equipment being repaired to prevent tools and components from being lost in the sand. The automotive-artillery grease possesses a significantly high-temperature capability. If not available, an alternative is general purpose wide-temperature range (WTR) aircraft grease.

Oil should be changed about twice as often under desert conditions as under US or European conditions, not only because grit accumulates in the oil pan, but also because noncombusted low-octane fuel seeps down the cylinder walls and dilutes the reservoir. Diluted lubricants cool less effectively, and evaporate at the higher temperatures generated during engine operation. Oil changes and lubrication of undercarriage points at more frequent intervals will prolong engine and vehicle life under desert conditions. Units employed in desert environments should reevaluate their engine oil requirements and plan accordingly.

Keeping sand out of maintenance areas is critical due to the strong possibility of sand or dust entering the cylinders or other moving parts when the equipment is stripped. Baggies, cloth, or plastic can be used to protect open or disassembled components from blowing sand and dust. The same applies for disconnected water, oil, or other fluid lines. Be sure to cover both ends of the connection if stored. It is essential to have screens against blowing sand (which also provides shade for mechanics). The surrounding ground may be soaked in used oil or covered with rocks to bind it down. Mechanics must keep their tools clean.

Dust and sand can easily cause failure of such items as radio and signal distribution panels, and circuit breakers, and cause small electrical motors to burn out. Wheel and flight control bearings may require daily cleaning and repacking, and engines should be flushed of contaminants daily.

Rotor heads have reduced life, requiring more frequent inspections than in temperate climates. Pay particular attention to sand-caused wear on rotor heads, leading edges of rotor blades, and exposed flight controls. Over 200 pounds of dirt has been known to accumulate in the fuselage area of helicopters operating in desert conditions. These areas must be routinely checked and cleaned to prevent a pound-for-pound reduction in aircraft-lift capability.

Filters must be used when refueling any type of vehicle, and the gap between the nozzle and the fuel tank filler pipe must be kept covered. It takes comparatively little dirt to block a fuel line. Fuel filters will require more frequent cleaning and will need to be checked and replaced often. Engine oil should be changed more often and oil filters replaced more frequently than in temperate climates.

Compression-ignition engines depend on clean air; therefore, examine and clean air cleaners of every type of equipment at frequent intervals. The exact interval depends on the operating conditions, but as a minimum, should be checked at least daily.

Air compressors are valuable pieces of equipment in the desert. They are essential for cleaning air filters and removing dust and sand from components. Intake filters require cleaning daily.

Windblown sand and grit, in addition to heat, will damage electrical wire insulation over a period of time. All cables that are likely to be damaged should be protected with tape before insulation becomes worn. Sand will also find its way into parts of items such as "spaghetti cord" plugs, either preventing electrical contact or making it impossible to join the plugs together. Use a brush, such as an old toothbrush, to brush out such items before they are joined. Electrical tape placed over the ends of spaghetti cords also works.

Radio is the primary means of communications in the desert. It can be employed effectively in desert climates and terrain to provide the reliable communications demanded by widely dispersed forces. However, desert terrain provides poor electrical ground, and a counterpoise (an artificial ground) is needed to improve the range of certain antennas.

Some receiver-transmitters have ventilating ports and channels that can get clogged with dust. These must be checked regularly and kept clean to prevent overheating. Mobile subscriber equipment may require the deployment of additional radio access units (RAU) AN/VRC-191. These assemblages are the primary link for the mobile subscriber radio telephone terminal (MSRT) AN/VRC-97s which are located down to battalion level. The normal operating range of the receiver-transmitter used with these radios may only be 10 kilometers in the desert.

Dust and sand adversely affect the performance of weapons. Weapons may jam or missiles lock on launching rails due to sand and dust accumulation. Sand- or dust-clogged barrels lead to in-bore detonations. Daily supervised cleaning of weapons is essential. Particular attention should be given to magazines which are often clogged, interrupting the feeding of weapons. Cover missiles on launchers until required for use. To avoid jamming due to the accumulation of sand, the working parts of weapons must have the absolute minimum amount of lubrication. It may even be preferable to have them totally dry, as any damage caused during firing will be less than that produced by the sand/oil abrasive paste. Paintbrushes are among the most useful tools to bring to the desert; they are extremely effective in cleaning weapons and optics.

Take precautions to prevent exposure of floppy disks and computers to dust or sand. Covering them in plastic bags is a technique that has worked for several different units. A number of units have successfully operated PLL computers in inflatable medical NBC shelters (MIS 1). This technique has obvious drawbacks since the shelter was not designed for this; however, until a materiel fix is developed, this sort of innovation may be necessary. Compressed air cans, locally purchased from computer vendors, will facilitate the cleaning of keyboards and other components of computer systems.

All optics are affected by blown sand, which gradually degrade their performance due to small pitting and scratching. It is necessary to guard against buildup of dust on optics, which may not be apparent until the low light optical performance has severely deteriorated. It may be advisable to keep optics covered with some form of cellophane film until operations are about to start, especially if the unit is in a sandstorm. A cover that has no sand on the underside should also be used and must be secured so it cannot vibrate against the wind screen. Both of these measures are equally important to tactical security as sun reflected from these optics will reveal positions.

Sand and dirt can easily accumulate in hull bottoms of armored vehicles. This accumulation, combined with condensation or oil, can cause jamming of control linkages. Sand accumulation at the air-bleeder valve can inhibit heat from escaping from the transmission and result in damage to the transmission. Park tactical wheeled vehicles with the rear facing the wind to keep sand out of the radiator. Tracked vehicles should park to protect the engine compartment (grille doors away from wind) from the same sort of damage. The operator's checks and services increase in importance in this environment.

HUMIDITY

Some deserts are humid. Where this is the case, humidity plus heat encourages rust on bare metal and mold in enclosed spaces such as optics. Bare metal surfaces on equipment not required for immediate use must be kept clean and very lightly lubricated.

Items such as optics must be stored in dry conditions; those in use should be kept where air can circulate around them, and should be purged at frequent intervals. Aircraft must be washed daily, particularly if there is salt in the air, using low-pressure sprays.

TEMPERATURE VARIATIONS

In deserts with relatively high-dew levels and high humidity, overnight condensation can occur wherever surfaces (such as metal exposed to air) are cooler than the air temperature. Condensation can affect such items as optics, fuel lines, and air tanks. Drain fuel lines both at night and in the morning (whenever necessary). Clean optics and weapons frequently. Weapons, even if not lubricated, accumulate sand and dirt due to condensation.

THERMAL BENDING

Weapon systems such as the tank cannon are affected in several ways by the desert. One is thermal bending, which is the uneven heating and cooling of a gun tube due to ambient temperature changes. Modem tanks, like the Ml, have been designed to compensate for these factors. The muzzle reference system (MRS) allows the crew to monitor any loss of gun sight relationship and to comet for any error using the MRS update at regular intervals. Ml-series tanks are equipped with a thermal shroud, allowing for more even heating and cooling of the gun tube. Both factors can greatly reduce the accuracy of a tank weapon system. By boresighting at regular intervals and constant monitoring of the fire control system, the tank crew can maximize its readiness. "Gun tube droop" can be countered using the MRS update at least four times in a 24-hour period: at dawn as part of stand-to; at noon to compensate for gun tube temperature chang: before EENT, for TIS reticle confirmation; and at 0100 hours to compensate for gun tube temperature changes.

OPTICAL PATH BENDING

The apparent illusion of target displacement is commonly called refraction. Under certain light and environmental conditions, the path of light (line of sight) may not appear to travel in a straight line. Refraction may cause problems for tank crews attempting engagements at ranges beyond 1,500 meters. Figure 1-17 shows an example of optical path bending in the desert. Refraction may occur in the following conditions:

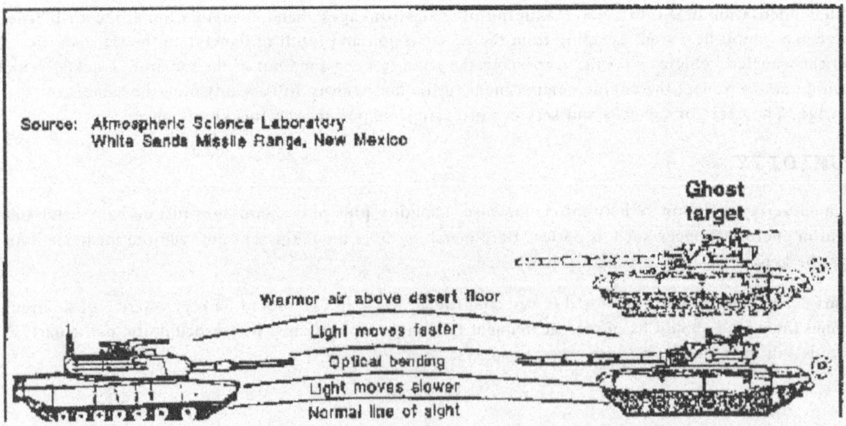

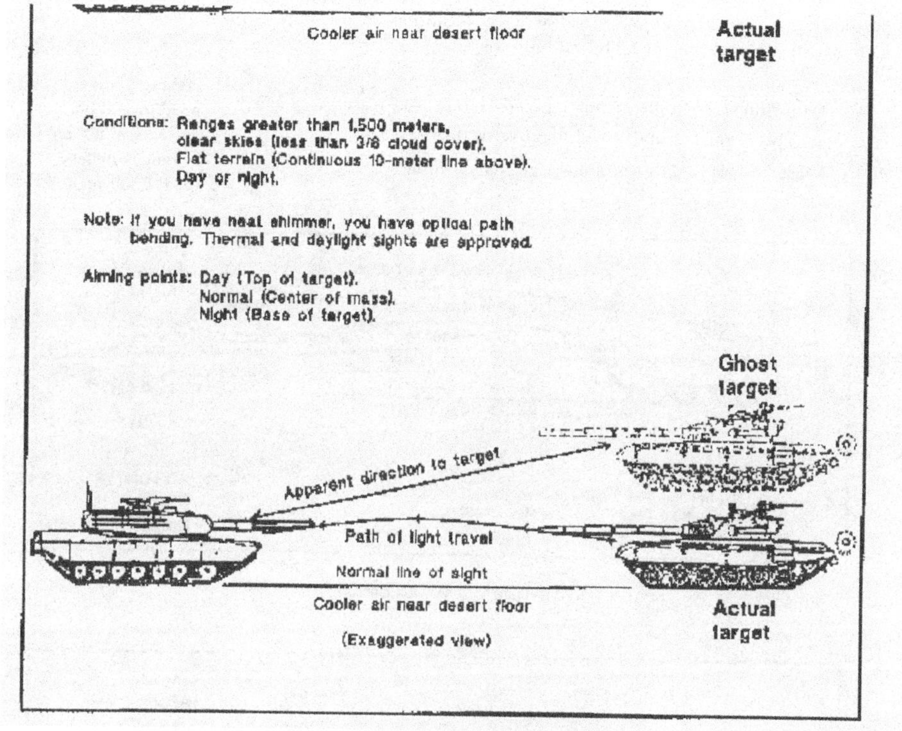

Figure: 1-17. Optical path bending in the desert.

- Day—Clear sky, flat terrain, winds less than 10 miles per hour,
- Night-Clear sky, flat terrain, winds under 4 miles per hour.

NOTE: Any time heat shimmer is present, refraction may also exist.

The effect of refraction is to make the target appear lower during the day; the sight picture, though it appears center of visible mass to the gunner, is actually below the target. This may result in a short round. At night, the effects are the opposite and may result in an over round. Crews must not be fooled by what appears to be a good range from their laser range finder (LRF); the laser beam will refract with other light rays and still hit the desired target.

The most effective measure available to the crew to minimize refraction is an elevated firing position. A position at least 10 meters above intervening terrain will generally negate any effects. If this type of position is not available, a crew operating under conditions favorable to refraction, and having missed with their first round, should apply the following:

- Day--Adjust sight picture up 1/2 target form. See Figures 1-18 and 1-19 for examples of day and night refraction, respectively.
- Night—Adjust sight picture down 1/2 target form.

NOTE:Crews do not need to make a correction for refraction at ranges of less than 1,500 meters.

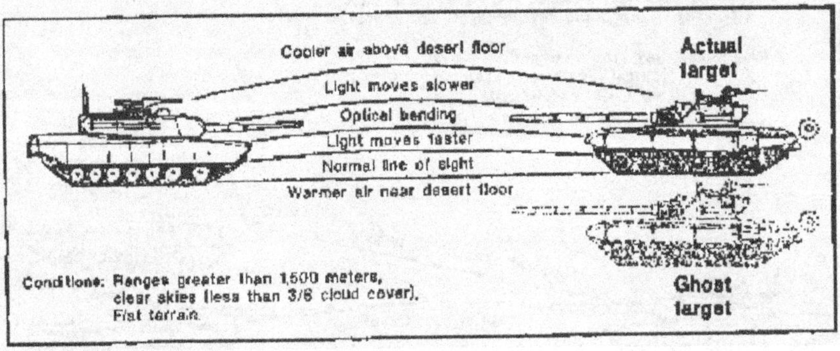

Figure 1-18. Day refraction.

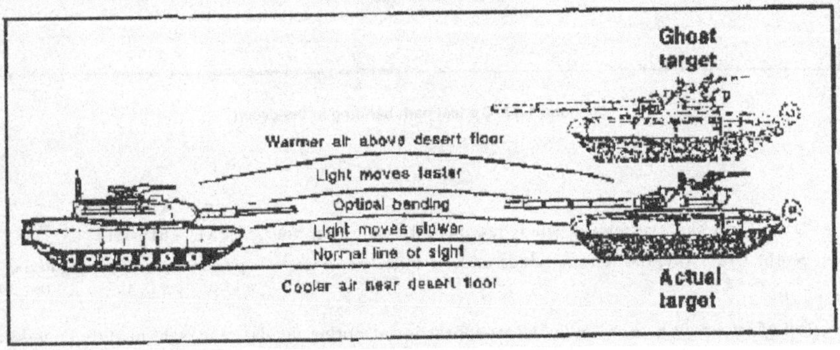

Figure 1-19. Night refraction

STATIC ELECTRICITY

Static electricity is prevalent and poses a danger in the desert. It is caused by atmospheric conditions coupled with an inability to ground out due to dryness of the terrain. It is particularly prevalent with aircraft or vehicles having no conductor contact with the soil. The difference of electrical potential between separate materials may cause an electrical discharge between them when contact is made, and if flammable gases are present, they may explode and cause a fire. Poor grounding conditions aggravate the problem. Be sure to tape all sharp edges (tips) of antennas to reduce wind-caused static electricity. If you are operating from a fixed position, ensure that equipment is properly grounded.

Establish a metal circuit between fuel tankers and vehicles before and during refueling. Ensure the fuel tankers and vehicles are grounded (for example, by a cable and picket or by a crowbar). Grounding of vehicles and equipment should be accomplished in accordance with appropriate operations manuals.

Static electricity is also a hazard with helicopter sling loads. Exercise care when handling and transporting unlike materials that might generate static electricity. Also turn off all switches, uncouple electrical connectors, and ground vehicle or aircraft electrically-operated weapon systems before reaming. Static electricity will also ruin circuit boards and other electronic equipment.

RADIANT LIGHT

Radiant light may be detrimental to plastics, lubricants, pressurized gases, some chemicals, and infrared tracking and guidance systems. Items like CO^2 fire extinguishers, M13 decontamination and reimpregnating kits, and Stinger missiles must be kept out of constant direct sunlight. Optics have been known to discolor under direct sunlight (although this is unusual), so it is wise to minimize their exposure to the sun's rays.

CHAPTER 2

PREPARATIONS FOR DESERT OPERATIONS

This chapter describes preparations for deployment and training for operations in a desert environment. A force sent to a desert theater will fight with the equipment it has in accordance with current doctrine. While equipment and doctrine can be modified to suit the new environment, much will depend on how well soldiers/marines and leaders have mastered their individual training. Units that have trained in Germany and the United States will have the basic technical and tactical skills that can be adapted for desert warfare. Well-trained troops and leaders can adapt quickly to the peculiar conditions of the new environment. If their individual and collective skills have been neglected, no amount of desert lore will remedy the situation.

Section I. Factors to be Considered When Preparing for Desert Operations

When a unit is alerted for operations in a desert environment the commander must first answer some or all of the following questions:

- To what country is the unit going?
- What are the climatic and terrain conditions of that country?
- Will the unit be taking its own equipment overseas?
- What is the tentative timeline for departure? Will there be a period of time where the unit has time for individual training while the vehicles are in transit?
- What unit equipment is being sent overseas and what items will it require for modification (including camouflage painting)?
- What special equipment does the unit require for desert operations?
- What special maintenance is required for weapons and equipment before deployment to or arrival in a desert environment?
- Are there personnel in the unit who—
    - Have desert experience as observers or controllers?
    - Have any experience in desert conditions?
- Are all personnel physically fit?
- How many soldiers/marines are nondeployable?
- What types of operations are expected?

Once these and other questions have been answered the commander must develop a program to bring his unit to a level where it is fully capable of successfully operating in harsh desert conditions. To do this, first set a list of priorities for both individual and unit training. The training priorities listed below are shown as a guide only. They can be modified as necessary depending on the state of readiness of the unit when it is first alerted for desert employment.

Section II. Individual Training

In order to fight and survive in desert operations troops must fully understand the desert environment. The objective of individual training is to prepare the individual for operations in a desert environment. This requires both mental and physical preparation.

To the extent practicable, troops should be acclimatized before arrival in the area of operations. The requirement for acclimatization will vary slightly between individuals, but physical conditioning (fit soldiers/marines acclimatize more easily) is a part of the acclimatization process. Acclimatization should take place in climatic conditions that are similar, or slightly more strenuous, than those of the prospective area of operations.

CAMOUFLAGE AND CONCEALMENT

Camouflage and concealment training may be divided into concealment from the ground and concealment from the air. Particular attention must be paid to movement, color, shadow, and deception. Units should practice erecting and disassembling camouflage netting in order to become more efficient. Well-trained crews can save time and headaches. Camouflage and concealment are equally important for combat service support troops. Appendix E contains information about desert camouflage and concealment techniques.

DRIVER TRAINING

Because of the absence of established roads in desert areas, driving requires experience, individual skill, and physical endurance on the part of the vehicle operator. Driver training exercises should be long and arduous to expose vehicle operators to the rigors of the desert as well as to the effects of fatigue. The need for dispersing and avoiding preceding vehicles is stressed when operating over crusted surfaces or when the trail deteriorates while operating over sand (except suspected mine areas). Training should be directed toward driver proficiency in dune areas, choice of the best ground, selection of proper gear ratios, and driver knowledge and appreciation of the exact capabilities of his vehicle. Driver skill should be developed in taking maximum advantage of momentum, gear shifting, estimating and utilizing proper speeds, and avoiding sudden driving or braking thrusts. Additional driving techniques are contained in Appendix C.

SURVIVAL, EVASION, AND ESCAPE

Convincing a soldier/marine that he is capable of surviving in the desert environment strengthens his Self-confidence, and thus his morale. FM 21-76 contains details on survival, evasion, and escape in the desert, but the following points concerning FM 21 -76 should be included in desert survival, evasion, and escape training due to their importance:

- It is unlikely that wells will be poisoned. However, some wells in the North African desert have such strong concentrations of mineral salts that water taken from them may lead to intestinal irritation and subsequent illness.

- Although water is undoubtedly the most important factor in survival, a soldier/marine should not discard his personal weapon or any navigational equipment except in extreme circumstances. Mirrors of any type should be retained for signaling aircraft or other ground forces.

FIELD TRAINING

Following minimum preliminary training in garrison, desert living can only be practiced in the field, often as part of unit training. Important aspects that should be covered include:

- The effects of heat, including possible dehydration and salt loss (the need to maintain the body fluid level).
- The effects of temperature variations.
- The effects of cold weather in the desert.
- First aid for heat illnesses. Each soldier/marine should be issued a memory aid card showing symptoms and immediate treatment.
- Maintenance of morale and the ability of the individual to accept the challenge of the desert. Self discipline and common sense.
- Environmental effects such as those of sand, wind, and light.
- Water discipline.
- Hygiene and sanitation.
- Comet clothing equipment, including how to wear and maintain and clothing.
- Precautions against snakes, scorpions, insects, and disease-bearing organisms.

To the extent possible, the commander should train his unit in terrain and environmental conditions similar to what he expects to find in the operational area. It would be both shortsighted and dangerous for example, to allow water for bathing if the expected operational area is totally waterless. To further accustom the troops to hardships, contact with garrison or other urban areas should be kept to the minimum except for medical or welfare reasons. Once field training has started, necessary supplies should be brought to field locations and items that are unlikely to be available in the operational area (commercial soft drinks and foods) should not be permitted. To gain the maximum value from this training, the unit's exposure to outside influences should be kept to a minimum.

ENEMY ORGANIZATIONS AND TACTICS

This can be taught in garrison on sand tables and map maneuvers, followed by tactical exercises without troops (TINT), and unit exercises in the field. If enemy equipment is available it should be brought to the unit so it can be studied firsthand.

DESERT NAVIGATION

Troops must be thoroughly briefed on the type of terrain and the general environment they will encounter, including-

- Water sources, if any.

- Landmarks or significant permanent terrain features.
- Friendly and enemy areas of operation.
- Prevailing winds.
- Whether or not the local populace is pro or con the US.

This information will assist navigation by reconnaissance units or individuals who become separated from their units. Although maps are the most obvious navigation aids, numerous types of equipment and techniques are available to assist soldiers/marines during desert operations.

Maps

Although maps used in field training will be those of the local area, sufficient maps of the operational area should be obtained to allow distribution for study and possible use during garrison training. This is particularly important if the operational maps use foreign words to describe terrain, such as sebhka, summan, hidiba, and dikaka.

In addition, the grid system on some maps differs from the universal transverse mercator grid system on US maps. In many Middle East countries that were previously under British influence, for example, the Palestine grid system is used on military maps. These maps, generally last surveyed during World War II or the following decade, are widely used, not only in the area of Palestine, but also in Egypt and much of Saudi Arabia. And since they are commonly produced in either 1:100,000 or 1:50,000 scale, they do not mesh with standard US maps. In some instances, accurate maps may not be available. An alternative is to draw the grid lines on attached blank sheets of paper. This method can be highly effective when used in conjunction with navigational aids such as the GPS and LORAN.

Latitude and Longitude Conversions

One of the oldest systematic methods of location is based upon the geographic coordinate system. By drawing a set of east-west rings around the globe (parallel to the equator), and a set of north-south rings crossing the equator at right angles and converging at the poles, a network of reference lines is formed from which any point on the earth's surface can be located. The distance of a point north or south of the equator is known as its latitude. The rings around the earth parallel to the equator are called parallels of latitude or simply parallels. Lines of latitude run east-west but north-south distances are measured between them. A second set of rings around the globe at right angles to the lines of latitude and passing through the poles are known as meridians of longitude or simply meridians. One meridian is designated as the prime meridian. The prime meridian of the system we use runs through Greenwich, England, and is known as the Greenwich meridian. The distance east or west of a prime meridian to a point is known as its longitude. Lines of longitude (meridians) run north-south but east-west distances are measured between them.

Geographic coordinates are expressed in angular measurement. Each circle is divided into 360 degrees, each degree into 60 minutes and each minute into 60 seconds. At any point on the earth, the ground distance covered by one degree of longitude is about 111 kilometers (69 miles); one second is equal to about 30 meters (100 feet). The ground distance covered by one degree of longitude at the equator is also about 111 kilometers, but decreases as one moves north or south, until it becomes zero at the poles. For example, one second of longitude represents about 30 meters at the equator but at the latitude of

Washington, D. C, one second of longitude is approximately 24 meters.

Geographic coordinates appear on all standard military maps, and on some they may be the only method of locating and referencing the location of a point. The four lines that enclose the body of the map (neatlines) are lines of latitude and longitude. Their values are given in degrees and minutes at each of the four comers. In addition to the latitude and longitude given for the four corners, there are, at regular intervals along the sides of the map, small tick marks extending into the the body of the map. Each of these tick marks is identified by its latitude and longitude value. Different methods exist for converting longitude/ latitudes to the military grid system. Special equipment such as the global positioning systems have the capability to convert longitude and latitudes to grid coordinates, or this may be accomplished through manual means outlined in FM 21-26. For more information on map reading and land navigation techniques consult FM 21-26.

Navigation Aids

Navigation aids vary in sophistication and complexity and may include the following:

Sun Compasses/Sextants

These systems can be used on moving vehicles and require accurate timekeeping.

Lensatic Compasses

Individual compass error and local deviation must be known before using the lensatic compass. The lensatic compass cannot be used with any accuracy on dense steel vehicles such as tanks. A crew member should dismount to obtain an azimuth. It is unreliable near large quantities of metal, and can also be affected by underground mineral deposits. Power lines also adversely affect the lensatic compass.

Gyro Compass

An efficient gun azimuth stabilizer (a gyroscope) used on fairly flat ground is useful for maintaining direction.

Fires

Planned tracer fire assists in maintaining bearings, and field artillery and mortar concentrations, preferably smoke (or illumination at night), are useful checks on estimated locations.

Distance Recorders

It is essential to record distance moved, which may be done by using a vehicle odometer.

Beacons

These are particularly useful for aircraft navigation, but can also permit the enemy to locate friendly forces. It may be necessary to place them in open desert with unit locations being marked at certain distances and bearings from them.

Radars

Provided the position of a radar is known, it can measure range and bearings and, therefore, the position of a vehicle.

Aerial Photographs

The advantage of aerial photographs, particularly to aviators, is their ability to show up-to-date views of the variations in color and texture of the desert soil.

Global Positioning Systems

The GPS is a space-based, radio-positioning navigation system that provides accurate passive position, speed, distance, and bearing of other locations to suitably equipped users. The system assists the user in performing such missions as siting, surveying, tactical reconnaissance, artillery forward observing, close air support, general navigation, maneuver, and ground-based forward air control. It can be operated in all types of weather, day or night, anywhere in the world; it may also be used during nuclear, biological, and chemical warfare. It is important to remember these types of devices are aids to navigation; therefore, users should continuously plot their positions. In the event of a GPS failure, you can revert to more traditional navigation and position determination methods.

Position and Azimuth Determining System

The PADS is a highly mobile, self-contained, passive, all-weather, survey-accurate position/navigation instrument used by field artillery and air defense artillery units for fire support missions. The system provides roil-time, three-dimensional coordinates in meters, and a grid azimuth in roils. It also gives direction and altitude. The PADS can be used by the land navigator to assist in giving accurate azimuth and distance between locations. A unit requiring accurate information as to its present location can use PADS to provide this information.

Position Location Reporting System

The position location reporting system (PLRS)/joint tactical information distribution system (JTIDS), hybrid (PJH), is a computer-based system. It provides near real time, secure data communications, identification, navigation, position location, and automatic reporting to support the need of commanders for information on the location, identification, and movement of friendly forces. The PLRS is based on synchronized radio transmissions in a network of users controlled by a master station. The major elements of a PLRS community include the airborne, surface vehicular, and man-pack users; the PLRS master station; and an alternate master station. The system can handle 370 user units in a division-size deployment per master station with a typical location accuracy at 15 meters. The man-pack unit weighs 23 pounds and includes the basic user unit, user readout, antenna, backpack, and two batteries.

Dead Reckoning

The simplest system of navigation is known as dead reckoning. This is a means of finding where an individual is located by a continuous plotting of where he has been. More exactly, dead reckoning consists of recording and plotting a series of courses, each measured as to the distance and direction horn a known point, to provide a plot from which the position can be determined at any time. In the desert, the direction traveled is determined with a compass and the distance is measured by counting paces or reading the odometer of a vehicle. Detailed information on navigation by dead reckoning is contained in FM 21-26.

Shadow-Tip Method

This method provides orientation by reading the way the sun casts shadows. To use the shadow-tip method, find a fairly straight stick about 1 meter long and follow these steps:

- Step 1. Push the stick into the ground at a fairly level, brush-free spot where a distinct shadow will be cast. The stick need not be vertical; inclining it to obtain a more convenient shadow, in size or direction, does not impair the accuracy of the shadow-tip method.
- Step 2. Mark the tip of the shadow with a small peg, stick, stone, twig, your finger, hole in the sand, or other means. Wait until the shadow's tip moves a few inches (if you use a 1-meter stick, 10 to 15 minutes should be enough time),
- Step 3. Mark the new position of the shadow's tip.
- Step 4. Draw a straight line from the first mark to the second mark and extend it about a foot past the second mark.
- Step 5. Stand with the toe of the left foot at the first mark and the toe of the right foot past the line you drew.

You are now facing true north. Find other directions by recalling their relation to north. To mark directions on the ground (to orient others), draw a line at right angles to the first line, forming a cross and mark the directions.

If you cannot remember which foot to place on the first rock (see step 5), remember this basic rule for telling east from west: the sun rises in the east andsets in the west (but rarely due east or due west). The shadow's tip moves just the opposite. Therefore, the first shadow-tip mark is always in the west direction and the second mark in the east direction, everywhere on earth. Figure 2-1 depicts finding your direction by using the shadow-tip method.

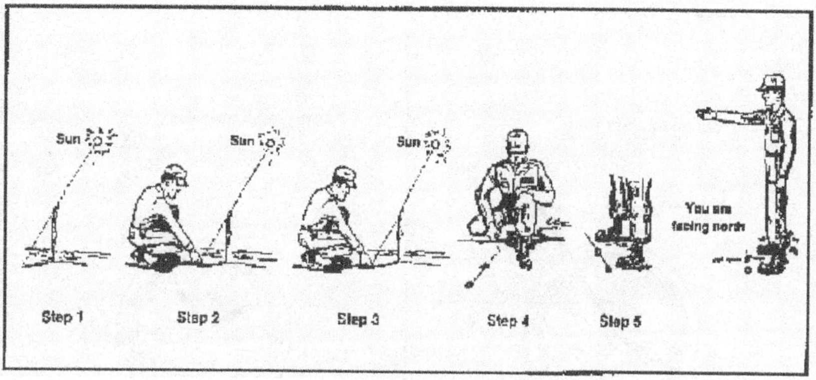

Figure 2-1.Finding direction Using shadow-up method.

Watch/Sun Method

An ordinary analog watch (with hands) can be used to determine the approximate true north in the North and South Temperate Zones. The North Temperate Zone is north of the equator and the South Temperate Zone is south of the equator. The temperate zones extend from latitude 23-1/2 degrees to 66-1/2 degrees in both hemispheres. In the North Temperate Zone only, the hour hand is pointed toward the sun (see Figure 2-2). A north-south line can be found midway between the hour hand and 12 o'clock. This applies to standard time. For daylight savings time, the north-south line is found midway between the hour hand and 1 o'clock. If there is any doubt as to which end of the line is north, remember that the sun is in the north, and remember that the sun is in the eastern part of the sky before noon and in the western part in the afternoon.

The watch may also be used to determine direction in the South Temperate Zone (see Figure 2-2). However, it is used a bit differently. Twelve o'clock is pointed toward the sun, and the north-south line will be halfway between 12 o'clock and the hour hand. If on daylight savings time, the north-south line lies midway between the hour hand and 1 o'clock.

On cloudy days, place a stick at the center of the watch and hold it so that the shadow of the stick falls along the hour hand in the North Temperate Zone. In the South Temperate Zone, the shadow falls along a line from the center of the watch through 12 o'clock. Direction is then determined using the appropriate technique.

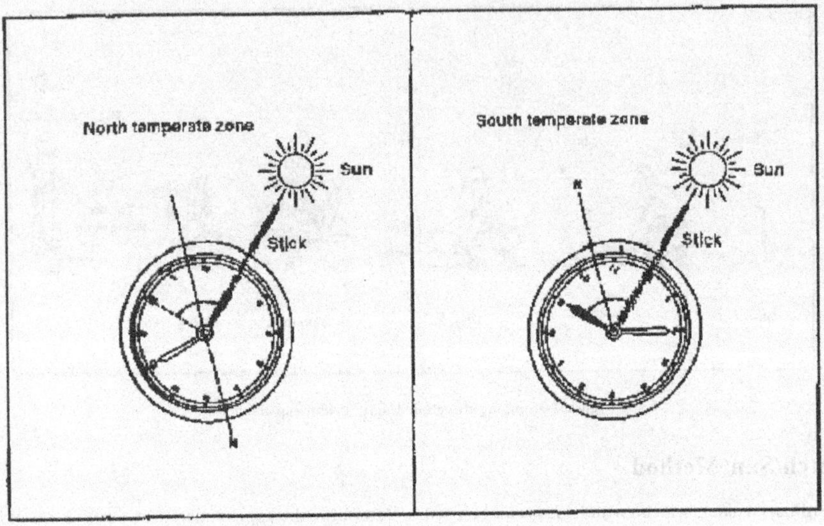

Figure 2-2. Using a watch to find north.

Celestial Navigation Method

North of the equator, locating your direction at night can be determined by locating the North Star. To find the North Star, look for the Big Dipper. The two stars at the end of the bowl are called pointers. In a straight line out from the pointers is the North Star (at about five times the distance between the pointer stars). The Big Dipper rotates slowly around the North Star and does not always appear in the same position (see Figure 2-3).

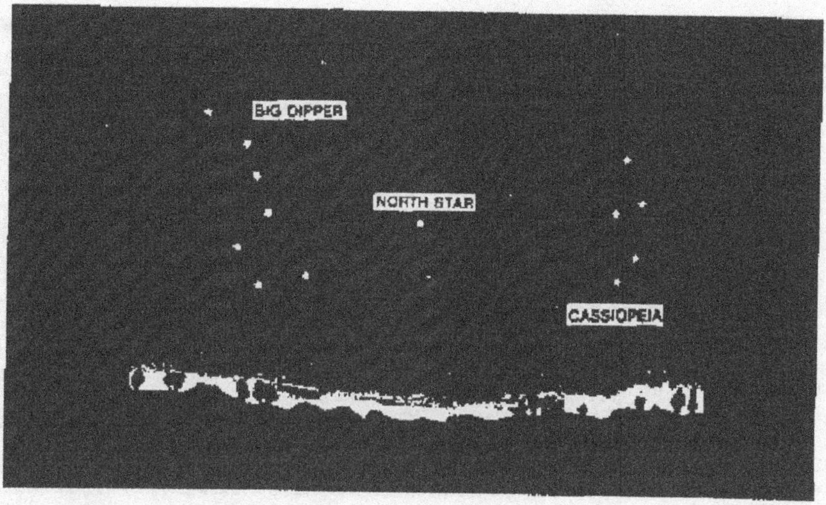

Figure 2-3. ***Big*** Dipper.

The constellation Cassiopeia can also be used. This group of five bright stars is shaped like a lopsided M (or W when it is low in the sky). The North Star is straight out from the center star, about the same distance as from the Big Dipper. Cassiopeia also rotates slowly around the North Star and is always almost directly opposite the Big Dipper. Its position, opposite the Big Dipper, makes it a valuable aid when the Big Dipper is low in the sky, possibly out of sight because of vegetation or high terrain features.

South of the equator, the constellation Southern Cross will help you locate the general direction of the south and, from this base, any other direction. This group of four bright stars is shaped like a cross that is tilted to one side. The two stars forming the long axis, or stern, of the cross are called the pointers. From the foot of the cross, extend the stem five times its length to an imaginary point (see Figure 2-4). This point is the general direction of south. From this point, look straight down to the horizon and select a landmark.

Figure 2-4. Southern Cross.

OPERATIONAL AREA (HOST COUNTRY)

A description of the host country should cover only those facts that apply to forthcoming operations, for example:

- Geographic description.
- Climate (throughout the year).
- Religious issues and constraints.
- Cultural differences and special considerations, important customs, and the behavior expected of US military personnel. (Such as speaking to a woman in some Arab countries, which can be offensive to the local inhabitants.)
- Population density.
- Industry and agriculture.
- Language(s) (phrase books may be issued).
- Communications and transportation network.
- The armed forces (and possibly police), including organization, equipment, and rank structure.
- The situation that has led to the introduction of US forces and reasons why US forces are being introduced. No soldier/marine should have to question why he is fighting for a country other than his own, if this is the case.

Treatment of these subjects will vary in degree according to category. Personnel who may require additional information such as the country's history, can find it in the appropriate DA 550-series pamphlet.

DESERT MANEUVER

Chapter 3 describes the influence of the desert environment on tactical operations. This subject should first be taught to a limited number of leaders and commanders as a theoretical subject, down to platoon level. Leaders should then train their units during unit training. The emphasis should be on small unit tactics, including combined arms operations. Additional subject matter that should be covered includes -

- Terrain in the operational area, emphasizing differences and similarities with the training areas the units will use.
- Application of concealment, using terrain and artificial means such as smoke, and the application of maneuver techniques.
- Mobility in the desert.
- Command and control techniques for desert operations.
- Navigation.
- Conduct of fire in desert operations.
- Resupply during desert operations.
- Special equipment techniques.

GENERAL TRAINING

Nearly all equipment will be affected in one way or another by the environment as described in Chapter L The purpose of this training is to train operators. Training should include-

- Likely effects on the equipment they operate.
- Efficient operations of the equipment within the limits imposed by the environment, including tactical limitations of the equipment, for example, helicopters may have difficulties flying NOE; and radios will normally be operated on reduced output due to the environment and enemy ECM.
- Preventive maintenance-employing any special techniques required by the desert environment. The appropriate equipment technical manual or lubrication order provides specific information concerning hot climate operations and maintenance.
- Basic desert recovery and repair techniques, including defensive measures, and camouflage required during recovery and repair operations.

Instruction must be oriented toward the expected operational area. For example, it is possible to keep radios cool by using ice packs, but if ice packs are not going to be available in the area of operations, then do not teach troops this technique as it will not be practiced.

SPECIAL MAINTENANCE AND SUPPLY TECHNIQUES FOR STAFF AND LEADERS

Special maintenance techniques that need to be addressed are the same as those taught to specialists, however, they only need to emphasize aspects that ordinarily require control or supervision, or affect the employment of equipment in desert terrain. This training should include any special handling techniques required in the operational area. The importance and difficulties of supply in desert operations are described in Chapter 4. Training should be modified according to-

- Modified tables of organization and equipment (MTOE) and mission of the unit.
- Supply situation expected in the area of operations.
- Capabilities of logistic units likely to support unit operations with special attention given to units not normally found in conventional operations, for example, well-drilling teams, refrigeration assets available in the theater, and transportation cargo carrier companies.

NUCLEAR, BIOLOGICAL, AND CHEMICAL (NBC) TRAINING

Wearing protective clothing, flak vests and masks in the desert environment will make a person extremely uncomfortable. Troops should not participate in strenuous activity while wearing protective clothing until they are acclimatized. Training in MOPP gear should become progressively more strenuous. Physical training sessions in field protective masks progressing to foot marches in MOPP 4 is a recommendation. Use of protective clothing in severe desert heat is described in Chapter 1 and Appendix D. Points that should be emphasized during training are-

- The value of being uncomfortable rather than dead.
- The need to avoid heat illness by¬

  - Reducing the labor rate to the minimum, and delaying work until cooler hours.

  - Maintaining proper body water and salt levels, particularly during a time of chemical threat.

  - Detecting the first symptoms of heat illness in others by constant vigilance.

  - Increasing the time factor of an operation as troops will move slowly when wearing protective clothing.

DESERT TERRAIN APPRECIATION

When training soldiers/marines to appreciate desert terrain, leaders should focus on the effects of the different types of desert terrain on the capabilities and limitations of unit equipment. Highlight the impact of the terrain on vehicular trafficability, fields of fire, and observation in the likely operational area. When possible, crews and small unit leaders should learn to appreciate desert terrain from practical experience in terrain as nearly similar as possible to that in the likely combat zone.

MEDICAL TRAINING CONSIDERATIONS

The unit surgeon can provide valuable information on the medical implications of operations in the desert environment. He can advise the unit commanders on measures to take to ensure training includes preventive medicine concepts essential to keeping nonbattle injuries to a minimum. Bum casualties should also be a medical training consideration as these will be the most likely casualties in a mechanized environment. Hydration and mouth-to-mouth resuscitation of injured personnel in a field protective mask should also be a training consideration. Nonbattle casualties, due to a lack of consideration of preventive medicine concepts, can far outnumber combat casualties.

Section III. Unit Training

When determining unit training requirements, the commander must first consider the training level of his unit when alerted for deployment. When time does not permit a comprehensive training program, the commander must concentrate on those areas where his unit is least proficient, considering the priorities previously described. In order to operate in the desert environment, the unit must, above all, be physically fit, so physical conditioning is of paramount importance.

COMBINED ARMS AND SERVICES TEAM

The greatest combat power results when leaders synchronize combat, combat support, and combat service support systems to complement and reinforce one another. The slice concept refers to CS and CSS units task organized to support a particular maneuver or combined arms unit. Leaders should routinely practice habitual relationship and cross attachment of units. Training as combined arms teams is critical to successful desert operations.

PHYSICAL CONDITIONING AND ACCLIMATIZATION

To the extent possible, physical conditioning and acclimatization should take place simultaneously. When a unit is training in a hot environment, begin physical training at night or during the cooler part of the day and work up to rigorous training, such as foot marches in open sand terrain at midday. Emphasis on mounted operations in desert warfare does not imply that foot marching can be totally disregarded. Physical conditioning must be continued after arrival in the area of operations.

Medical advice should always be available during periods of physical training in hot weather. Training several days prior to deployment to combat the effects ofjet lag works well, if performed as a command directive. Avoiding coffee and alcohol, and drinking plenty of water will also assist in overcoming the effects ofjet lag. Planning for the change in time zones and developing a sleep plan are recommended and will facilitate readiness.

WEAPONS TRAINING

Soldiers/marines must train to proficiency at all ranges but accuracy at maximum effective range both in daytime and nighttime must be emphasized. Firing should also be practiced during the heat of day to condition troops to heat haze and mirages. Emphasis must also be placed on maintenance of individual weapons in view of sandy desert conditions.

NBC TRAINING

The purpose of unit NBC training is to train individual soldiers/ marines to become proficient as a team while wearing protective clothing and masks, and when in combat vehicles-while buttoned up. This training should be conducted both in daytime and nighttime, until the unit can operate effectively under these conditions. Additional information and guidance on the effects of the environment on NBC weapons are contained in Appendix D.

MARCH DISCIPLINE

Although of particular importance to combat support and combat service support units, all units should be trained in tactical road marches. Training should emphasize-

- Off-road movement over open terrain. Irregular spacing when moving in convoy.
- The need to maintain sufficient distances between vehicles to preclude "dust blindness."
- Actions to be taken when stuck in sand, and when a vehicle breaks down.
- Vehicle camouflage.
- Removal of tracks which would reveal friendly locations.
- The need for dispersion when halted.
- Air defense drills.
- Practicing incoming artillery drills.

OBSTACLES AND BARRIERS

In some desert areas, natural obstacles such as wadis or other terrain features can be found. Often, however, it will be necessary to use artificial obstacles if enemy movement must be slowed. A minefield, to be of any tactical value in the desert, must usually cover a relatively large area, so mechanical means and engineer support are required. Since there are often too many avenues of approach to be covered with mines, it is usually best to employ mines to cover any gaps between units, especially at night. Minefields are most effective when they can be covered by observation or fire. During unit training, soldiers/marines should be trained to lay mines wearing gloves, since human scent attracts desert animals who may attempt to dig them up. Emphasis should be on antitank minefield since combat vehicles are the most dangerous threat.

SCOUTING, SURVEILLANCE, AND PATROLLING

Effects of the environment on scouting, observation, and surveillance techniques are described in Chapter 3. Effects of the environment on surveillance, target acquisition, and night observation devices are described in Chapter 1.

ADJUSTMENT AND CONDUCT OF FIRE

The principles for adjustment and conduct of fire in the desert are the same as for operations in more temperate climates. However, the following considerations, somewhat peculiar to desert operations, should be kept in mind:

- Obscuration from sand, dust, smoke, or a combination of these, can affect direct-fire adjustment.
- There may be major inaccuracies of initial rounds from indirect-fire weapons due to misjudgment of target location.
- The target may be concealed by sand or dust if rounds land short, on, or near the observer target line.
- Heat haze and mirages can mislead gunners and observers as to target location. This condition can particularly affect antitank guided missile gunners.

Direct-fire gunners may have to depend on flank observers, who may be any individual on the battlefield equipped with a radio. If observation is lost, subsequent corrections are very unlikely to cause a second round hit. The following considerations can help to overcome the obscuration or sending problem:

- An observer requesting indirect fires needs to ensure that initial rounds land beyond the target to preclude short rounds obscuring the target, and then adjust accordingly.
- By remembering the greatest impact of heat haze (which varies throughout the day) is on ATGM gunners-when both gunner and target are within 2-3 feet of the desert surface.

AIR DEFENSE

In desert operations any type of unit, be it tank, infantry, trains, tactical operations centers, or supply points, can expect to be a target for air attack. Air attacks may be from fighter bomber aircraft using cannons, missiles, bombs, napalm, and machine guns, or from attack helicopters using machine guns, rockets, or missiles.

Enemy air superiority should be assumed during all field training and simulated fighter bomber attacks and attack helicopter missions should be flown against the unit whenever possible. When practical, aerial photographs of positions should be taken, and pilots interviewed to assist in the critique of air defense, both passive and active. Points that should be emphasized during training are mentioned in the following paragraphs.

Passive air defense measures should be taken routinely. When stopped for any period of time, take every advantage of whatever cover and concealment are available. As previously described, natural cover and concealment will be difficult to find in many desert areas. Nevertheless, vehicles, particularly unarmored vehicles, should be irregularly dispersed and dug-in, or revetments provided. When appropriate, air guards, trained in aircraft recognition, should be posted, with clear instructions on actions to be taken when aircraft are sighted. Units not being attacked by aircraft, but in close proximity to the attack, may desire to remain stationary in order to avoid detection. Reducing infrared signatures is also a passive anti-air consideration. Artificial camouflage can be used as described in Appendix E.

Active air defense techniques used in desert operations are the same as those described in other doctrinal manuals, appropriate to the level of command. However, at small unit level, additional emphasis should be given to air defense using small arms. When combat vehicles on the move are engaged by enemy aircraft, their immediate action will depend on whether or not they are maneuvering in contact with the enemy. If they are in contact they should continue to maneuver, relying on overwatch elements and air defense artillery to engage attacking aircraft.

Vehicles about to be engaged by enemy aircraft in open desert where cover is not available, should move perpendicular to the attacking aircraft to evade rocket or machine gun fire. Engage the aircraft with small arms fires, if possible. Meanwhile, the remainder of the unit should mass small arms fire to the aircraft's front. Sudden variations in course may also distract the pilot.

COMMUNICATIONS

Good communications in desert terrain will often depend on the state of mind of the operators. They must be enthusiastic, persistent, and determined to make and maintain contact. Unit training should concentrate on ECCM techniques. When conducting field training, higher headquarters can provide assistance in the form of small teams to jam unit nets. Practice actions to be taken when radio contact is lost due to heat (described in Appendix C).

FRATRICIDE CONSIDERATIONS

The following are fratricide considerations when operating in a desert environment and may be topics for fratricide awareness training:

- Because of the absence of easily identifiable terrain features in the desert, knowing your exact location can be especially important. Be in the right place at the right time. Use position location/navigation (GPS) devices; know your location and locations of adjacent units (left, right, leading and follow on). Synchronize tactical movement.
- Ensure positive target identification. Review vehicle/weapons ID cards; know at what ranges and under what conditions positive ID of friendly vehicles/weapons is possible. This is especially important in the desert due to the likelihood of weapons being able to fire at their maximum ranges.
- Maintain situational awareness-be aware of current intelligence, unit locations/dispositions, denial areas (minefields/FASCAM), contaminated areas (e.g., ICM and NBC), SITREPs, and METT-T. This can be more difficult in desert environments because navigation is more difficult.
- Conduct individual and collective (unit) fratricide awareness training. This includes target identification/recognition; fire discipline; and leader training.
- Use common language/vocabulary. Use doctrinally correct, standard terminology and control measures (e.g., fire support coordination line (FSCL), zone of engagement, and restrictive fire line (RFL)).
- Consider the effects of key elements of terrain analysis on fratricide. These include observation and fields of fires, cover and concealment, obstacles and movement, key terrain, and avenues of approach.
- Gun tube orientation can also assist in avoiding/preventing fratricide.

CHAPTER 3

OPERATIONS IN DESERT CONDITIONS

In the two previous chapters the environment and its effects on personnel and equipment and preparations for desert operations were described. This chapter describes desert operations and is divided into three sections.

Section I. How the Desert Environment Affects Tactical Operations

The key to success in desert operations is mobility, and this was clearly evident in ground operations in Desert Storm. The tactics employed to achieve victory over Iraq were wide and rapid flanking movements similar to those Rommel and Montgomery executed in North Africa.

Trafficability and cross-country movement become critical to desert operations when using these tactics. Trafficability is generally good in the desert and cross-country movement is a lesser problem, but not always. Salt marshes can create NO-GO conditions during the rainy season. Sand can also bog down traffic and make foot movement slow and exhausting. The steep slopes of dunes and rocky mountains can make vehicular movement a NO-GO. The wadis create cross-compartmented terrain. The banks of these stream beds can be steep and unconsolidated. Then, when it rains, it becomes a torrent of dangerously rushing water, leading to flat lake beds that can create NO-GO mud conditions. Rock quarries and mining areas can also adversely affect mobility and trafficability. Often these areas are not reflected on maps. Satellite imagery can be helpful in identifying these areas, as was the case in Kuwait during Operation Desert Storm. In rocky terrain, tires can easily be punctured by sharp angular debris; however, overall movement is mostly uninhibited. Given ample fuel, water, and other resources, units can go around natural and man-made obstacles.

Movement can easily be detected because of sand and dust signatures left due to the loose surface material. In an actual engagement, this may not be all that bad because a unit is obscured from direct fire while advancing, but the element of surprise may be lost. Moving at night becomes the logical choice. The dust is still there, and vehicles (which should be widely spaced) can get separated. But at night, reflection of the sun's rays from glass, mirrors, or metal, which can give away movement and positions up to 20 kilometers away, is not a concern.

Using the ability to make fast and wide flanking movements, a unit can encircle and cut off enemy forces. The Israeli forces under General Ariel Sharon did just that to the Egyptian Third Army in the 1973 War, and the British did the same to the Italians in North Africa in January 1941. In Desert Storm, the night-fighting AH-64 helicopters, combined with field artillery fires, made for an unbeatable team in this regard.

Land navigation is a challenge during movement in the wide expanses of many arid regions. There are few landmarks to key on, and maps and even photos can become dated quickly, especially in the sandy deserts where dunes migrate. The global positioning system (GPS) with the small lightweight GPS receivers (SLGRs) is a major aid for desert operations.

Refuel and resupply operations require periods in which forces assume the defense, but only temporarily. The flat sandy desert topography that is characteristic of Saudi Arabia is not conducive to defense, compared to rocky plateau topography. In mountains and canyons, a defensive posture can be favorable. Controlling the passes, as mentioned earlier, can essentially close off vast areas to an attacker and make it extremely costly for him.

While a unit is in the defense, it needs both ground and air reconnaissance to detect enemy movement at long range. Obstacles must be placed in all types of topography, primarily to slow advances and channel columns. Neglecting these security measures in the flat sandy regions can lead to disaster.

MILITARY ASPECTS OF THE TERRAIN

The following paragraphs describe how terrain affects tactical operations in the desert. This discussion follows the outline of the terrain analysis process summarized by the factors of OCOKA.

Observation and Fields of Fire

Observation and fields of fire are generally excellent in most desert areas. The atmosphere is stable and dry, allowing unrestricted view over vast distances, but this can also be a problem. Range estimation by "gut feeling" is subject to error. The effective ranges of weapons can easily be reached, and a correct estimation of maximum ranges is critical for all weapons, especially for wire-guided munitions.

Flat desert terrain permits direct-fire weapons to be used to their maximum range. Open terrain and a predominantly clear atmosphere generally offer excellent long-range visibility; but at certain times of the day visibility may be limited or distorted by heat.

Two primary considerations in the desert environment are longer range observation, and fields of fire at the maximum effective ranges for weapons. However, rapid heating and cooling of the atmosphere hinder these factors and cause distortion of ranges to the aided and unaided eye. Mechanical and electronic means must be used to verify estimated ranges such as GSR and laser range finders. Boresight and zero more frequently at standard ranges.

The desert is not absolutely flat, so weapons are sited to provide mutual support. Dead space is a problem. Even though the landscape appears flat, upon closer inspection it can be undulating with relatively deep wadis and depressions. These areas must be covered by indirect fire.

When on the offense, attacks should be initiated with the sun at or near one's back whenever possible. This eliminates most shadows that degrade optical weapon guidance and makes visual target acquisition difficult.

When there is no usable dominant terrain available, the only means of observation may be from an aeroscout, or limited to short-range observation by the vehicle commander. Other visibility problems are caused by heat distortion. Heat waves at the desert surface cause images to shimmer making positive identification difficult and degrade depth perception. Ranges to targets may be distorted from heat rising from the desert surface. Use range finders to verify correct distances. Be prepared to use bracketing techniques with large adjustments to hit an enemy target with artillery.

Radars are unlikely to be affected by heat haze so they could be valuable on flat terrain during midday heat if optical vision is hopelessly distorted; however, they arc almost useless in sandstorms. Image intensification is of limited value in sandstorms, and depends on the phase of the moon at night. If there is no moon, use artificial illumination outside the field of view of the system.

Since thermal imagery devices depend on the difference between ambient temperature and equipment temperature, they are more useful at night than in the day. Because of the distinct advantages of thermal sights, they should be used as the primary sighting systems for vehicles so equipped.

Correction of field artillery fires, especially those of larger pieces, may be complicated by dust hanging in the air following the impact of ranging rounds. Forward observers should consider placing initial rounds beyond a target rather than short of the target. Observation of fires, especially direct fires by tanks, may be difficult due to dust clouds, so wingmen may have to observe direct fires.

Cover and Concealment

Cover and concealment are generally scarce in the desert. The flat sandy deserts provide little if any natural cover or concealment, especially from aerial attack or reconnaissance. Ground concealment and protection from fire can be found behind dunes or in wadis. Troops must be aware of the potential for flash floods when using wadis for ground concealment.

Some arid regions have vegetation that can provide limited concealment from ground observation. In rocky, mountainous deserts, cover and concealment are best found behind boulders and in crevices. Daytime vehicular movement eliminates nearly any possibility of surprise, as dust trails created by the traffic can be spotted for miles. At night noise and light discipline is critical, as both sound and light travel great distances because of the unobstructed flatness of the terrain and atmospheric stability. Camouflage can be effectively employed to improve on natural cover and concealment. See Appendix E for additional information on concealment and camouflage.

Obstacles

Natural obstacles do exist in the desert, and arid regions are well suited for man-made obstacles. The wadis and steep slopes of escarpments, mountains, hills, and dunes hinder cross-country movement. Sand dunes may stretch for miles and prevent direct movement across their length. These sand dunes are often more than 100 feet in elevation and consist of loose sand with high, steep downwind faces that make vehicular traversing next to impossible. Aerial reconnaissance immediately before any large movement is advisable because sand dunes migrate with shifting winds and they may not be where maps or even photographs show them.

In the Desert Storm area, the salt marshes have a crust on the top that can deceive a vehicle driver. These dry lake beds can become obstacles, especially in the wetter seasons when the water table is higher. A top crust forms on the surface, but below the crust the soil is moist, similar to marsh conditions. The surface may look like it has good trafficability, but the crust will collapse with the weight of a vehicle, and the vehicle becomes mired. The high premium on fuel and time makes it costly to go around these natural obstacles.

Sandy deserts are ideal for employing minefield. Although windstorms can reveal previously buried mines, these mines can still channel movement and deny access to certain areas. The battles of the Bi'R Hacheim Line and El Alamein were influenced by minefield. Other obstacles include ditches, revetments, and barriers, such as the Bar Lev Line along the Suez Canal, made by bulldozing sand mounds or by blasting in rocky mountainous areas to close passes.

Key Terrain

Key terrain in the desert can be any man-made feature, mountain pass, or source of water, and of course, high ground. Because there are few man-made features in the desert, those that do exist can become important, perhaps even key.

Passes through steep topography are also likely to be key, again because they are so few. The North African campaigns of World War II focused on the control of passes, specifically the Solium and Halfaya. In the Sinai Wars between Egypt and Israel, the Mitla, Giddi, and Sudar passes were key. In Afghanistan, control of the mountain passes provided the Mujahideen safe haven from the Soviets. Oases, where wells exist, become important for water resupply. The high ground in desert terrain is usually key terrain. The relative flatness and great distances of some deserts, such as in Iraq, make even large sand dunes dominant features.

Avenues of Approach

Avenues of approach are not clearly defined in arid regions. The vast, relatively flat areas permit maneuver from virtually any direction. This point became obvious to units establishing defensive positions in Desert Storm. Wide envelopments are possible, as demonstrated in the Desert Storm ground campaign. Modem sensor technology, limited natural concealment, and improved observation make the element of surprise a challenge. Yet, surprise was achieved during Desert Storm-Iraqi commanders were shocked when they discovered US tanks in their perimeters.

The major limitation with respect to avenues of approach may be fuel. The great distances a unit must travel to outflank enemy positions require significant amounts of fuel and complicate resupply. In mountainous and canyon topography avenues are much more limited, and the wadis and valleys are likely to be the only possible access routes. Any roads that do exist are probably in the valleys. Nevertheless, none of the considerations outlined above are reasons to disregard flanking movements.

MANEUVER

Army operations are ideally suited to desert environments. Its thrust of securing and retaining the initiative can be optimized in the open terrain associated with the desert environments of the world. In that environment, the terrain aspect of METT-T offers the potential to capitalize on the four basic tenets of the doctrine initiative, agility, depth, and synchronization.

Initiative

Israeli efforts in 1967 and initial Egyptian assaults in 1973 clearly illustrate the effects of initiative in the desert environment.

Agility

The Egyptian success in 1973 was negated by their failure to ensure agility. Conversely, the Israeli actions on the flanks of the Egyptian force demonstrated the effects of a force capable of rapid and bold maneuver.

Depth

Depth does not necessarily relate to distance. In the nonlinear battlefield offered by the desert, depth often equates to an agile reserve force of sufficient size to counter enemy efforts into flanks and rear areas. Depth is also a concept of all-around defense for forces-the ability to fight in any direction.

Synchronization

To a large measure, the German successes against the British in the Western Desert were due to their ability to synchronize their operating systems. More recent events illustrate this tenet between and internal to, operating systems. Heavy/light operations have demonstrated that light forces can be key to achieving tactical and operational momentum. The Israeli airmobile assault against supporting artillery in the 1967 battle of Abu Ageila is a good example of the effective use of light forces in this type of environment.

Maneuver must be at the maximum tactical speed permitted by the terrain, dust conditions, and rate of march of the slowest vehicle, using whatever cover is available. Even a 10-foot sand dune will cover and conceal a combat vehicle. Air defense coverage is always necessary as aircraft can spot movement very easily due to accompanying dust clouds. In some situations movement may be slowed to reduce dust signatures. Rapid movement causes dramatic dust signatures and can reveal tactical movements.

Another consideration during maneuver is dust from NOE flight, which can be seen as far as 30 kilometers. This is especially true when the enemy is stationary. Aeroscouts must use caution to avoid blundering into enemy air defense weapons.

To achieve surprise, maneuver in conditions that preclude observation, such as at night, behind smoke, or during sandstorms. In certain circumstances, there may be no alternative to maneuvering in terrain where the enemy has long-range observation. Then it is necessary to move at the best speed possible while indirect fires are placed on suspected enemy positions. Speed, suppressive fires, close air support, and every other available combat multiplier must be brought to bear on the enemy.

Tactical mobility is the key to successful desert operations. Most deserts permit good to excellent movement by ground troops similar to that of a naval task force at sea. Use of natural obstacles may permit a force to establish a defensive position that theoretically cannot be turned from either flank; however, these are rare. Desert terrain facilitates bypassing enemy positions and obstacles, but detailed reconnaissance must be performed first to determine if bypassing is feasible and will provide an advantage to friendly forces.

Dismounted infantry may be used to clear passes and defiles to eliminate enemy ATGM positions prior to the mounted elements moving through.

Avenues of approach of large forces may be constrained due to limited cross-country capability of supply vehicles coupled with longer lines of communications. The limited hard-surface routes that do exist are necessary for resupply.

RECONNAISSANCE

Reconnaissance is especially important in desert environments. Reconnaissance is a mission undertaken to obtain information by visual observation, or other detection methods, about the activities and resources of an enemy, or about the meteorologic, hydrographic, or geographic characteristics of a particular area. The desert environment may influence any or all of these techniques. The environmental effects on troops and their equipment may also influence observation techniques, or the frequency of vehicle and equipment maintenance that is required. Reconnaissance produces combat information. Combat information is a by-product of all operations, acquired as they are in progress. Reconnaissance, however, is a focused collection effort. It is performed prior to or in advance of other combat operations, as well as during that operation, to provide information used by the commander to confirm or modify his concept. Cavalry is the Army corps or division commander's principal reconnaissance organization.

Surveillance is a primary task of Army cavalry during reconnaissance operations. Surveillance is the systematic observation of airspace or surface areas by visual, aural, electronic. photographic, or other means. Scouts, ground and air, are the principal collectors of information. Scouts and their associated equipment are particularly affected by the environmental aspects of deserts. They require equipment that enhances their senses allowing them to conduct mounted and dismounted surveillance with stealth, at long-range, and in limited visibility, all of which can be adversely influenced by the desert environment.

SECURITY

Security operations obtain information about the enemy and provide reaction time, maneuver space, and protection to the main body. Security operations are characterized by aggressive reconnaissance to reduce terrain and enemy unknowns, gaining and maintaining contact with the enemy to ensure continuous information, and providing early and accurate reporting of information to the protected force. Security operations may be affected by various aspects of the desert environment including the sun, wind, sand, vegetation, sandstorms, terrain, and heat. Security operations include-

- Screen.
- Guard.
- Cover.

Counterreconnaissance is an inherent task in all security operations. Counterreconnaissance is the sum of actions taken at all echelons to counter enemy reconnaissance and surveillance efforts through the depth of the area of operation. It is active and passive and includes combat action to destroy or repel enemy reconnaissance elements. It also denies the enemy information about friendly units.

COMMAND, CONTROL, AND COMMUNICATIONS

The following paragraphs describe command, control, and communications considerations when operating in a desert environment.

Command

The effort to synchronize battlefield operating systems during the planning process can be negated by the failure to continue the synchronization effort during the preparation phase of a mission. This is especially true in the construction of engagement areas for defensive operations. Direct fire, indirect fire, and obstacles are linked, and the adjustment of one requires the adjustment of all. The commander must know and have a feel for what his unit can do, how long his unit takes to accomplish a mission, and what he really wants his unit to accomplish.

Adjustment of the elements of the battlefield operating systems can unravel the focus of a commander's intent. This is especially true in open terrain. Tactical commanders should personally direct the synchronization of engagement areas. Obstacles should be positioned, indirect fires adjusted, and direct fires rehearsed under the personal supervision of the commander. The commander controls operations using a highly mobile command group located well forward. He personally directs the battle, but must not be drawn into personally commanding an isolated segment of the force to the detriment of the remainder of the command. As previously mentioned, dry desert conditions can sometimes reduce radio signal strength and create unforeseen blind spots, even in aircraft operating nap of the earth.

Units may employ either a jump TOC or retransmission stations to facilitate communications with rear areas, as maneuver units are unlikely to be in one place very long. (If wire is used it should be buried to a minimum depth of 12 inches to avoid damage from track vehicles or shell fire.) There must be plenty of slack in the line to allow for sand shift and accurate map plots of buried wire should be kept. If overhead wire must be used, it should be mounted on posts erected in the form of tripods to avoid falling during severe weather.

Air or vehicle mounted liaison officers can be used if units are stationary or under listening silence. They should be proficient in navigation and sufficiently equipped to facilitate parallel planning. Liaison officers are highly effective and should be employed at every opportunity.

Continuous Operations

Continuous operations are affected by a number of factors in a desert environment. Fatigue is probably the foremost degrader of performance. Performance and efficiency begin to deteriorate after 14 to 18 hours of continuous work and reach a low point after 22 to 24 hours. Most tasks involving perceptual skills begin to show a performance degradation after 36 to 48 hours without sleep. Soldiers/marines cease to be effective after 72 hours without sleep. Performance decreases dramatically in an NBC environment and sleep becomes more difficult in MOPP gear. Sleep deprivation coupled with the environmental factors of the desert and the stresses of combat can significantly affect mission accomplishment.

The two categories of personnel who can be expected to show signs of fatigue first are young immature soldiers/marines who are not sure of themselves and seasoned old soldiers/marines upon whom others have relied and who have sustained them at cost to themselves. Commanders and leaders often regard themselves as being the least vulnerable to fatigue. Tasks requiring quick reaction, complex reasoning, and detailed planning make leaders the most vulnerable to sleep deprivation. Leaders denying themselves sleep as an example of self-control is extremely counterproductive. These factors are complicated by the environmental aspects of desert operations and should be considerations for operational planning.

Control

Clear identification of engagement areas is necessary to facilitate the massing and distribution of fires. In the absence of identifiable terrain, target reference points (TRPs) can be created with damaged/destroyed vehicles that are moved into required locations at the direction of commanders invested with the responsibility for specific engagement areas. Other types of TRPs could be used. For example, marker panels, visible and infrared chemical lights, flags, and white phosphorus/illumination rounds could be used. The construction or fabrication of TRPs must be resourced and well planned in order to be effective. For example, how will TRPs be replaced for subsequent defensive operations? Another common problem is TRP proliferation, which makes TRPs difficult to identify when each echelon of command has allocated too many TRPs.

Pyrotechnics are usually more effective in desert climates than in temperate climates; however, heat mirages and duststorms may impair or restrict their use. Even heliographs (signal mirrors) may be useful as they are directional and therefore can aid security. Sound communications are usually impractical due to distance, vehicular noise, and storms, but can be used for local alarm systems.

Colored flags with prearranged meanings can be used as a means of communication in the flat open terrain of the desert. Colored flags tied to antennas may also assist in vehicle/unit recognition during limited visibility operations and offensive operations.

As previously described, the desert offers excellent fields of fire. Tanks and heavy antitank weapons should be sited to take advantage of their long range and accuracy. Firing first and accurately are the most important considerations in desert operations.

Target identification is the recognition of a potential military target as being a particular object (such as a specific vehicle, by type). At a minimum, this identification must determine the potential target as friendly or threat (identify friend, foe, or neutral [IFFN]). Because it is easy to become disoriented, it is often necessary to mark sectors of fire on the ground with poles or rocks, if available.

Communications

Communications support is also adversely affected by high temperatures. The heat causes anomalies in radio and other electrical transmissions, and radio battery life is reduced. Radio range is shorter during the day than at night. At night, range improves but static electricity may cause interference. FM communications range can be reduced by as much as 50 percent because of high temperatures. HF ground wave propagation over the dry sandy soil is reduced.

Night communications make communications security a concern, as it always should be. Experience in Desert Shield and Desert Storm indicates vastly expanded ranges of FM radios. Communications between units 40 to 50 kilometers apart was not unusual. Communications obviously affect command and control as well as intelligence collection and dissemination, and their importance must not be underestimated.

COMBAT SUPPORT

A force operating in the desert must be a balanced force with combat support and combat service support-it must be a combined arms team. While principles of combat support operations are found in doctrinal manuals dealing with a specific arm of service, there are some techniques that must be modified or emphasized in the desert.

INTELLIGENCE

The relative importance of intelligence sources may vary from that expected in more conventional areas. Enemy prisoners of war require immediate interrogation as the flexibility of operations will rapidly make their information outdated. Information given by civilians encountered in desert operations should be treated with caution unless corroborated. Military intelligence teams located in the area of operations can determine if these EPWs and civilians are in fact what they say they are, or infiltrators sent to harass the rear area and commit acts of sabotage. Electronic support measures are a major source of intelligence in desert warfare. Enemy activity, or the lack of it, is a good source of information; so punctual, accurate reports by all sources, both positive and negative, are necessary.

FIRE SUPPORT

The Allies in North Africa in 1942 found that placing small field artillery units in support of small maneuver units gave the units a sense of security, but produced limited results. Field artillery was effective only when massed (battalion or higher) and only when continued for some time because of the protective posture and mobility of the target. Typically, the control of massed fires was the responsibility of the division artillery.

The Allies in North Africa in 1942 experienced heavy casualties from Axis units overrunning the artillery positions after penetrating the armor and infantry positions. Often, the Axis units would attack from the east at one time, from the west later, and from several directions simultaneously. At first, the Allies simply emphasized direct fire. Later, the Allies attached antitank gun units to the artillery battalions to increase the artillery's antitank ability.

When armor and infantry units move, the artillery must move with them. The most useful technique is for the artillery to move in a formation with a lead vehicle so that, immediately upon stopping, the artillery is in a position or formation to deliver fire in any direction and simultaneously defend the position from any direction. The Allies in North Africa in 1942 and units in Desert Shield/Storm found that the armor and infantry units would outdistance the artillery unless the artillery moved with them. The artillery moved within 2-3 kilometers of the leading troops to provide responsive fires. The armor and infantry provided protection for the artillery. The whole group moved in one cohesive formation, sometimes in a large box or diamond formation.

Due to the fluid nature of desert operations and the possibilities for excellent enemy observation, close and continuous field-artillery support for all levels of the force is necessary. Field-artillery pieces should be at least as mobile as the force they are supporting. Crews must be proficient in direct fire and prepared to defend against a ground attack.

Due to the threat of immediate counterbattery fire, field artillery units must be prepared to move into position, fire, and rapidly displace to another position. A battery should be prepared to displace several times a day.

Field artillery units employed in desert operations should be equipped with the most sophisticated survey devices available. Manual systems are slower and not necessarily as accurate, thus affecting tactical employment and reducing response time.

Aerial observation may often be extremely difficult due to enemy air defense, so most adjustment is by ground observers. How the environment affects observation of fires was described previously in this chapter in the paragraph, "Observation and Fields of Fire." Recompute weather conditions frequently as weather conditions can change rapidly from the morning to the evening, and thus affect the accuracy of fires.

Fires are planned as in temperate climates. When there are no significant terrain features along a route of advance, targets are planned using coordinates.

A moving force in a desert is at a disadvantage in comparison with a stationary unit due to lack of concealment and the presence of dust clouds. The defender may engage with missiles from an unexpected direction or from terrain features of no apparent significance. The attacker must be prepared to rapidly shift fires to suppress unforeseen targets. Tactical aircraft may be used to suppress or destroy targets. Targets for aircraft can be marked with indirect- or direct-fire smoke. White phosphorus or illuminating rounds set for low-air burst are also effective.

Indirect fires are used to slow the enemy advance, to suppress enemy weapons and observers, and to conceal movement between positions using smoke. Defensive operations in deserts are characterized by long-range engagement with tanks and ATGMs.

AIR DEFENSE

Identification of friend or foe is difficult. Throughout the entire theater of operations there will be numerous weapon systems that are common to both sides of the conflict. The individual soldier/marine is going to be faced with the monumental task of separating friend from foe by more than just from the recognition of the manufacturer or silhouette of a piece of equipment. This will be true of both air and ground systems. This identification problem will be compounded by the nonlinear battlefield where the focus of operations will not be separated by a line.

The desert is an outstanding environment for employing aircraft. Every unit must be extremely proficient at passive and active air defense. The Allies in North Africa and the Israelis in the Middle East found that dispersion limited the effects of air attacks, and small arms air-defense techniques were effective. Almost every weapon in North Africa had a secondary antiaircraft or antitank mission.

Emphasize to each unit that, when in position, units must disperse very widely making a less lucrative target. When moving in column and under air attack, units must move at least 40 to 50 meters off the road because aircraft normally have nose guns trained on troth sides of the road. A vehicle on the road or on both sides of the road will die.

Because of the wide open spaces characteristic of many deserts and the relatively large areas associated with desert operations, forces fighting in the desert should be reinforced with additional air-defense weapons. Still, there may not be sufficient dedicated air-defense systems to fully cover the force. When this is the case, commanders must be especially careful when establishing air-defense priorities in view of relatively long lines of communication and the tendency to maneuver over relatively large areas. In any event, all units must include a scheme for countering air attacks in their battle plans using both active and passive measures.

Although Army armored and mechanized infantry division air-defense weapons are tracked, this does not necessarily apply to corps medium-altitude air defense units. However, Army corps surface-to-air missile (SAM) units have considerably greater ranges and are equipped with more sophisticated early warning and control systems. Some corps units should be employed well forward. These weapons will have to displace by section to ensure continuous coverage.

Air-defense units should be located close to elements of supported units to provide for ground defense. When the supported unit moves, the air defense unit must also move, which requires careful coordination to ensure that movement of the supported unit is not delayed. Airspace management difficulties are compounded in the multinational environment. SOPS should be exchanged among multinational forces to lessen the confusion of airspace management.

ENGINEERS

Engineer operations in the desert are similar to those in temperate climates although there are fewer natural terrain obstacles to be crossed. Depending on the terrain anticipated in the operations area, a dry-gap crossing capability may have to be obtained from corps support units. Important tasks for engineers in desert operations include-

- Mobility/countermobility/survivability support, including construction of obstacles, logistics facilities and routes, field fortifications, airfields, and helicopter landing pads.
- Water supply.
- Topographic support (map-making).

Mobility

The vastness of the desert makes mobility a prime concern. Roads are usually scarce and primitive. Cross-country mobility is possible in some areas, but trafficability is poor in soft sand, rocky areas, and salt flats. Engineers assist maneuver by reducing slopes, smoothing rock steps, and bridging dry gaps.

Expanded engineer reconnaissance capability will be needed to identify routes, existing obstacles, and minefield locations. Flat, open areas provide good sites for aircraft landing strips; however, in most cases the soil must be stabilized. Normally, desert soil produces extensive dust and has limited weight-bearing capacity.

Engineers use various agents to alleviate severe dust conditions (diesel, JP4, or oil mixtures for example). This is particularly critical in reducing engine wear in areas supporting rotary wing aircraft. It is also important along heavily traveled roads and in cantonment areas. Engineers also use soil-stabilization techniques to increase soil-bearing capacity for airstrips and MSRs.

The application of the fundamentals of breaching-suppress, obscure, secure, and reduce-and the organization of the force in terms of supporting, breaching, and assaulting elements, are even more important in the desert due to the enhanced observation and fields of fire. However, the desert does offer greater opportunities to bypass enemy obstacles because of the greater range of mobility afforded by desert terrain. Caution must be exercised when choosing to bypass enemy obstacles since the bypass may lead the force to the enemy's kill sack.

The increased mobility in the desert makes it easier for the enemy to counterattack exposed flanks of attacking forces. Plan obstacles to protect flanks during offensive operations. Beyond conventionally emplaced minefield, FASCAM, which includes artillery-delivered mines, GEMSS, and air-delivered Gator munitions, are all systems that lend themselves to situational development. FASCAM and conventional minefield may be appropriate, but consider the time required to employ FASCAM when selecting this option. Artillery-delivered FASCAM does not deploy well in soft sand and removes a majority of your indirect-fire assets from the fight.

Countermobility

Due to the mobility inherent in desert operations, obstacles must be extensive and used in conjunction with each other and with any natural obstacles, and covered by direct and indirect fires. Isolated obstacles are easily bypassed.

Mines are easily emplaced in a sand desert, and blowing sand will effectively conceal evidence of emplacement. However, the following potential problem areas must be considered when emplacing mines:

- Large quantities of mines are required for effectiveness.
- Sand can cause malfunctioning.
- Shifting sand can cause mine drift.
- An excessive accumulation of sand over the mines can degrade performance.
- Sand may be blown away and expose the mines.

In suitable terrain, antitank ditches that exceed the vertical step of enemy main battle tanks may be used. Because antitank ditches cannot be conceded, they must be dug so they do not outline a defensive front or flank. They have the advantage of not requiring as much logistic support as minefield. They must be covered by observation and fire to prohibit enemy infantry using them as ready-made trenches.

Because of limited off-road mobility of most combat service support vehicles, considerable engineer efforts may be necessary to construct and maintain routes forward to maneuver units. Local resources, such as salt-marsh mud laid on sand can be used. Track vehicles should not use these routes since they could easily ruin them.

Most desert regions have a natural terrain structure that restricts maneuver such as sandy dunes, rocky plateaus, mountains, and wadis. These structures must be interpreted rapidly and correctly, and then reinforced with obstacles to fix, turn, or disrupt enemy movement, according to the commander's plan.

Minefield and antitank ditches are the primary means of creating obstacles in the desert. Antitank ditches require extensive preparations, but they are effective when adequate preparation time is available. Many desert villages have irrigation ditches that can be used tactically. Other countermobility methods are generally not effective. Road craters, for example, are usually easy to bypass. In sandy areas, ditches can easily be filled in, so they are not good obstacles. Opportunities for bridge destruction are rare, and local materials for expedient obstacles are scarce.

Engineers and combat forces should coordinate the siting of planned obstacles to support the defensive concept. In defensive operations the effectiveness of obstacles requires synchronization.

Survivability

Desert terrain varies from region to region. Generally, however, observation is excellent and concealment is difficult. Deserts provide little cover and concealment from ground-based observers and even less from aircraft. These conditions make modem weapon systems more lethal in deserts than in any other environment.

In the desert, hull and turret defilades for tactical vehicles are essential. This allows the defending force to take advantage of their long-range weapon systems in the face of enemy fires. Dispersion and frequent moves are other survivability techniques that can be used.

The preparation of fortifications in the desert is difficult. Fortifications in sandy soil often require revetments. In rocky plains or plateaus it may be impossible to dig. To counter this problem, build up emplacements with rocks and use depressions.

Camouflage is very effective when properly employed; however, patterns and techniques must be carefully selected to match the local desert environment. Camouflage nets should be provided for all equipment. See Appendix E for additional comments on desert concealment and camouflage.

Desert Survivability Positions

Defensive positions are very vulnerable to offensive fire due to long-range observation and fields of fire in the desert. This, coupled with a lack of natural obstacles, may lead the commander to invest the bulk of his engineer effort into survivability positions. Survivability positions enhance the ability of all direct-fire elements to survive indirect-fire and to return fire on the enemy. Survivability positions are normally more important than antitank ditches, especially in open terrain. See Figures 3-1 through 3-6, for examples of survivability positions. The following are some things you should or should not do when preparing survivability positions:

DO-

- Ensure adequate material is available.
- Dig down as much as possible.
- Maintain, repair, and improve positions continuously.
- Inspect and test position safety daily, after heavy rain, and after receiving direct and indirect fires.
- Revet excavations in sandy soil.
- Interlock sandbags for double-walled constructions and corners.

DO NOT-

- Fail to supervise.
- Use sand for structural support.
- Forget to camouflage.
- Drive vehicles within 6 feet of a position.
- Overfill sandbags.
- Put troops in marginally safe bunkers.
- Take shortcuts.
- Build above ground unless absolutely necessary.
- Forget lateral bracing on stringers.

- Check stabilization of wall bases.
- Fill sandbags approximately 75 percent.
- Construct to standard.
- Use common sense.

The commander's responsibilities during construction of survivability positions are to—

- Protect troops.
- Continuously improve and maintain unit survivability.
- Provide materials.
- Periodically inspect.
- Plan and select fighting position sites.
- Get technical advice from engineers, as required.

In a combat situation, it may be necessary to improvise construction of a survivability position by using materials not normally associated with the construction. Some examples of field-expedient materiel are-

Wall Revetment

- Sheet metal.
- Corrugated sheet metal.
- Plastic sheeting.
- Plywood.
- Air mat panels.
- Air Force air-load pallets.

Overhead Cover Stringers

- Single pickets.
- Double pickets.
- Railroad rails.
- "I" beams.
- Two-inch diameter pipe or larger.
- Timbers 2" x 4", 4 " x 4 ", and larger.
- Reinforced concrete beams.
- 55-gallon drums cut in half longitudinally.
- Large diameter pipe/culvert, cut in half.
- Precast concrete panels, 6-8 inches thick.
- Airfield panels.
- Air Force air-load pallets.

Wall Construction (Building up)

- Sand-grid material.
- 55-gallon drums filled with sand.
- Expended artillery shells filled with sand.
- Shipping boxes/packing material.
- Prefabricated concrete panels.
- Prefabricated concrete traffic barriers.

Stand-Alone Positions

- Prefab concrete catch basins, valve pits, and utility boxes.
- Military vans.
- Connexes or shipping containers.
- Large diameter pipe/culvert.
- Steel water tanks.
- Other storage tanks (cleaned and ventilated).
- Vehicle hulks.

- Shipping pallets.

The following is a suggested inspection checklist to follow when preparing survivability positions:

- Location is sited tactically sound.
- Low profile is maintained.
- Materials are of structural quality (standard construction material).
- Excavation-walls are sloped.
- The setback for overhead is a minimum of 1 foot or 1/4 the depth of cut.
- Stringers¬
  - Are firmly on a structural support.
  - Have lateral bracing emplaced along supports.
- 2" x 4" or 2" x 6" stringers are used on the edge; the strength is on the depth of the lumber.
- Supports¬
  - Stringers are firmly on supports.
  - Supports extend past the excavation by 1/2 the depth of cut.
- Revetments—
  - Quality of construction is checked.
  - Sheeting is supported by pickets.
  - Pickets are tied back.
- Overhead cover¬
  - Quality of structural layer is inspected.
  - Quality of dust layer-plywood or panels-is inspected.
  - Layer is cushioned at least 18 inches deep.

The one-man fighting position is the individual's basic defensive position. The one-man fighting position with overhead cover (see Figure 3-1) provides protection from airburst weapon fragments. A good position has overhead cover that allows the soldier/marine to fire from beneath it. Stringers extend at least 1 foot on each side of the position to provide a good load-bearing surface for overhead cover.

Figure 3-1. One-man Fighting position with overhead cover.

Generally, the two-man fighting position is preferred over a one-man position since one soldier/marine can provide security while the other is digging or resting. The position can be effectively manned for longer periods of time; if one soldier/marine becomes a casualty, the position is still occupied. Further, the psychological effect of two men working together permits occupation of the position for longer periods. Overhead cover also improves the position's effectiveness; it is made as described for the one-man position (see Figure 3-2).

Figure 3-2. Two-man fighting position wiih overhead cover.

Fighting positions for machine guns are constructed so the fires are to the front or oblique; the primary sector of fire is usually oblique so the gun can fire across the unit's front. The position is shaped so the gunner and assistant gunner can get to the gun and fire it to either side of the frontal direction. Overhead cover is built over the middle of the position (see Figure 3-3). It is constructed as described for the one-man position.

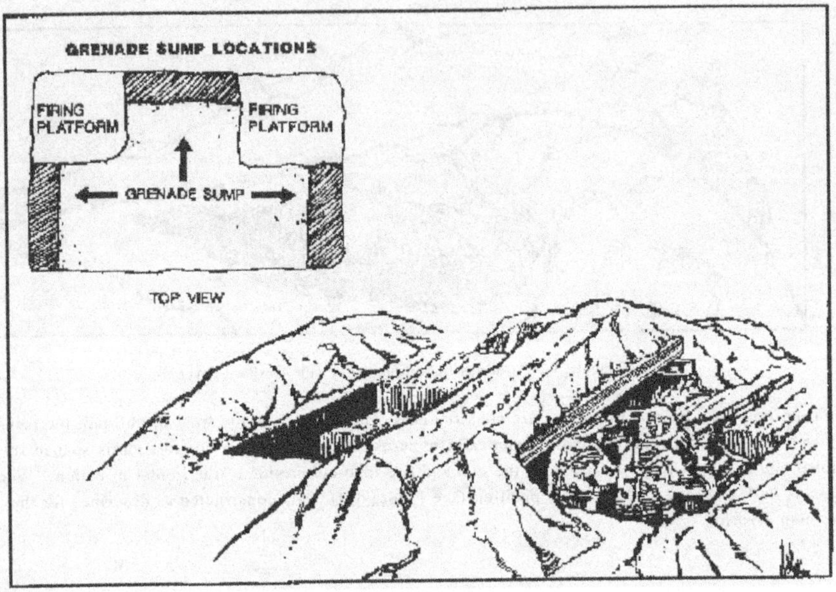

Figure 5-3. Machine-gun position with overhead cover.

Protective shelters and fighting bunkers are usually constructed using a combination of the components of positions mentioned thus far. Protective shelters are primarily used as command posts, observation posts, medical aid stations, supply and ammunition shelters, and sleeping or resting shelters. Figure 3-4 shows an example of a command bunker.

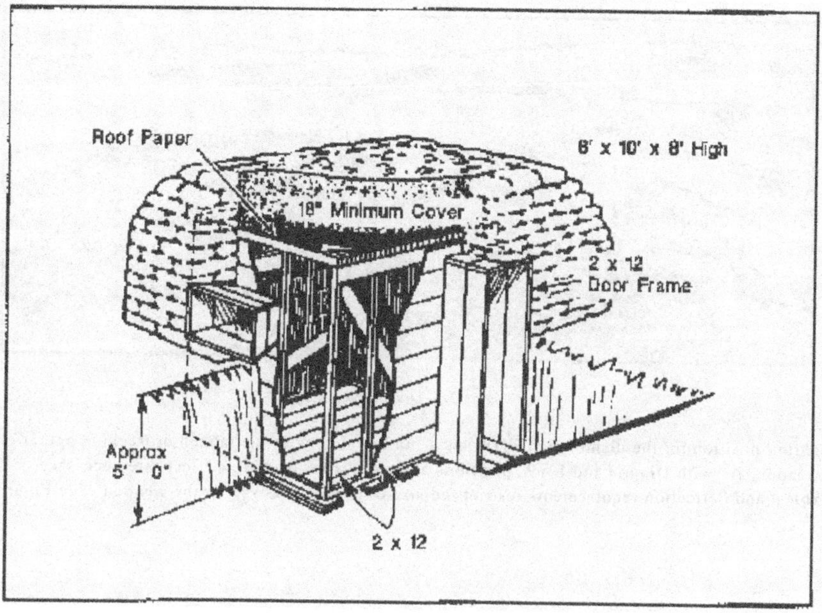

Figure 3-4. Command Bunker .

The Dragon position requires some unique considerations. The soldier/marine must consider the Dragon's extensive backblast and muzzle blast, as well as cleared fields of fire. When a Dragon is fired, the muzzle extends 6 inches beyond the front of the position, and the rear of the launcher extends out over the rear of the position. As the missile leaves the launcher, stabilizing fins unfold. Therefore, the soldier keeps the weapon at least 6 inches above the ground when firing to leave room for the fins. A waist-deep position will allow the gunner to move while tracking a target. Because of the Dragon's above ground height, soldiers/marines should construct frontal cover high enough to hide the soldier's/marine's head and, if possible, the Dragon's backblast. The soldier/marine must dig a hole in the front of the position for the biped legs. If cover is built on the flanks of a Dragon position, it must cover the tracker, missiles, and the gunner. Overhead cover that would allow firing from beneath it is usually built if the backblast area is clear (see Figure 3-5).

Figure 3-5, Dragon Position.

A fighting position for the dismounted TOW must not interfere with the launch or tracking operations of the weapon. As with Dragon and LAW positions allowances for backblast effects are necessary. Backblast and deflection requirements restrict the size of overhead cover for the weapon. See Figure 3-6.

Figure 3-6.Dismounted TOW position.

Designers of fighting positions and protective positions in desert areas must consider the lack of available cover and concealment. Fighting positions should have the lowest profile possible, but mountain and plateau deserts have rocky soil or "surface chalk" soil which makes digging difficult. In these areas, rocks and boulders are used for cover. Because target acquisition and observation are relatively easy in desert terrain, camouflage and concealment, as well as light and noise discipline, are important considerations during position construction.

Indigenous materials are usually used in desert position construction. However, prefabricated structures and revetments, if available, are ideal for excavations. Metal culvert revetments can be quickly emplaced in easily excavated sand. Sandbags and sand-filled ammunition boxes are also used to prevent side walk of positions from collapsing.

FM 5-103 discusses vehicle fighting positions in detail and should be consulted for more information. Figure 3-7 provides specifications for vehicle survivability defilade positions that can be dug by the D-7 dozer.

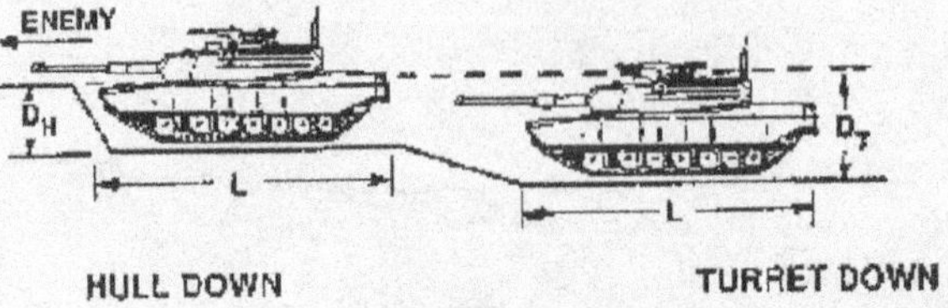

VEHICLE	L	D_H	D_T	WIDTH
M1 Tank	2 Dozer lengths	Top of battery box	Top of your head	1 1/2Blade widths
M60 Tank	2 Dozer lengths	One fist above battery box	Top of smoke stack	1 1/2 Blade widths
M2 BFV M3 CFV	1 1/2 Dozer lengths	Top of engine housing	Top of smoke stack	1 1/2 Blade widths
M113 APC	1 1/2 Dozer lengths	One fist above battery box	Top of air intake cover	1 1/4 Blade widths
M901 ITV /FISTV	1 1/2 Dozer lengths	Top of engine housing	Top of your head	1 1/4 Blade widths
M109 155 Howitzer	6 1/2 Dozer lengths	One fist above battery box	Not required	1 1/2 Blade widths
M110 8" Howitzer	6 Dozer lengths	Top of blade tip crown	Not required	1 1/2 Blade widths
M577 Command Post Vehicle	1 1/2 Dozer lengths	Top of your head	Not required	1 1/4 Blade widths

These dimensions are described with respect to you and your dozer. Techniques for construction of survivability positions are described in FM5-34, ENGINEER FIELD DATA.

Figure 3-7. Survivability defilade positions.

Logistics areas (BSA/DSA) require additional survivability support. Desert operations require that logistics concentrations such as BSAs and DSAs be given additional considerations for survivability support. These sites are large, relatively static, and difficult to camouflage. As a result, these support areas are vulnerable to enemy interdiction. Military vans or connexes should be covered with sandbags to improve protection. Additionally, if they are covered with heavy plastic, with plastic drapes over the entrances, protection against NBC effects can be improved.

Engineer digging assets, such as bulldozers, should be tasked to provide survivability support to these sites with particular emphasis placed on hardening ammunition and fuel storage locations. Caution should be used when digging foxholes and tank hide positions since some areas have a tendency to cave in.

Water Supply

Water supply is the most important mission of engineers in the desert. The search for water sources requires continuous, intensive reconnaissance. Water may be obtained by drilling beds of dry water courses, or by deepening dry wells. Once found, water must be made potable and stored or transported. Since water purification trucks may be high-priority targets and barely sufficient for the task, any force operating in the desert must be augmented with water supply units (including well drilling), water purification and water distillation teams, and transportation teams. Another possible water source is the reverse osmosis water purification unit (ROWPU). This unit is an ISO frame-mounted, portable water purification system capable of purifying water from almost any shallow well, deep well, and surface water or raw water source. The ROWPU is capable of removing NBC contaminants, minerals, and biological impurities. The single greatest benefit of the reverse osmosis process is the ability to desalinate sea water. The ROWPU is capable of producing potable water at a rate of 600 gph. The ROWPU is powered by a 30-kilowatt generator set.

Topographic Support (Map-making)

Large areas of the world's deserts are not covered by maps of any useful tactical scale. Existing maps are frequently inaccurate and increase the difficulties of navigation. Therefore, engineer topographic companies must augment the force by preparing, printing, and distributing up-to-date maps of the operational area. USAF, Army, and Marine aviation support can be used to produce gridded maps from aerial photography of the area forward of the line of contact.

MILITARY POLICE

Combat support provided well forward by military police will continue in desert operations, although over increasingly extended distances. MP tactical and physical security will be of special importance over extended lines of communication, such as petroleum pipelines and viaducts transporting water over long distances. Protection of these critical items demands both active and passive measures, including overflight by returning aircraft or overwatch by convoy movements. The storage sites for water, food, POL, and ammunition have historically been principal targets for enemy action, and consequently must receive augmented security.

The indefinite conditions and number of roadways will require increased circulation control points to direct traffic, redirect stragglers, and provide information so that throughput forward to the fighting elements will be expedited. Military police are especially valuable when the combat commander must employ concentration or economy of force in the face of the enemy to gain a favorable combat ratio. MPs can secure the roadways, enforce priority movement, and prevent any delay of the elements undertaking passage of lines to blocking or defensive positions. MPs can also assist in the handling of EPWs.

US AIR FORCE SUPPORT

US Army and Marine forces fighting in the desert can expect to be supported by USAF tactical fighter-bomber and airlift aircraft. Close air support by USAF tactical fighter bombers is more important in desert warfare in view of lack of concealment, relatively large areas of operations, and mobility of forces employed by each side. Air support in a desert environment has advantages over more temperate areas of operations. For example, it is easier to locate targets; visual observation is normally far superior to that in temperate climates; and ground movement is more readily apparent.

Air attacks can be handicapped by lack of covered approaches, but increased visibility permits engagement from standoff ranges. When flying close air support missions it is important for pilots to be able to differentiate between enemy and friendly forces. Use panels or other visual or electronic identification means to assist in identification.

Because of the extended lines of communication likely in desert operations, USAF tactical airlift should be used whenever possible. This is particularly true of resupply operations conducted from a lodgement area to forward trains areas when considerable distances are involved.

Planning for air support must be as detailed as time permits to determine mission and armament requirements, time over target, and method of control. The joint air-ground operations system (AGOS) used to request and coordinate the use of US Air Force tactical air support is described in FM 100-26.

US NAVY SUPPORT

When the force is being supported by US Navy gunfire, or Navy or Marine aircraft, elements of a Marine air and naval gunfire liaison company (ANGLICO) are attached to Army ground forces. The mission of the company is to support an Army division by providing control and liaison agencies for employment of this support.

ANGLICO platoons and teams can advise commanders on the capabilities, limitations, and employment of naval gunfire, and USN or USMC air support. Platoons are normally placed with brigades or higher headquarters, and air and gunfire support teams placed with battalion task forces. Although the company has organic vehicles and some combat services support capability, its elements generally require additional administrative and logistical assistance from the supported unit. In order to communicate with Army/Marine units, additional communications equipment may also have to be provided. Additional information on ANGLICO employment can be found in FM 31-12.

COMBAT SERVICE SUPPORT

Combat service support for desert operations is described in detail in Chapter 4. When planning a desert operation it is necessary to consider the following factors:

- The speed of supply may be slowed and lines of communication can be vulnerable due to the distances between units. Except for Class V and sometimes Class III, resupply should beat night for reasons of security.
- A great demand for water can tie down large quantities of transport and may involve laying pipelines. Water is vital, so consider the water situation during every operation estimate.
- Increased maintenance is required due to heat, sand, and dust damage to equipment. This not only increases the repair work load, but also increases demand for replacement items due to increased wear.

Supply

Offensive operations in this environment may involve considerable expenditure of ammunition and high POL consumption. Units must carry maximum combat supplies, and plans for resupply must be widely disseminated and clearly understood. Use every opportunity for resupply.

Due to the importance of combat service support, attacking the rear will be more immediate y effective at lower levels of command in the desert than in temperate climates. In the case of water, for example, the enemy must be able to obtain resupply. Degradation of the enemy's trains places him in a situation where his troops must maneuver against the attacker regardless of the planned scheme of defense.

Combat service support must be reliable and timely, using vehicles that can travel over difficult terrain to reach combat units. In the desert, more than anywhere else, the commander must ensure that he has support that is capable of maintaining his unit for a specified period of time, even if the logistic line of communication is temporarily broken.

The mobility and freedom of tactical maneuver are tied to the ability of the logistic chain to supply maneuver units. Two alternatives are available: increase the rate of supply, probably requiring more vehicles, or prestock, which ties units to the stocked area. Some important supply considerations are outlined below:

- Class I. It is often impractical to supply hot rations from mess trucks, especially when the unit is subject to enemy air reconnaissance or target acquisition devices. T rations and B rations are the usual method of troop feeding.
- Class III. Daily requirements for POL in desert operations can be expected to be high. Estimates for POL requirements should take into consideration large-scale maneuver inherent in desert operations.
- Class V. Estimates of ammunition requirements should reflect the high level of commitment that can be anticipated in desert operations.

Maintenance

Disabled vehicles are vulnerable targets. Both disabled vehicles and the maintenance vehicles used in working on them must be concealed during the day, and strict light and sound discipline imposed at night. Maintenance contact teams should carry Class IX supplies that have a quick turnover.

Section II. Offensive Operations

This section discusses offensive operations as they are modified by desert terrain.

GENERAL

The main purpose of offensive operations in desert terrain is to destroy the enemy, Operations may be undertaken to secure key or decisive terrain, to deprive the enemy of resources or decisive terrain, to deceive and divert the enemy, to develop intelligence, and to hold the enemy in position. Destruction of the enemy can be accomplished by concentrating friendly forces at a weak point in the enemy's defense and destroying enemy combat units, or by driving deep into the enemy's rear to destroy his combat service support and cut his lines of communication. No force can survive in the desert for long without combat service support.

An imaginative commander is not bound by terrain constraints in seeking and destroying the enemy. Due to the scarcity of key terrain in the desert, normally the only constraints placed upon a maneuvering force is its ability to maintain the only constraints placed upon a maneuvering force is its ability to maintain responsive combat service support and to protect its combat service support from enemy attack. The longer the lines of communication become, the more susceptible they are to being cut.

In most deserts, the scarcity of large areas of defensible terrain means that an enemy force has at least one flank open to attack, The attacking force must seek this flank and attempt to maneuver around it into the enemy's rear before the enemy can react and block the envelopment with mobile reserves.

Successful offensive operations depend on rapid, responsive, and violent maneuver, seeking a vulnerable enemy flank while exposing none to the enemy. The enemy, realizing the danger of remaining stationary in this terrain, may choose to conduct spoiling attacks or to counterattack. The resulting meeting engagement between the two attacking forces will often be a series of flanking actions and reactions with success going to the one who can find the other's unguarded flank first.

Attacking forces may conduct or participate in movement to contacts or hasty or deliberate attacks. Within a division, lead elements of forward units may be conducting a deliberate attack on the enemy's weak point or flank to open a gap for following units to move through and exploit success. Lead units of the exploiting force will be conducting a movement to contact and hasty attacks to overcome pockets of enemy resistance. Regardless of the type of operation being conducted, attacking units use the fundamentals for offensive operations described in FM 100-5/FMFM 6-1 and other tactical manuals, modified to suit the terrain.

FUNDAMENTALS OF THE OFFENSE

The attacker must conduct active and aggressive reconnaissance to the front, flanks, and rear, not only to locate and identify enemy obstacles, units, weak points, and flanks, but also to give early warning of threats to his flanks and combat service support elements. A moving force is at a disadvantage in the desert due to a lack of concealment. Therefore, it is necessary to push reconnaissance units as far out from the main body as possible to allow early warning and to deny the enemy close-in observation.

Information gathered by this reconnaissance must be passed promptly to all units. In the desert, a negative report may be as important as an enemy sighting. Commanders and staffs must avoid the two extremes of either passing too little information or overwhelming their subordinates with useless trivia. Similarly, reconnaissance units must also avoid extremes. There is a very real possibility that extensive reconnaissance in one area will alert the enemy of intended operations in that area. Therefore, the need for reconnaissance must be tempered with the need for deception. In fact, reconnaissance may even serve as a deceptive measure to draw the enemy's attention away from the real objective or area of operations.

Concentrate on overwhelming combat power. Mass is achieved in both time and space. Units must be able to rapidly concentrate at a given time and place, and then disperse just as rapidly to avoid offering a lucrative target to the enemy. Concentration does not necessarily mean that vehicles and men are massed in a small area, but that units have the ability to place an overwhelming concentration of fires on the enemy.

Mutual support is as important in the desert as in temperate climates. Due to the large distances covered by maneuver in the desert, mutual support does not mean that any one unit is always in position to fire against an enemy threatening another unit. However, units must be capable of maneuvering in support of one another without disrupting the scheme of maneuver.

Concentration requires movement, and possibly weakening of forces facing the enemy in another part of the zone. Due to the enemy's observation capabilities, movement should take place at night or in conditions of limited visibility whenever possible. Deception measures play an important part in concentration, either to mislead the enemy as to the strength or true intentions of the opposing forces, or their avenues of approach. In this environment of negligible concealment, deception cannot be overemphasized.

The enemy's objective is to stop and destroy the attacking force by direct and indirect fires, obstacles, and counterattacks. The attacker must in turn suppress enemy weapon and surveillance systems to degrade their effects and their intelligence-gathering capability.

Attack helicopters and high-performance aircraft are extremely useful due to their ability to maneuver and apply firepower over a large battlefield in a short time. So, suppression of enemy air defense has a high priority during offensive operations. The destruction of enemy antitank capabilities must also have a high priority due to the shock potential of armor in the desert. No target that has a long-range antitank capability should be disregarded. Good gunnery and well-planned fire distribution are preeminent.

In featureless desert terrain, the requirement to shock, overwhelm, and destroy the enemy demands accurate reconnaissance to identify actual positions from false positions, and excellent navigation so that a commander may be certain of the deployment of his forces. Reconnoiter to find a gap or assailable flank (without alerting the enemy that the area is being reconnoitered) and concentrate the main body to go through or around it with suppressive fires on the flank(s). A gap must be wide enough to allow one unit to bypass another unit that could be stalled. Obstacles are likely to be placed so that attempts to go around them will often lead the attacker into a tire sack. Equipment capable of breaching obstacles must be located well forward.

ENVIRONMENTAL CONSIDERATIONS

As a general rule, a force attacking in daylight should try to wait until the sun is comparatively low and position behind it. This enables enemy targets to be plainly seen without their shadows, while the defenders are handicapped by glare, mirages, and haze. It is not always possible (nor essential) for the sun to be directly behind the attackers. To rely on this leads to a stereotyped method of attack which could become evident to the defenders. The commander of a maneuver force should attempt to keep the sun somewhere on a 3,200-mil arc to his flanks or rear, giving a wide choice of angle of attack.

Dust is an observational hazard to a maneuvering force, especially where there is little or no wind. Teams should move in echelon with overmatching elements on the upwind side, and observers and attack helicopters should operate well to the flank. Since it is impossible to disguise movement during daylight, the assault should be as rapid as possible to minimize enemy reaction time.

The decision to move through a sandstorm will depend on the unit's distance from the enemy, trafficability, the presence of minefield, and the direction and density of the storm. If the advancing unit is caught in a storm blowing from the enemy's direction, the safest alternative is to halt until it abates, although this may not always be possible. In some situations it may be possible for platoons to form close column, using taillights only, and continue movement. When the storm is blowing toward the enemy it is possible (and extremely effective) to conduct an attack immediately behind the storm.

In certain circumstances equipment or positions that are camouflaged and are less than 1 meter from the ground are invisible to an observer at the same height out to approximately 2,000 meters. At the same time, mirages allow observation of objects below the horizon, although these may be distorted, enlarged, or fuzzy to the point of being unrecognizable. These effects often depend entirely on the angle of the sun to the observer and are best combated by-

- Maintaining observers as high above the desert floor as possible, even if only in hull-down positions behind sand dunes.
- Allowing a vehicle's crew on one side of a position to warn a crew on the other side of a possible threat to his front by crews observing over wide areas.

Many offensive operations take place at night. Observation in these conditions varies according to the amount of ambient light. During nights when the moon is full or almost full, the clear desert sky and ample ambient light allow good observation, both with the naked eye and with night observation devices. Maneuvering units using night-vision devices must continually scan the surrounding terrain to pickup enemy activity that normally would be acquired by peripheral vision in the daylight.

The desert night is extremely dark when there is little or no moon. Under these conditions passive-vision devices, with the exception of thermal imagery, are of little value unless artificial light is used. Active light sources will have to be relied upon. Employment of artificial light must be strictly controlled by the headquarters directing the operation to maintain surprise. As a general rule, direct-fire weapons should not illuminate their target themselves, as their vision will be obscured by debris kicked up due to muzzle blast. Following contact, when some targets should be on fire, passive devices can be used.

MANEUVER

If the terrain permits masking of maneuvering units, and trafficability is good, normal fundamentals of fire and maneuver are used. Trafficability may be restricted by rocky terrain as in the Golan Heights, or the ground may be so flat that the defender has total observation of the area. Movement in these circumstances requires speed of maneuver, deception, and considerable suppression to degrade enemy observation and fires. Frontal attacks should be avoided, especially in conditions of restricted trafficability. It is preferable to maintain pressure on enemy units in unfavorable terrain, while other forces find enemy weaknesses in terrain that is more favorable for an attack.

Lack of clearly defined terrain features complicates navigation and phased operations. Units conducting an enveloping maneuver are apt to lose direction unless routes have been carefully reconnoitered by the maximum number of leaders.

Section 111. Defensive Operations

This section discusses defensive operations as they are modified by desert terrain.

GENERAL

It is unlikely that a US force will be fully deployed in a desert country before an enemy attack. The more probable situation, assuming a secure lodgement area, will be that part of the force will be in position supporting an allied force, while the remainder is moving in by air and sea. Tactically, the allied force will be outnumbered, so the initial mission will be to gain time until the entire force is present in the operational area. This will require a defensive posture initially, but a defense undertaken so aggressively as to convince the enemy that his offensive action will be too costly in personnel and equipment to be worth maintaining. The enemy will be well aware that US forces are arriving in the area, and will make every effort to conclude his operation successfully before the force is fully prepared for combat operations.

The force may conduct defensive operations during subsequent stages of the operation for any of the reasons described in FM 100-5/FMFM 6-1. Portions of the force may be required to defend the important types of terrain described below:

- Man-made features such as ports, key logistic installations, roads, railroads, water pumping stations, airfields, and wells.
- Natural features, such as mountain passes, or dominating ground, such as Mount Hermon on the border of Syria and Israel, or the Solium escarpment near the sea between Libya and Egypt.
- Key or decisive terrain that need not necessarily be a major feature, but one whose loss will inhibit the force in some manner. For example, the loss of terrain relatively close to a lodgement area may hinder the planned rate of buildup.

With the exception of the above cases, the retention of desert terrain normally makes little difference to the final outcome of battle. This does not mean that a commander has complete discretion to move his force wherever and whenever he wishes, as this movement will affect the dispositions of other US forces or allies. It means that possession of terrain is less important than the destruction of enemy forces. Although it will be necessary to dominate certain terrain or retain freedom to maneuver in large areas of the desert, there is no more sense in permanently occupying such areas than occupying a patch of sea. Assuming equal equipment capabilities for both opposing forces, the critical factor in defense will be the force ratios involved and the state of morale and training of the opposing forces.

A defense using aggressive maneuver at all levels is the best way to destroy large numbers of enemy without being destroyed in the process. If the defending force fails to remain mobile and active, the enemy will easily outflank it and strike directly at vital targets, such as the lodgement area. It is almost certain that one flank or the other will be open as were the south flanks of the British and German forces in Egypt and Libya in 1940-43. Since it will not be possible to maintain an unbroken line between strategic obstacles, air and ground security forces must be positioned in width and depth to guard against an enemy trying to outflank the defender.

Obstacles, both natural and artificial, are used to slow, contain, or isolate enemy units in order to defeat and destroy his units one at a time. Forward units block the enemy and canalize him into one or two avenues where he can be engaged from the flank. A reserve can then counterattack to destroy any remaining enemy.

Mutual support is normally a factor of time rather than weapon range due to the large areas to be covered. Gaps in initial positions may have to be accepted between and within task forces; although the ideal is to site units in such a manner that forces in at least two positions can engage an enemy maneuvering on any one of them. This greatly reduces any possibility of defeat in detail. When gaps exist they must be kept under surveillance. The defensive plan must include provisions for maneuvering to fire on any part of a gap before the enemy can move through it. A unit's area of responsibility must be defined by higher headquarters and should be clearly identifiable on the ground, which, due to the absence of significant terrain features, may require marking by artificial means.

FUNDAMENTALS OF THE DEFENSE

The fundamentals of defense are described in FM 100-5/FMFM 6-1 and in doctrinal manuals appropriate to each level of command. The following paragraphs discuss some points to remember in desert operations as they apply to the fundamentals of defense.

Reconnaissance and security units and force surveillance systems must focus on-

- What is the enemy's short-term objective?
- What are the enemy's avenues of approach, and what force is employed on each of them?
- Are the apparent movements real or feints?

As soon as these questions have been answered the commander will be able to maneuver to destroy the enemy. Until they are confirmed he can do nothing more than react to enemy initiatives, This is dangerous in any circumstance and doubly so in the desert as the side with the greatest potential for maneuver is more likely to win.

Direct-fire weapons must be used to their maximum effective range both by day and night. Limitations in night-vision equipment cannot be allowed to reduce depth or frontages; so plans for field artillery or mortar illumination are made for defense during limited visibility.

It is essential that all elements of a force retain their tactical mobility and efficient communications so that they can immediately react to changes in the commander's plans. Each individual weapon must be sited in a number of firing positions, even though vehicular movement may be exposed to air attack. Infantry fighting vehicles must remain in positions where they are concealed, capable of giving fire support to the dismounted squad, and available for immediate remounting.

Combined arms teams are essential to give the commander the capability he requires to fight the defensive battle. Defending forces orient on primary enemy approaches but units must also be prepared for attack from any other direction. Itis neither possible nor necessary to have maximum firepower in all directions, provided weapons can be moved to threatened areas before the enemy reaches them. Air cover or an air defense umbrella is necessary for a successful defense.

It is rare to find positions where any substantial part of the unit area of operations can be protected by natural obstacles. This require extensive use of artificial obstacles, depending on time, personnel, and combat service support available. Obstacles are used to divide the enemy force to improve local force ratios, and to slow the enemy's advance, thus permitting a flank attack. Conventional minefield must be clearly marked on the friendly side and recorded to avoid unnecessary losses if friendly forces later maneuver over the area.

STRONGPOINTS

Strongpoints are rare in desert warfare; however, they may be necessary to defend an oasis, mountain pass, or other key terrain essential to the defender's scheme of maneuver. When it is necessary to deny terrain to an enemy force, it is far better to initiate the defense well forward of the terrain feature, conduct the defense in depth, and destroy the enemy or force him to break off his attack before he reaches the critical feature.

In some cases the level of fortification and the deployment of the enemy may be a function of time, or the enemy's intention and his understanding of what our forces are intending to do. The effectiveness of these strongpoints depends on the range of fires, the level of fortifications, and the decision of the opponent to attack them.

Deeply dug and well-prepared strongpoints surrounded by a minefield and having underground accommodations are usually used in the desert. Although these strongpoints may be neutralized by air or artillery fire and bypassed, eventually they will have to be assaulted. If they have been carefully sited and are well defended they can be quite effective. Variations of the strongpoint defense are used in rear operations. Combat service support units will use this method in perimeter defenses or base-cluster defenses. See Figure 3-8 for an example of a strongpoint and Figure 3-9 for an example of a strongpoint holding key terrain.

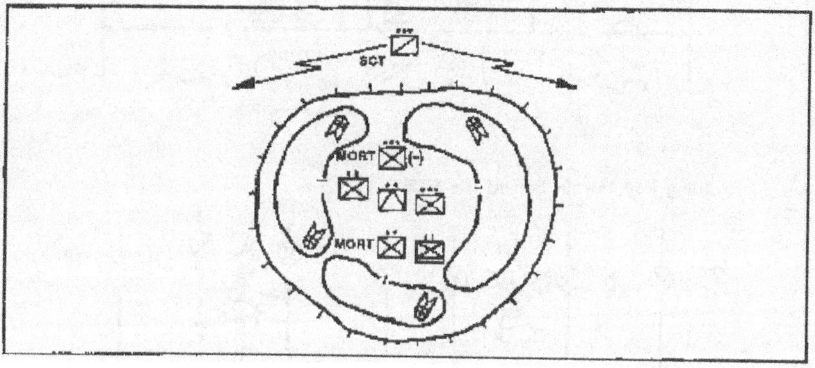

Figure 3-8, Strongpoint

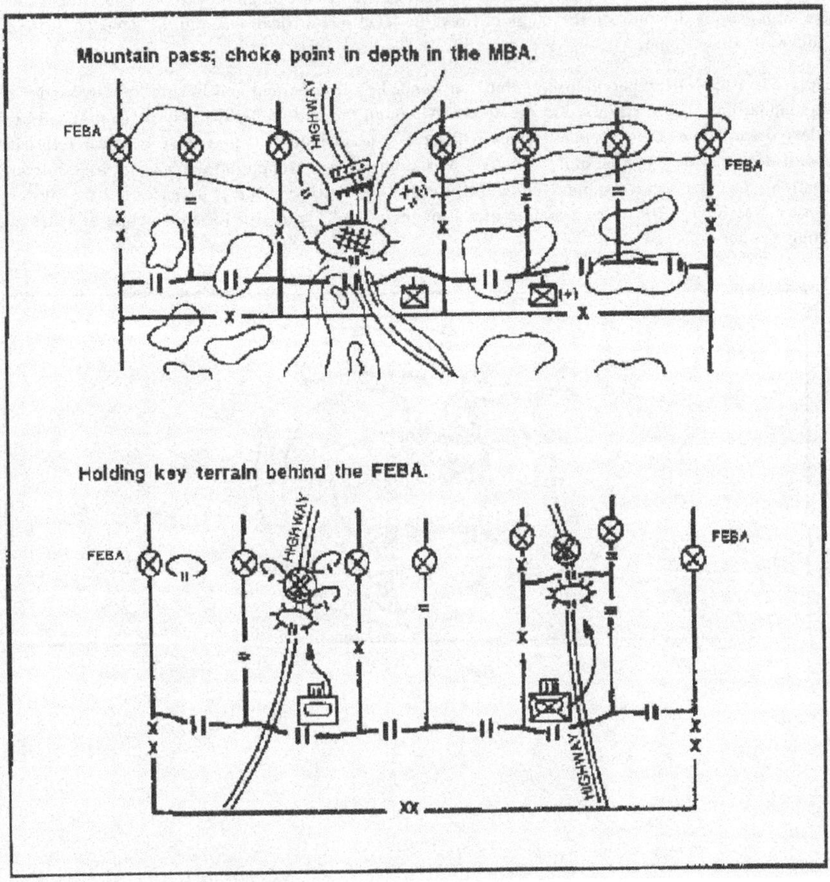

Figure 3-9, Strongpoints holding key terrain.

REVERSE SLOPE DEFENSE

The use of the reverse slope defense takes on added importance in the desert. Concealment is hard to achieve in the open desert. Detection of a unit's location invites both direct and indirect fires in abundance. The use of reverse slope positions will deny the enemy direct observation of positions until he is within the range of direct-fire weapons. Reverse slopes can even be found on seemingly flat desert floors; an intervisibility line will provide the reference for the establishment of engagement areas to support a reverse slope defense. A common misconception is that the desert is flat, when in fact, deserts are normally very uneven, with large breaks in the terrain.

Desert environments give special significance to the terrain aspect of METT-T. Commanders at all levels should place emphasis on the impact of desert terrain as it relates to the other factors of METT-T. The reverse slope defense in desert terrain warrants special considerations.

Direct-fire positions should be placed at the maximum effective ranges from the intervisibility line. This is where the enemy cannot see or engage a force with direct fire until he is within its engagement area. He can only deploy limited forces at a time. This allows the defender to mass fires on one portion of the enemy force at a time. The attacking force will have difficulty in observing and adjusting indirect fires. Obstacles may not be seen by the enemy until he is upon them and force him to breach under massed frees. Observation posts (OPs) positioned forward to see the advance of the enemy can influence the fight through indirect fires. The OPs can direct indirect fires on enemy forces that are slowed or stopped outside direct-fire ranges.

This defensive technique may be used in all defensive missions. Light infantry units use the reverse slope for protection against enemy long-range fires and to reduce the effects of massive indirect fires (artillery and close air support). The reverse slope defense brings the battle into the defender's weapons' ranges. Use of the reverse slope provides an opportunity to gain surprise.

The goal is to cause the enemy to commit his forces against the forward slope of the defense, resulting in his force attacking in an uncoordinated fashion across the crest. A reverse slope defense is organized on the portion of a terrain feature or slope that is masked from enemy direct fires and observation by the topographical crest, and extends rearward from the crest to maximize the range of the defender's weapon systems. See Figure 3-10 for an example of a reverse slope defense and Figure 3-11 for the organization of the reverse slope defense.

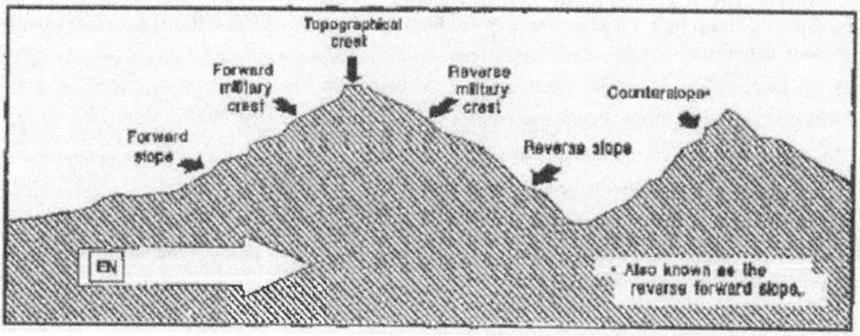

Figure 3-10. Reverse slope defense.

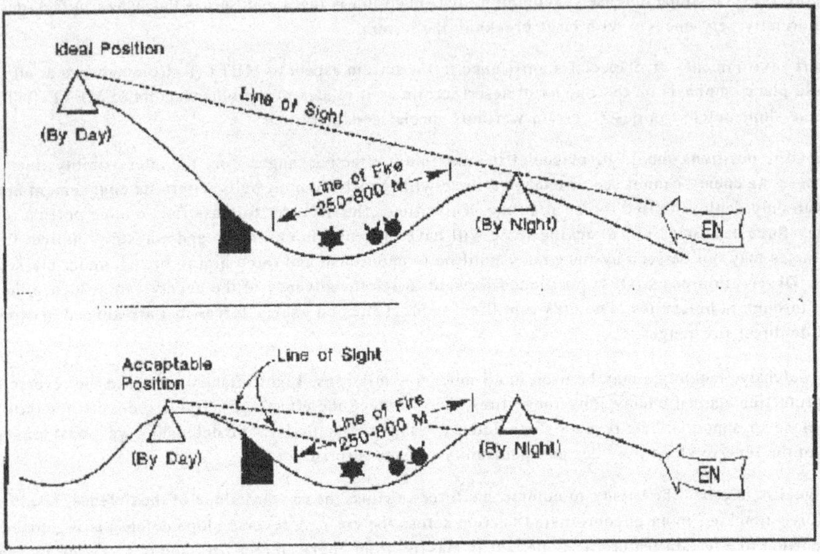

Figure 3-11. Organization of the reverse slope defence.

A disadvantage may be that the maximum ranges of weapon systems may not be employed due to the terrain available. The desert may be the best environment for the reverse slope defense. It may allow the use of weapons at maximum ranges as well as facilitating advantages. The following are advantages of a reverse slope defense:

- It hinders or prevents enemy observation of the defensive position.
- Attacking forces will not be able to receive direct-fire support from following forces.
- Enemy long-range antitank fires will be degraded.
- Attacking enemy forces will be silhouetted on the crest of the hill.
- Engineer work can be conducted away from direct-fire and observation from the enemy.

Reverse slope defense is not one concept, but a series of concepts that produce the potential for success. The concepts are-

- Pursue offensive opportunities through surprise and deceptive actions, with the intent of stealing the initiative, imposing the commander's will on the enemy, and breaking the enemy's morale.
- Afford the defender a variety of options in positioning his troops, with each option designed to draw the enemy into unfamiliar terrain.
- Enhance light infantry effectiveness and survivability.

A hasty or deliberate reverse slope defense may be considered when any of the following conditions exist:

- When the forward slope lacks cover and concealment, and effective enemy fire makes that position untenable.
- When the terrain on the reverse slope affords appreciably better fields of fire than those available on the forward slope.
- When it is desirable to avoid creating a distortion or dangerous salient in friendly lines by relying on forward slope positions.
- When it is essential to surprise and deceive the enemy as to the unit's true defensive positions or main effort.
- When seeking to gain protection from the enemy as he is massing fires.

DELAY OR WITHDRAWAL

When it is necessary to delay or withdraw, a desert offers many advantages to the defender. Long-range fields of fire allow engagements at maximum effective range of direct-fire weapon systems, and disengagement before the defender's position. However, dust clouds created by a moving force make it necessary to disengage under cover of smoke or darkness. Even a sandstorm can be used to the advantage. Field artillery, US Air Force fighter bombers, and attack helicopters can also be used to allow a ground maneuver unit to disengage and move rapidly to the next position.

When it is necessary to trade space for time, often a counterattack to destroy enemy advance units will do more good than trying to defend longer from an intermediate position.

Commanders at all levels should clearly understand the scheme of maneuver concept of the operation, and what it is they are expected to do, especially if communications should fail. Plans must include provisions for alternate means of communication. Routes should be clearly marked and reconnoitered to the maximum extent practical.

Due to the distances involved and constantly changing task organization, passage of lines is more difficult to coordinate and control. Pay extra attention to the identification of vehicles, routes of passage, signals, and coordination of movements.

Deception should be a part of all desert retrograde operations. The object of deception is to conceal the fact that a retrograde operation is taking place and that units are thinning out. Smoke and dummy positions can be used, false radio messages transmitted, and even dust clouds used to deceive the enemy.

ENVIRONMENTAL CONSIDERATIONS

In the desert it is necessary to modify the techniques of defense as described in doctrinal manuals applicable to each level of command and according to the mission, the fundamentals described in the preceding paragraph, and to the environmental considerations that are described in the following paragraphs.

Observation

The enemy will try to attack when the sun is low and behind him so as to dazzle the defender. The defender's observers must be as high as possible above the desert floor to see the advancing enemy as soon as possible.

Active light sources can be detected from great distances, especially during nights with low ambient light. Positive control of active light sources must be maintained until the battle is joined. Even then, the force equipped with passive devices will have the advantage over the force that is not equipped with these devices.

Heat from combat vehicles can give an enemy using thermal imagery devices a complete picture of the defensive scheme. So, combat vehicles should not prematurely occupy battle positions at night.

Sandstorms

Sandstorms may be used by the enemy to hide an offensive operation especially if the storm is blowing from the enemy's direction. When this is the case, units should immediately occupy their battle positions before the storm arrives. The unit should remain there until it ends, ready to fire and maneuver against the attacker after the storm abates. If vehicle patrolling is possible, a scout platoon or similar unit should cover all gaps, preferably moving in pairs, and on straight lines in view of navigational difficulties.

Terrain

From the point of view of a defending brigade or battalion task force commander, avenues of approach will often seem unlimited. Long-range observation must be maximized and scouts employed well forward to offset this problem. Radars should also be used extensively to provide early warning. It is necessary to identify the enemy's main effort early in order to move to concentrate.

Lack of concealment, especially from the aerial detection, prohibits units from occupying firing positions until just before engaging the enemy. Combat vehicles must displace immediately after engagement or risk destruction. Because of frequent displacement, mutes between battle positions should be reconnoitered and marked when possible, without revealing the scheme of defense. Smoke must be used frequently to conceal movement.

TACTICAL DECEPTION OPERATIONS

Analysis of desert operations from World War II to the present day indicates that tactical deception and surprise are clearly linked to the ability to move and mass forces during periods of limited visibility.

Operational planning should emphasize night movement of units. To minimize the problems of dust and to enhance deception, movement should be accomplished using multiple routes. Place priority on training to support this requirement. Associated with night movement is the requirement for night passage through lanes in minefield and forward passage through friendly forces.

In every modem desert war, deception has played a major role. The lack of concealment leads commanders to believe that with a reasonable reconnaissance effort they can gain an accurate picture of the enemy's dispositions. Reconnaissance by German, British, Israeli, Egyptian, and Syrian forces in modern desert warfare has been sufficient to detect the presence of combat forces in the desert. Deception has been successfully used in each of the modem desert conflicts to mislead commanders.

Since the desert environment makes it difficult to hide forces, the alternative is to make them look like something else-trucks and plywood made to look like tanks, and tanks made to look like trucks.

The movement of personnel and equipment and the placement of logistic support installations are normally indicators of a force's intent. The movement of empty boxes or pallets of ammunition and the establishment of fuel storage areas with real or dummy assets can deceive the enemy as to planned offensive actions. Use minimal actual transportation assets and make numerous, visible trips to simulate a large effort.

There are many examples of successful deception efforts by US forces from World War II In September 1944, the 43rd Cavalry Reconnaissance Squadron (Reinforced) occupied a 23-mile fronton the left flank of XX (US) Corps on the Metz Front. This squadron portrayed an armored division for several weeks and was so successful that the German Order of Battle Maps showed the 14th (US) Armored Division (AD) to be in the area. The 14th AD was not even in Europe at the time. Expertise in deception operations is critical to success.

Deception plays a key part in offensive operations and has two objectives: the first objective is to weaken the local defense by drawing reserves to another part of the battlefield. This may be done by making a small force seem larger than it is. The second objective is to conceal the avenue of approach and timing of the main attack. Some deception methods that can be used in offensive operations are-

- Using dummy units and installations.
- Using phony radio traffic.
- Using movement and suppressive fires in other areas timed to coincide with the real attack.
- Using small convoys to generate dust clouds.
- Filling ration boxes with sand and stacking them at landfills.
- Moving trucks into and out of the area giving it the appearance of being a storage facility or logistic base.
- Emulating damage to induce the enemy to leave important targets alone. For example, ragged patterns can be painted on the walls and roof of a building with tar and coal dust, and covers placed over them.
- Stacking debris nearby and wiring any unused portions for demolition. During an attack, covers are removed under cover of smoke generators, debris scattered, and demolitions blown. Subsequent enemy air photography will disclose a building that is too badly damaged to be used. Troops using the building after an attack must guard against heat emissions after dark and care must be taken to control electromagnetic emissions.
- Using phony minefield to simulate live minefields. For example, disturb the ground so that it appears that mines have been emplaced and mark boundaries with appropriate warnings.
- Making a real minefield to appear as phony or camouflaging it. For example, once a real minefield is settled, a wheel or a specially made circular wooden tank track marker can be run through the field, leaving track or tire marks to lure the enemy onto live mines. Antipersonnel mines should not be sown in such a field until the track marks have been laid. Another method is to leave gaps in the mechanically laid field, run vehicles through the gaps, and then close them with hand-laid mines without disturbing the track marks.
- Using decoys to confuse the enemy as to the strength of friendly forces and the unit's identity, or to conceal unit movement by being sited in a position after the real unit has moved. For more information on deception operations, see FM 90-2.

LONG-RANGE SURVEILLANCE OPERATIONS

Desert characteristics affecting LRS operations are: lack of water (a major problem), scarcity of vegetation, extensive sand areas, extreme temperature ranges, brilliant sunlight, and usually excellent observation. Movement using animals, vehicles, or by foot may be considered and is generally restricted to darkness. More training in land or air navigation and terrain orientation procedures may be necessary.

AIR ASSAULT OPERATIONS

An air assault task force provides commanders with truly unique capabilities. They can extend the battlefield, move, and rapidly concentrate combat power like no other forces.

An air assault task force uses the helicopter to move to and close with the enemy. Initial assault elements must be light and mobile. They are often separated from weapon systems, equipment, and materiel that provide protection and survivability on the battlefield. Thus, an air assault task force may be particularly vulnerable in a desert environment to enemy-

- Attack by aircraft and air defense weapon systems during the movement phase due to differences in desert effects on observation and fields of fire.
- Attacks (ground, air, artillery) during the loading and unloading phases and at other times when the infantry is not dug in.
- Small arms fire that presents a significant threat to helicopters.
- Artillery or other fires that may destroy helicopters and air assault forces during PZ (pickup zone) or LZ (landing zone) operations.

Air assault operations are discussed in greater detail in FM 90-4. Marine Corps assault support ensures the rapid buildup of combat power and facilitates the quick maneuver of ground forces. See FMFM 5-35 for more information.

AIRBORNE OPERATIONS

The airborne division is organized to rapidly deploy anywhere in the world. It is the only US Army division with a rapid, strategic, combined arms, forced-entry capability. It will most likely be the initial force deployed for contingency operations. It is ideally suited and primarily designed to seize, secure, and repair airfields in order to provide an airhead for follow-on forces and to delay, disrupt, and reduce enemy forces.

Airborne operations can be adversely affected by various environmental considerations of the desert. High winds, sandstorms, and heavy rainfall or thunderstorms may impact on mission accomplishment. Planning considerations for combat service support are complicated by the fact that airborne forces will be the first Army forces in an immature and austere desert theater. For more information on airborne operations, see FM 90-26.

MARINE AIR GROUND TASK FORCE OPERATIONS

Marine operating forces are organized for combat as Marine air ground task forces (MAGTFs) composed of command, ground combat, aviation combat, and combat service support elements. The MAGTFs are closely integrated combat forces capable of rapid response to any crisis or contingency. Their Naval/Marine Corps expeditionary nature makes them ideal for immature and austere environments as was seen in Operations Desert Shield/Storm (Southwest Asia) and Restore Hope (Somalia).

A variety of types of MAGTFs may be formed in support of national strategy and rapid crisis response. The Marine expeditionary force (MEF) is the Corps' principal organization for combat and peacetime readiness, and is formed from the legislated division and aircraft wing teams. These MEFs provide a reservoir of integrated combined arms combat power that can be task organized to simultaneously execute a wide range of global missions. The MAGTFs are mission tailored and range in size from very powerful MEFs, capable of prosecuting operational campaigns against the most capable potential threat through rapidly deployable and employable Marine expeditionary units (MEUs), to small special purpose forces (SPMAGTFs) formed for specific missions or crises. In the early moments of Operation Desert Shield, a MEF provided the nation a powerful combined arms combat force to stand against aggression, while US forces and equipment and supplies were being assembled in Southwest Asia. For more information on MAGTF operations, see FMFRP 2-12.

MARITIME PRE-POSITIONING FORCE OPERATIONS

The maritime pre-positioning force (MPF) gives the nation an added dimension in mobility, readiness, and global responsiveness. The MPF program involves 13 ships, organized in three squadrons. Maritime pre-positioning squadron one (MPSRON-1) operates in the eastern Atlantic Ocean, MPSRON-2 in the Indian Ocean, and MPSRON-3 in the western Pacific. The MPF, when called upon, provides equipment and 30 (days of supplies for a 16,000-man Marine expeditionary brigade (MEB). The MEB's personnel and selected equipment can be airlifted quickly using roughly 250 airlift sorties to an objective area to join with its equipment at a secure site. Equipment and supplies can also be selectively off-loaded to support smaller MAGTFs. During Operation Desert Shield/Storm, all three MPFs were off-loaded in Southwest Asia providing immediate support to deploying forces. During Restore Hope, one MPF supported operations. For more information on MPF operations, see FMFM 1-5.

ARMY SPECIAL OPERATIONS FORCES

To meet our nation's global commitments, the Army maintains a balanced force of armored, light, special operations, and support forces for use across the operational continuum. Army special operation forces (ARSOF) are an integral part of the total Army force. ARSOF have five elements: special forces, rangers, Army special operations aviation (ARSOA), PSYOP, and civil affairs. These forces offer significant capabilities to the desert theater of operations. Details of ARSOF capabilities are discussed in FM 100-25.

MILITARY OPERATIONS ON URBANIZED TERRAIN (MOUT)

Because there are few man-made features throughout the expanse of the desert, those that do exist can become important, perhaps even key. Key terrain in the desert can be any man-made feature. Settlements (where a logistics base may be established), road junctions, shelters, and airfields, all become important, simply because they are so few in number. Growing villages and settlements straddle these lines of communication, and small villages may exist near water sources and other key terrain. It may be necessary to conduct MOUT operations to control these areas. In areas involved with Desert Storm, paved roads and even dirt roads were considered key terrain for both high-speed movement and for providing clearly defined directions and locations. Commanders must be prepared to fight on terrain that is constantly being modified by man. More information on conducting MOUT can be found in FM 90-10.

CHAPTER 4

COMBAT SERVICE SUPPORT

There is a lack of every kind of resource in the desert, especially of the sophisticated infrastructure of ports and railways for their high capacity for moving combat supplies. Logistical support is always a challenge, and an arid environment burdens all types-supply, aviation, communications, and maintenance. Commanders must be sensitive to the constraints, and those providing support must work to overcome them. A unit's tactical effectiveness in the desert depends to a large degree on the combat service support available. Equally, its vulnerability lies in its exposed lines of communications and the immobility of its bases of supply and support.

Inherent to the success of any tactical operation is continuous, sound, logistical planning for adequate supply, medical, and maintenance support. This is especially important in the desert because the greater distances used in maneuver and deployment complicate supply procedures. Other reasons are the shortage of locally available water and the increased maintenance requirements due to sand and dust damage. The effects of the environment on equipment are severe, requiring increased levels of support to maintain a standard level of efficiency. The extended supply lines required for expanded frontages call for special considerations and procedures to ensure adequate and timely supplies arrive to sustain combat in the desert.

US forces in the desert operate at the end of a long, perhaps tenuous line of communication. Cargo space must not be wasted to provide all the comforts of home. A significant difference in living standards between rear area support personnel and those in forward combat areas must be avoided since this can affect morale and weaken the ability of combat units to resist psychological warfare. Transportation priority must be given to minimum essential materials and the support base should be austere.

Section I. Base Development Plan

US forces deploying for operations in a desert environment should expect to begin operations from a lodgement area. When this is the case, it is necessary for the headquarters deploying the force to prepare a detailed base development plan. How the plan is developed will depend on a number of factors that are described below:

- The mission and size of the force. The size of the force depends on its mission and the operations it is expected to conduct. The size of US forces deployed for desert operations could vary from a small force conducting a show of force, to a joint task force capable of full-scale operations.
- Security of the lodgement area. A lodgement area will probably be secured by allied forces or US Marines before deployment of US Army forces into the operational area. However, it may be necessary to use US Army forces, either air-dropped or air-landed, to secure a lodgement area.

- Transportation of US forces into the lodgement area. It is probable that initial forces will be transported by air and follow-on forces by sea. Another possibility is that initial forces will be transported by sea with follow-on personnel being transported by air.
- Strategic lines of communication (LOC). The initial strategic LOC will probably be an air LOC. However, at some point in the operation a sea LOC will be established to convey the bulk of the supplies, supplemented by an air LOC to haul time-sensitive items.
- Theater lines of communication. Lines of communication within the operations area should be analyzed before selecting the lodgement area. The analysis should include ports available, airfields throughout the operational area, road nets, and railroads. It may be necessary to stage engineer construction units into the operational area early to improve existing facilities and LOCs or to construct new ones. In a single or multi-corps theater, a theater army headquarters provides overall management of CSS operations. It establishes priorities, assigns missions, and allocates resources in accordance with the theater army commander's concept of operations.
- Local resources. These are extremely important as they will affect logistics planning. Typical information about resources in the operational area that should be obtained before base development planning includes-

  -Airfields.

  -Water sources.

  -Fresh rations.

  -Labor supply.

  -Construction materials and available equipment.

  -Material-handling equipment at ports and airfields.

  -Local hospitals, maintenance capability, and storage sites.

  -Local power supply to include types and equipment.

  -Railroad rolling stock and gauge of tracks in local areas.

After consideration of the factors listed above, the lodgement area is selected Ideally, a lodgement area should have a deep water port and airfield suitable for heavy strategic airlift, located at the end of an adequate road or rail system suitable for an intratheater LOC. Once the lodgement area has been selected, then LOC-port units can be specially tailored for early deployment to the operational area.

Section II. Theater Support

THEATER HEADQUARTERS COMMAND

Should a friendly nation ask for military assistance either following an invasion or in anticipation of one, the first combative units to arrive in the theater will probably be a force designed to secure entry points. The assembly and movement of armor and mechanized forces will take time, time that the logistics staffs can put to good use preparing for the reception of these forces. Elements of the theater staff can assess the assets and facilities available in the following areas and make an estimate of the work, labor, equipment, and other resources required to support the buildup of the force.

Logistics Reconnaissance

Port capacity, cranes and off-loading equipment, storage facilities, the availability of stevedores, and local shipping assets at the main port facilities should be determined. Contract and requisition of host nation support assets can be arranged by initial theater staff personnel. The theater staff should also determine airfield facilities and capacities that are available. Railway assets that can be made available for logistics support and the movement of troops should also be determined.

Local Resources

The host nation in a Middle Eastern theater will probably be an oil producer and may be able to provide the bulk of our fuel requirements. However, if the host nation exports its oil in bulk crude, it may have only a limited refinery capacity to meet a local domestic demand for diesel, kerosene, high-octane fuel, and lubricants.

Fresh Rations

The host nation may be able to produce a limited supply of fresh rations but the bulk may have to be imported from neighboring countries. If there is arable land within the host nation, it may be possible to start farms to relieve the burden on the local economy and on that of neighboring friendly states.

Hospitals

The extent to which hospitals in the host nation can accept long-term patients until our own base hospitals can be established is also an issue of concern.

Power Supply

The share of the host nation's electric power supplies which can be offered to our forces must be determined. Initially, there is unlikely to be any shortage in power supply, but as a large force (including other allies) builds up there maybe a generating capacity problem. At the beginning of the campaign the difficulty may be in distributing electricity where it is required throughout the theater. It may be necessary to construct overhead or underground cables together with transformers.

Transient Camps

Sites should be located for transient camps for troops arriving in the theater and convalescent camps for the recovery of the sick and wounded.

Local Currency

Local currency may also be required for extended operations, both to pay troops and to locally purchase supplies.

Maps

Maps are a critical item that maybe more readily available through local survey teams or oil companies.

Interpreters

Interpreters will be required to communicate with the host-nation troops, contractors, and labor forces.

Stockpiles

Higher stock levels are required in the desert due to the following factors: use of the limited life of many perishable items in a harsh environment; the enormous distances stockpiles must be lifted and over inadequate transport systems; the loss of supplies due to sudden changes in the fortunes of war, and the time it takes to replace items from the US. The levels of each commodity to be held in theater and the proportion of the totals will be decided during the staff planning process. The distribution will depend on the tactical situation and the vulnerability of the lines of communication to enemy action.

Expenditure Rates

While the requirement for rations and water remain relatively constant, the expenditure of fuel and ammunition will vary far more, not just because of the fluctuation between quiet periods and intense operations, but because of the environment. The amount of driving in soft sand and the longer distances to be traversed combine to increase consumption beyond central European rates. Similarly, the expenditure of tank and artillery ammunition may be increased because of the open terrain.

Host-Nation Support

In a desert environment where resources, CSS personnel, and equipment are limited, the use of host-nation support assets can be vital to the success of an operation. Host-nation support assists in the accomplishment of missions and functions in support of US forces and enhances their capability to perform their wartime role.

All forms of peacetime transition to wartime, and wartime host-nation support should be included in the planning process.

Host-nation support includes-

- Government agency support such as police, fire companies, and border patrols, may be available to support US forces.
- Contractor support such as supplies and services, including laundry, bath, bakery, transportation, labor, and construction.
- Host-nation civilians may be able to provide needed skills for laborers, stevedores, truck drivers, managers, and technicians.
- Host-nation military units may provide traffic control, convoy escort, installation security, cargo and troop transport, POL storage and distribution, and rear operations.
- Host-nation facilities may be contracted and used for hospitals, headquarters, billets, maintenance shops, or other activities.

- Functional or area support maybe provided in the form of rail operations, convoy scheduling, air traffic control, and harbor pilot services.
- Services may be provided by the host nation for gymnasiums, recreation facilities, and other morale and welfare demands.
- Supplies and equipment needed for missions may be acquired locally, precluding or reducing materiel shipments from CONUS.

Section III. Corps Support Command

An Army corps support command (COSCOM) deploying to support desert operations must be carefully tailored to meet the needs of combat forces operating in a harsh environment. Requirements for long-haul truck companies, engineer construction battalions, water production units, and LOC-port units previously described, must be carefully weighed. A shortfall of these units could significantly impair combat operations. Organization of the COSCOM should be planned based on the factors described in the previous paragraphs, with particular attention given to-

- Number of troops to be supported.
- Quantity and types of equipment to be maintained.
- Tonnage to be handled.
- Available local resources and labor force.
- **Types** of units to be deployed to the theater of operations.

The organization of the COSCOM and a description of its tasks are provided in FM 63-3J. Initial corps forces entering the theater can be supported by a forward support battalion (FSB) of a division support command and a corps support battalion (CSB) of a corps. Once initial forces have arrived in the theater, additional tailored elements from the COSCOM must immediately follow, or even arrive first to minimize the requirement for the DISCOM cooperate such activities as ports or airheads.

Section IV. Division Support Command

As previously mentioned, combat service support units are high-priority targets for any desert enemy. In most cases, Army division support command (DISCOM) units will not be able to provide for their own security, considering the many ways in which they could be attacked. Air defense protection must be provided. It may even be necessary to provide a maneuver unit or additional MP units to secure DISCOM elements. Nearby maneuver units can also be designated to move to their defense-attack helicopters are especially suited for this purpose-and on-call fires should be planned by nearby field artillery units. Any pipelines in the division area must also be secured by any means at hand. Observation helicopters can be used to patrol pipelines.

Stocks should be kept as mobile as possible in the event rapid displacement is necessary. Stockpiling off vehicles must be held to a minimum, as should stockage levels. To the extent practical, supplies located forward of the division support area should be stored aboard vehicles to minimize the Possibility of having to leave them behind. For this purpose, a force operating in the desert should be augmented with additional transportation assets.

DISCOM organizations of the lead divisions in austere and immature theaters may be called upon to establish forward logistic bases. In these situations division assets may have to assume other support or transportation responsibilities temporarily until area support groups can establish support operations and transportation.

Section V. Combat Service Support Element

The Marine Corps' combat service support element (CSSE) is a task organized service support element of the MAGTF. Its composition is based on many factors, to include-

- MAGTF size (MEF, MEB, MEU, or SPMAGTF).
- MAGTF mission.
- Type of operation.
- Area of operation.

The considerations listed under Sections III and IV are also true for a CSSE. For more information on CSSE operations, see FMFM 4-1.

Section VI. Support Operations

Listed below are some of the factors that make support operations complicated.

SUPPLY

Consumption rates must often be developed after the force has operated for some time in the area. Water has to be found, purified, stored, and transported.

MAINTENANCE

There will be a greater demand for such items as filters, oils, and lubricants. More Class IX stores are required than normal, and the work load on maintenance units is much greater. Supply items and spare parts should be packed or wrapped as if to be air and water tight to prevent blowing sand from contaminating or damaging them. All echelons that request supplies and repair parts should be using the same or compatible equipment for requisitioning, with alternative means in place as a backup so there is little or no slow down in the reorder process.

CONCEALMENT

It is difficult to conceal trains areas. However, trains areas must be concealed to the best extent possible. These are soft targets in any environment and are high-priority enemy targets as their destruction (especially water, HETTS, and fuel supplies) effectively cripples the force.

DISTRIBUTION

Maneuver units may be farther apart, both in width and depth, than in temperate environments. They move more frequently and faster. Lines of communication are longer. Terrain away from the main supply mute (MSR) maybe such that it is only trafficked by cross-country vehicles, and then only with reduced payloads. Lack of significant terrain features may increase navigational problems, requiring local guides.

COMBAT SERVICE SUPPORT PLANNING

The commander's intent and METT-T analysis must dictate the CSS plan to support the tactical mission. However, CSS planners must not become locked into rigid CSS plans. The situation will dictate how trains are configured, echeloned, and controlled. Commanders and their staffs must use a logical and fast means to evaluate the battlefield and reach decisions. The military decision-making process described in FM 101-5 provides the framework within which the commander and staff interact to arrive at and execute a decision. Battlefield support must be planned to satisfy requirements during the following operational phases:

- Prior to D-day (before).
- Commitment to battle (during).
- Future mission (after).

All areas of CSS (man, arm, fuel, fix, move, and protect) must be considered during each operational phase to ensure an integrated, responsive plan of support. Support requirements must be projected and plans developed to satisfy these projected requirements. Supporting CSS plans should be as detailed as planning time permits.

CSS commanders and planners must thoroughly understand the tactical mission and plans and the commander's intent. They must know-

- What each of the supported elements will be doing.
- When they will do it.
- How they will do it.
- Where they will do it.
- What the priority of support is.
- Density of personnel and equipment being supported.

After analyzing the concept of the operation, CSS commanders and planners must be able to accurately predict support requirements. They must determine-

- What type of support is required.
- What quantities of support are required.
- The operational commander's priorities, by type and unit.

Using the support requirement of the tactical plan as a base, the support capabilities of the CSS structure are assessed. The staff must determine-

- What CSS resources are available (organic, lateral, and higher headquarters).
- Where the CSS resources are.
- When CSS resources will be available to maneuver units.
- How the FSB will make these resources available.

Based on this information, the staff must then develop support plans that apply resources against requirements in a manner that results in the most responsive support possible. Communications links must be established and maintained.

Orders that clearly describe tasks to be accomplished must be issued. Continuous follow-up must ensure tasks are being accomplished as planned.

CSS functions should be performed as far forward as the tactical situation and available resources will permit. They should be performed at or close to the site where the weapon system is located to lessen evacuation requirements. Support must be continuous, using immediately available assets. This will involve bringing ammunition, fuels, parts, end items, maintenance personnel, and occasionally replacement crews or individuals, to the forward elements such as battalion field trains, combat trains, and equipment downsites. Planning and execution emphasize the concept of providing support to forces in the forward areas.

CSS planners must know priorities for support. This is necessary to ensure that units with the highest tactical priority receive required support first. The commander and his staff provide mission directives, determine CSS requirements, and establish priorities within the unit.

Section VII. Security of Supply Routes

Long lines of communications require convoys from the support base to the combat forces, and convoys are subject to air attacks (as learned during World War II when convoys from Casablanca to Al Guettar Tunisia, were frequently targeted by Luftwaffe raids), Enemy ambushes on main supply routes (MSR) are always a threat in desert operations. Enemy patrols may also place nuisance mines on routes, especially at critical points such as defiles. Actions must be taken to minimize the threat to supply routes.

The MSR will be considerably longer in the desert. The logistical assets utilizing the MSR are extremely vulnerable and must be protected. It maybe necessary to allocate maneuver forces to maintain an open and relatively safe MSR. This will allow supplies to be pushed forward and casualties to be moved to the rear. Coordination must be accomplished between maneuver elements as to where responsibility will end and begin on the MSR during each phase of an operation. The MSR requires constant patrolling to ensure safest operations and continuance of supplies to the maneuver forces. Marking of the MSR facilitates security. Different techniques for marking MSRs range from chemical lights, to spray paint, to signs. MPs can also position themselves along the MSR to help guide units.

Section VIII. Supply

Time and distance factors developed on different terrain by experience are of little value in the desert. The absence of roads in forward areas, navigation problems, vulnerability of trains and supply installations to attack by ground forces or aircraft, sandstorms, and wide dispersion, all require a different appreciation of time for resupply operations.

CLASSES OF SUPPLY

Requirements for supplies vary from that of temperate climates according to the classification of supply. Differences that may be expected in any desert are described in the following paragraphs.

Class I

Until the theater is fully developed and ration requests can be implemented, ensure enough MRE rations for three to five days are stored on combat vehicles. Meals from this combat load are eaten only when daily Class I resupply cannot be accomplished. Frequency of unit feeding and use of A or B rations depends on the tactical situations, If possible, troops should receive at least one hot meal per day. Hot rations should be packed in platoon-size portions rather than consolidating company-size packages.

It is critical to plan for the cooling of water supplies. Troops will drink any potable water available to them; however, they would prefer chilled water. Commanders and staffs must plan for water coding systems, ice, and individual soldier/marine field-expedient devices. Troops fighting in the desert will likely be wearing the battle dress overgarment and body armor. This fact will impact on the planning for water consumption.

The key to having enough water in a task force conducting desert operations is the capacity of that force to store and transport it. Current water trailers are inadequate. The potential for water consumption is high when you consider personal use and consumption, decontamination, medical needs, messing operations, and maintenance uses. Possible solutions include converting fuel tankers to water tankers, the use of blivets, and local purchase of civilian water-holding tanks through host-nation support. Water is vital, yet local supplies may be scarce or nonexistent in a desert combat zone. If water is plentiful, as it is in mess around Tripoli and Benghasi in Libya, water supply should not be a problem, provided that normal water supply procedures are followed.

This paragraph deals with situations where local water is difficult to obtain. All units must maintain a continuous watch for possible sites such as oases, dry wells, dry water courses, open water (even marshes), or captured enemy dumps. These should be reported to the next higher headquarters, giving the location and quantity and flow, if possible. It is not the responsibility of these units to test the water for potability, which could be dangerous for untrained personnel. This task should be left to specialists.

Since distances between water points may be long, it may be desirable to augment the division with additional 5,000- and 2,500-gallon bulk water tankers, processed to haul water.

Priorities for water use should be established. See Appendix G for a suggested list of water priorities and additional information concerning drinking water needs, water requirements, and water heating rates. If vehicle decontamination is necessary, it will take a high priority. NonPotable water should be used for this task.

Class II

There is little change in Class II consumption. However, clothing variations, from tropical clothing to sweaters and sleeping bags, must be anticipated Requirements for items such as neck scarves and canteens will be increased as well as those for hand tools, since tools tend to get lost in the sand.

Class ill

There is a marked increase in oils and lubricants used in preventive maintenance however, the actual quantities depend on operating conditions. Some types of desert terrain can lead to greatly increased fuel consumption per mile moved or hours that equipment is used. Use of cans or fuel bladders in certain circumstances should also be considered as they allow fuel to be spread more evenly among cargo vehicles since a loaded fuel tanker's cross-country capability may be degraded in desert sand. HEMTTS may be a suitable replacement vehicle to solve the cross-country mobility problem.

Antifreeze requirements remain roughly the same as in temperate climates as antifreeze increases the boiling point of coolant and decreases wear on liquid-cooled engines. Various oils and lubricants are required in smaller user containers. This assists in preventing sand from contaminating larger containers since they would have to be moved from site to site, and opened and closed numerous times.

Class IV

The requirement for Class IV stores can be significantly more than in other theaters, and consumption of some items such as sandbags is greatly increased. Maximum use must be made of local materials. An engineer reconnaissance unit should be present in the theater from the initial buildup to establish what resources are available. All possible Class IV should be carried and incorporated into vehicle load plans when deploying. The construction of airstrips, minor port facilities, and rehabilitation of major port facilities and railways, are all engineer missions of particular importance in desert warfare.

Class V

Due to excellent firing conditions, and the need for extensive suppressive fires, ammunition consumption may be high. It may be necessary to restrict firing of certain types of ammunition once they have reached predesignated levels unless command approval is obtained. Battalion task force trains should contain a one-day supply of ammunition and missiles for all vehicles in the task force. Ammunition should be divided between combat and field trains when trains are echeloned. Units should keep ammunition as packaged until it can be uploaded on combat vehicles. This will protect the ammunition from sand that could cause weapon malfunction.

If artificial obstacles are to be employed, considerable quantities of mines will be required as minefield must be long and deep to be effective. Since extensive minefield will be preplanned, relatively few antitank mines need to be held in ammunition supply points forward of the division support area. When required, the quantities needed should be moved as close to minefield locations as possible. Only mines necessary to replenish unit basic loads used for local defense need be stocked forward of the division.

Class VI

The demand for Class VI supplies, especially beverages, is high. They are not, however, essential and if transportation is limited they are given a low priority, especially if refrigeration space is certain to be in short supply. Sundries packs can also be used.

Class VII

The demand for Class VII supplies depends greatly on maneuver and the intensity of the battle. The only variation that can be forecast is for refrigeration equipment, especially if it is necessary to move deceased personnel to the United States for burial.

Class VIII

Class VIII supplies may vary in type, but is unlikely that the overall quantity will vary significantly from that required in temperate climates.

Class IX

There is a large increase in demand for Class IX supplies due to environmental effects on equipment and the extra maintenance effort required. Small items with high usage rates should be held as far forward as team trains and may also be kept on fighting vehicles. Typical high consumption items are-

- Tires for wheel vehicles.
- Water pumps, gaskets, fan belts, water hoses, and clamps.
- All parts for ignition systems.
- Wheel and sprocket nuts, and wedge bolts.
- Spare caps for all liquid containers.
- Speedometers and cables (due to dead-reckoning navigation these are critical items).
- Filter elements.

FORECASTING

Due to extended lines of communication, consumption forecasts are very important in desert operations. Forecasts should be provided once a day and should include.

- A POL forecast for the next 24 hours.
- Status of the unit's basic ammunition loads.
- Equipment losses in the past 24 hours not previously reported.
- Status of reserve water and rations.
- Special supply shortages or maintenance problems not previously reported.

Section IX . Maintenance

In order to return equipment to battle as quickly as possible, repair of disabled equipment must be accomplished as close to the site of damage as possible. Evacuation should be limited whenever possible.

Due to unit dispersion, organizational maintenance personnel and direct-support contact teams will be thinly spread, so vehicle crews must be trained to make as many adjustments and repairs as they can. General guidelines for desert repair are-

- Repair only what is necessary to make the equipment combat effective.
- Recover and then evacuate to the nearest reasonably secure site, followed by on-the-spot repair.

An SOP for recovery and repair must be established either before or immediately upon arrival in the theater. The SOP should include-

- Guidelines for crew-level recovery and expedient repair.
- Recovery by organizational maintenance.

- Recovery by direct-support maintenance.
- Priorities for recovery by vehicle type.
- Limitations on field expedients. For example, the distance or time over which one tank is allowed to tow another tank considering the heat buildup in the transmission in this environment.
- Recovery of classified equipment such as crypto.
- Security and guides for recovery teams.

Section X. Personnel Support

POSTAL SERVICES

Mail is the soldier's/marine's link to family and friends. Inefficient distribution of mail can quickly undermine morale, regardless of the theater. Mail may be particularly affected by longer lines of communications in a desert theater of operations. Mail is important to the soldier/marine in the desert as it assists in defeating the sense of isolation caused by the environment and the necessary dispersion of units, It is especially important in the first few weeks to counter the shock of entering totally new terrain. Transportation of mail should be given a high priority on arrival in the theater of operations.

FINANCE SERVICES

The mission of finance support organizations during conflict is to provide high-priority support to the soldier/marine on an area basis. Mobile pay teams from corps-level finance organizations provide support to brigade-size units. Generally, finance support will not change in the desert environment.

LEGAL SERVICES

Legal service support will be provided to the commander and to troops by personnel of the division staff judge advocate section. This support will be on an as-required basis coordinated through personnel support channels. Legal advice will be available for the following areas:

- International law.
- Operational law.
- Foreign law.
- Status of forces agreements (SOFA).
- Rules of engagement.
- Claims and compensation payments.
- Public affairs.

RELIGIOUS SUPPORT

The chaplain is the staff officer responsible for implementation of the unit religious program. Included in this program are worship opportunities, administration of sacraments, rites and ordinances, pastoral care and counseling, development and mangement of the unit ministry team (UMT), advice to the commander and staff on matters of morals, morale as affected by religion, and ministry in support of combat shock casualty treatment. Many of the above elements may be affected by the religion of the host nation. With many of the deserts being in predominantly Muslim cultures, religious support may be affected and should be a consideration prior to deployment.

Section XI. Health Services

Medical unit requirements for desert operations are essentially the same as for temperate climates. It is essential that each brigade has an environmental sanitation team attached. When planning for medical support the following factors should be considered:

Increased dispersion and large areas over which battles are fought increases vehicle evacuation time. This problem can be further complicated if the enemy does not recognize the protection of the Red Cross, thereby inhibiting air evacuation within the range of enemy air defense weapons. The importance of units having trained combat lifesavers is critical to overcoming this. The reduction of the number of deaths due to slow evacuation time can be directly affected by the combat lifesavers available. One combat lifesaver per combat vehicle is an adequate number.

The comparatively long distances between units may limit the availability of medical aidmen to adequately support combat troops. Reinforcements may be required from the division medical battalion or from supporting corps level medical units. Augmentation should include vehicles as well as personnel.

The incidence of illness from heat injuries and diseases are higher than in temperate climates. Fevers, diarrhea, and vomiting, for example, cause loss of water and salt, which can culminate in heat illnesses. Cold weather injuries can also occur during a desert winter.

The mobility required of maneuver units will be inhibited if movement of any part of these units, including trains, is restricted by having to hold a number of casualties; therefore, the wounded and sick must be evacuated immediately.

In order to properly treat patients, all medical treatment facilities should be provided additional supplies of water. Medical personnel at all levels must assist tactical commanders in preventing or reducing heat casualties within their units.

Divisional medical units should be augmented with extra field ambulances from corps units. In an emergency, empty cargo trucks moving to the rear can be used for medical evacuation.

Evacuation from the battalion combat trains back to the brigade ambulance exchange point (AXP) or clearing station, will be performed by ground or air transportation. METT-T, availability of equipment, and the patient's condition, will be the determining factors on what method of transportation will be utilized.

The effects of nuclear weapons can be expected to be greater in desert terrain. Introduction of nuclear weapons by the enemy will greatly increase casualties and severely strain available medical resources. The same effects can be expected if the enemy introduces chemical weapons against unprepared troops.

Section XII. Naval and Air Force Assistance

NAVAL

During the initial stages of an operation it may be necessary to request logistic support from the US Navy. Ships, with the exception of a few special types, are neither designed nor equipped to give logistic support to ground forces. Limited support may be available if it is adequately coordinated in advance. A cruiser, for example, may have more than 20,000 gallons of water per day available beyond the requirements of the crew. Limited supplies of items such as bread may be available. Limited surgical and medical assistance may also be available.

AIR FORCE

The military air command (MAC) provides tactical airlift in support of the force. Air Force assistance must be coordinated with MAC to deliver personnel, supplies, and equipment forward to brigades and farther forward when necessary. Delivery is made by the most suitable means available, air landing, extraction, or airdrop. MAC also makes aircraft available for rearward movement of wounded persons or prisoners of war.

Section XIII. Other Combat Service Support Issues

MORTUARY AFFAIRS

The mortuary affairs program provides peacetime and wartime support to search, recover, identify, evacuate, and, when required, temporarily inter, disinter and re-inter deceased US military or civilian personnel, and allied and enemy dead. In addition, the program provides support to collect, inventory, store, and process personal effects (PE) of deceased, missing, captured, and medically evacuated US personnel and deceased allied and enemy personnel.

The goal of the mortuary affairs program is to search, recover, identify, and evacuate the remains of US military and certain civilian personnel from the theater of operations as long as feasible using available US Air Force aircraft.

The longer lines of communication required in desert environments may affect the evacuation of the deceased and therefore must be a planning consideration. Transport remains in palletized transfer cases when the tactical and logistical situation permits.

When the situation prohibits immediate evacuation, remains may have to be temporarily interred within the theater. When possible, the use of temporary cemeteries will be confined to echelons above corps. However, emergency war burial (mass burial) sites, as authorized by the theater commander, may be required as far forward as the brigade area. Desert environmental factors should be considered when establishing temporary interment sites, as the desert environment can significantly affect burial sites (strong winds, flash flooding and so forth. Remains will be buried with their personal effects in these burial sites to assist in the identification of remains when they are disinterred.

This concept will permit the use of host nation support (HNS) to dig and fill mass burial sites. Host nation support laborers will not actually handle or process remains or personal effects of US personnel, but will generally provide labor for interment operations. The mortuary affairs company commander has overall technical responsibility for the layout and survey accuracy of the cemetery. During World War II HNS/EPW labor was used successfully by mortuary affairs units.

CAPTURED MATERIEL AND PERSONNEL

In a desert theater of operations, where resources are scarce to begin with, the innovative use of captured materiel can be critical. This materiel can contribute to the retention of momentum by maneuver forces and decreases the need to consume our own supply stocks and to transport them to using units. Obvious sources are captured or overrun enemy fuel supply points, and materiel which may be used for barrier and fortification construction. Food and medical supplies may be used to feed and treat EPWs and civilians. Commanders and staffs must have a workable plan for handling EPWs. These potentially overwhelming requirements include health services, transportation, security forces, and so forth.

CLOTHING EXCHANGE AND BATH SERVICES

Clothing exchange and bath (CEB) services are provided by the supply and service company, when augmented. CEB services are requested through the brigade S4. The request must specify the location of the unit making the request, desired time for service, and range of clothing sizes for unit members. The requesting unit must be prepared to assist troops in setting up the CEB point.

NOTE: These activities may not be possible in the early stages of an operation, or not at all in forward areas. Additional effort will be required to provide these services in the desert.

APPENDIX A

DESERT COUNTRIES OF THE WORLD

This appendix provides brief descriptions of notable deserts of the world. Detailed information can be obtained from a terrain analysis from the Defense Intelligence Agency and the relevant volume of the DA Pamphlet 550-series. See Fi gure A-1 for desert locations of the world.

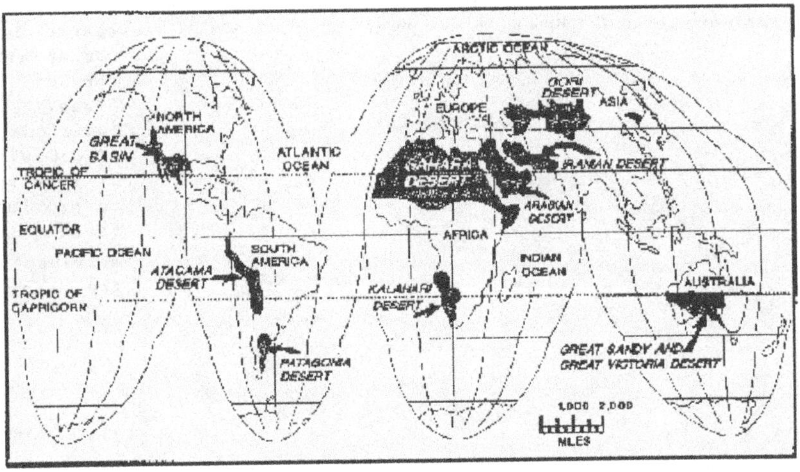

Figure A-1. Desens of the world

NOTABLE DESERTS OF THE WORLD

ARABIAN (EASTERN)	70,000 square miles	In Egypt between the Nile River and Red Sea, extending southward into the Sudan
ATACAMA	600 miles long	In northern Chile
CHIHUAHUA	140,000 square miles	In New Mexico, Texas, Arizona and Mexico

COLORADO	7,500 square miles	South of the Mojave desert, along the Colorado River in southern California
DEATH VALLEY	3,300 square miles	In eastern California and southwest Nevada
GIBSON	120,000 square miles	In the interior of western Australia
GOBI	500,000 square miles	In Mongolia and China
GREAT SANDY	150,000 square miles	In western Australia
GREAT VICTORIA	150,000 square miles	In western and southern Australia
KALAHARI	225,000 square miles	In southern Africa
KARA-KUM	120,000 square miles	In Turkmenia (former Soviet Union)
KAVIR	400 miles long	In central Iran
KYZLKUM	100,000 square miles	Kazakhstan and Uzbekistan (former Soviet Union)
LUT	20,000 square miles	In eastern Iran
MOJAVE	15,000 square miles	In southern California
NAFUD	40,000 square miles	Near Jawf in Saudi Arabia
NAMIB	800 miles long	Along the southwest coast of Africa
NUBIAN	100,000 square miles	In the Sahara in northeast Sudan
PAINTED DESERT	150 miles long In northern Arizona	In northern Arizona
SAHARA	3,500,000 square miles	In Northern Africa extending westward to the Atlantic (largest desert in the world)
SIMPSON	40,000 square miles	In central Australia
SONORAN	70,000 square miles	In southwest Arizona and southeast California
SYRIAN	100,000 square miles	In northern Saudi Arabia eastern Jordan, southern Syria,and western Iraq

Table 1-**1.** Desert counties of the world.

COUNTRY	AREA SQ MILES	POPULATION (MILLIONS)	LANGUAGE(S)	ADJACENT COUNTRIES
Algeria	919,487	24.2	Arabic French Berber	Morocco Western Sahara Mauritania Mali Niger Libya Tunisia
Bahrain	255	.480	Arabic English Farsi Urdu	Saudi Arabia Qatar
Egypt	386,660	53.3	Arabic English French Arabic	Libya Sudan Israel
Iran	636,293	51.9	Farsi Turkish Kurdish Arabic English	Iraq Turkey Former Soviet Union Afghanistan Pakistan
Iraq	167,924	17.6	Arabic Kurdish Assyrian Armenian	Syria Jordan Saudi Arabia Kuwait Iran Turkey
Israel	8,019	4.3	Hebrew Arabic English French	Egypt Jordan Syria Lebanon
Jordan	35,475	2.9	Arabic English	Israel Syria Iraq Saudi Arabia
Kuwait	6,880	1.9	Arabic English	Iraq Saudi Arabia
Lebanon	4,015	2.7	Arabic French Armenian English	Syria Israel

Table 1-1- Desert countries of the world (cont).

COUNTRY	AREA SQ MILES	POPULATION (MILLIONS)	LANGUAGES	ADJACENT COUNTRIES
Libya	679,359	4.0	Arabic Italian English	Egypt Sudan Chad Niger Algeria Tunisia
Oman	82,031	1.3	Arabic English Baluchi Urdu	UAE Saudi Arabia Yemen
Qatar	4,247	.328	Arabic English	UAE Saudi Arabia Bahrain
Saudi Arabia	829,996	15.5	Arabic	Bahrain Iraq Jordan Yemen Oman UAE Qatar Kuwait
Syria	71,498	11.6	Arabic Kurdish	Lebanon Israel Turkey Iraq Jordan
Tunisia	63,170	7.7	Arabic French	Algeria Libya
United Arab Emirates	32,278	2.0	Arabic English Farsi Hindi Urdu	Qatar Saudi Arabia Oman

AUSTRALIA

DESERTS: GIBSON, GREAT SANDY, GREAT VICTORIA, SIMPSON

AUSTRALIA - GENERAL INFORMATION

Area	2,967,895 sq mi
Population Growth	1.17 %
Population Density	5/sq mi
Capital City	CANBERRA

AUSTRALIA - MAJOR CITIES

	Population	Latitude, Longitude
CANBERRA	274,000	35.18S, 149.08E
Sydney	3,441,000	33.53S, 151.10E
Melbourne	2,949,000	37.45S, 144.58E
Brisbane	1,169,000	27.30S, 153.00E
Perth	1,001,000	31.57S, 115.52E
Adelaide	987,000	34.56S, 138.36E
Newcastle	423,000	17.20S, 133.21E
Wollongong	238,000	17.13s, 137.57E
Gold Coast	208,000	27.59S, 153.22E
Hobart	180,000	42.54S, 147.18E
Geelong	150,000	38.10S, 144.26E

AUSTRALIA - LANGUAGES, ETHNIC GROUPS AND RELIGIONS

Languages	English	
	Aborigine	
Ethnic Groups	White	95%
	Asian	4%
	Aborigine and Other	1%
Religions	Anglican	26%
	Catholic	26%
	Other Christian	24%
	Other	24%

SOUTH AMERICA

CHILE

DESERTS: ATACAMA

CHILE - GENERAL INFORMATION

Area	756,945 sq km
Population Growth	1.49 %
Population Density	17 /sq km

CHILE - MAJOR CITIES

	Population	Latitude, Longitude
SANTIAGO	4,320,000	33.30s, 70.40W
Vina del Mar	320,000	33.02S, 71.35W
Valparaiso	273,000	33.05s, 71.40W
Talcahuano	225,000	36.40S, 73.10W
Conception	281,000	36.50S, 73.03W
Antofagasta	203,000	23.40S, 70.23W
Temuco	175,000	38.45S, 72.40W
Rancagua	157,000	34.10s, 70.45W
Talca	145,000	35.28S, 71.40W
Chillan	130,000	36.37S, 72.10W
Arica	158,000	18.30S, 70.20W

CHILE - LANGUAGES, ETHNIC GROUPS AND RELIGIONS

Languages	Spanish	
Ethnic Groups	European and Mestizo	95 %
	Native American	3%
	Other	2%
Religions	catholic	89%
	Protestant and Other	11 %

ASIA

CHINA

DESERTS: GOBI

CHINA - GENERAL INFORMATION

Area	9,600,000 Sq km
Population Growth	1.24 %
Population Density	113 /sq km
Capital City	BEIJING

CHINA - MAJOR CITIES

	Population	Latitude, Longitude
BEIJING	6,000,000	39.55N, 116 26E
Shanghai	7,000,000	31.06N, 121 .22E
Tianjin	5,400,000	39.08N, 117 .12E
Shenyang	4,200,000	41.50N, 123 .26E
Wuhan	3,400,000	36.45N, 114 .15E
Guangzhou	3,300,000	23.08N, 113 .20E
Chongqing	2,800,000	29.30N, 106 .35E
Harbin	2,630,000	45.45N, 126 .41E
Chengdu	2,580,000	30.37N, 104 .06E
Nanjing	2,250,000	32.03N, 118.47E

CHINA - LANGUAGES, ETHNIC GROUPS AND RELIGIONS

Languages	Mandarin	
	Cantonese	
	Shanghainese	
	Fuzhou	
	Minnan and Other	
Ethnic Groups	Han Chinese	**94%**
	Other	**6%**
Religions	Atheist and Eclectic	97 %
	Other	3%

MONGOLIA

DESERTS: GOBI

MONGOLIA - GENERAL INFORMATION

Area	1,564,619 sq km
Population Growth	2.76 %
Population Density	1 /sq km
Capital City	ULAN-BATOR

MONGOLIA - MAJOR CITIES

	Population	Latitude, Longitude
ULAN-BATOR	511,000	47.54N, 106.52E
Darhan	75,000	46.29N, 109.24E
Erdene	46,000	46.18N, 100.35E
Choybalsan	23,000	48.02N, 114.32E
Nalayh	15,000	47.40N, 107.12E
Ulangom	15,000	49.59N, 92.00E
Uliastay	14,000	47.42N, 96.52E

MONGOLIA - LANGUAGES, ETHNIC GROUPS AND RELIGIONS

Languages	Khalkha Mongol	
	Turkish	
	Russian	
	Chinese	
Ethnic Groups	Mongol	90%
	Kazakh	4%
	Chinese	2%
	Russian	2%
	Other	
Religions	Tibetan Buddhist	94%
	Muslim	4%
	Other	1%

PAKISTAN

DESERTS: THAR

PAKISTAN - GENERAL INFORMATION

Area	803,943 sq km
Population Growth	2.70 %
Population Density	134 /sq km
Capital City	ISLAMABAD

PAKISTAN - MAJOR CITIES

	Population	Latitude, Longitude
ISLAMABAD	270,000	33.40N, 73.08E
Karachi	5,400,000	24.51N, 67.02E
Lahore	3,132,000	31.34N, 74.22E
Faisalabad	888,000	31.25N, 73.09E
Rawalpindi	810,000	33.40N, 73.08E
Hyderabad	770,000	25.23N, 68.24E
Multan	750,000	30.10N, 71.36E
Gujranwala	700,000	32.06N, 74.11E
Peshawar	600,000	34.01N, 71.40E
Sialkot	310,000	32.29N, 74.35E
Sargodha	300,000	32.0IN, 72.40E

PAKISTAN - LANGUAGES, ETHNIC GROUPS AND RELIGIONS

Languages	Urdu	
	English	
	Punjabi	
	Sindhi	
	Pashtu and Other	
Ethnic Groups	Punjabi	66%
	Sindhi	13 %
	Pashtun	9%
	Baluchi and Other	12%
Religions	Sunni Muslim	77 %
	Shi'a Muslim	20 %
	Other	3%

AFGHANISTAN

DESERTS: KAR KUM

AFGHANISTAN - GENERAL INFORMATION

Area	249,999 sq mi
Population Growth	2.35 %
Population Density	58 /sq mi
Capital City	KABUL

AFGHANISTAN - MAJOR CITIES

	Population	Latitude, Longitude
KABUL	1,127,000	34.31N, 69.12E
Qandahar	195,000	31.35N, 65.45E
Herat	156,000	34.20N, 62.12E
Mazar-i-Sharif	115,000	36.42N, 67.06E
Jalalabad	111,000	34.26N, 70.28E
Qonduz	110,000	36,45N, 68.51E
Baghlan	109,000	36.11N, 68.44E
Maymana	40,000	35.53N, 64.38E
Pul-i-Khomri	35,000	35.55N, 68.45E
Ghazni	32,000	33.33N, 68.26E

AFGHANISTAN - LANGUAGES, ETHNIC GROUPS AND RELIGIONS

Languages	Pashtu	
	Dari	
	Turkish Languages	
	Baluchi	
	Pashai	
Ethnic Groups	Pashtun	50 %
	Tajik	25 %
	Uzbek	9%
	Hazara	9%
	Other	7 %
Religions	Sunni Muslim	74%
	Shi'a Muslim	15 %
	Other	11 %

SOUTHWEST ASIA

EGYPT

DESERTS: SAHARA

EGYPT - GENERAL INFORMATION

Area	1,001,449 sq km
Population Growth	2.65 %
Population Density	53 /sq km
Capital City	CAIRO

EGYPT - MAJOR CITIES

	Population	Latitude, Longitude
CAIRO	11,000,000	30.03N, 31.1

Alexandria	2,905,000	31.13N, 29.55E
Giza	1,640,000	30.01N,31.10E
Shoubra El-Kheima	497,000	30.06N, 31.15E
El Mahalla El Koubra	355,000	30.59N, 31.10E
Tanta	344,000	30.48N, 31.00E
Port Said	364,000	31.17N, 32.18E
El-Mansoura	260,000	31,03N, 31.23E
Asyut	257,000	27.14N, 31.07E
Zagazig	256,000	30.35N, 31.30E
Suez	274,000	29.59N, 32.33E

EGYPT - LANGUAGES, ETHNIC GROUPS AND RELIGIONS

Languages	Arabic	
	English	
	French	
Ethnic Groups	Eastern Hamitic	90%
	Other	10%
Religions	Sunni Muslim	94%
	Other	6%

IRAN

DESERTS: KAVIR, LUT

IRAN - GENERAL INFORMATION

Area	1,648,000 sq km
Population Growth	3.23 %
Population Density	32 /sq km
Capital City	TEHRAN

IRAN - MAJOR CITIES

	Population	Latitude, Longitude
TEHRAN	6,028,000	**35.40N, 51.2 6E**
Mashad	1,120,000	**36.41N, 52.39E**
Isfahan	930,000	**32.41N, 51.41E**
Tabriz	855,000	**38.05N, 46.18E**
Shiraz	801,000	**29.38N, 52.34E**
Bakhtaran	535,000	**31.28N, 54.54E**
Karaj	527,000	**35.48N, 50.58E**
Ahwaz	471,000	**31.17N, 48.43E**
Qom	425,000	**34.39N, 50.57E**
Abadan	329,000	**30.20N, 48.15E**

IRAN - LANGUAGES, ETHNIC GROUPS AND RELIGIONS

Languages	Farsi
	Turkish

	Kurdish	
	Arabic	
	English	
Ethnic Groups	Persian	**63 %**
	Turkish	**18%**
	Kurdish	**3%**
	Arab and Other	16%
Religions	Shi'a Muslim	**93%**
	Sunni Muslim	**5%**
	Bahai and Other	**2%**

IRAQ

DESERTS: ARABIAN

IRAQ - GENERAL INFORMATION

Area	434,924 sq km
Population Growth	3.53 %
Population Density	40 /sq km
Capital City	BAGHDAD

IRAQ - MAJOR CITIES

	Population	**Latitude, Longitude**
BAGHDAD	3,250,000	**33,20N, 44.26E**
Basra	1,600,000	3030N,47.50E
Mosul	1,250,000	**36,21N, 43.08E**
Kirkuk	535,000	**35.28N, 44.26E**
An Najaf	141,000	**31.59N, 44.19E**
Erbil	99,000	**36.12N, 44.01E**

IRAQ - LANGUAGES, ETHNIC GROUPS AND RELIGIONS

Languages	Arabic	
	Kurdish	
	Assyrian	
	Armenian	
Ethnic Groups	Arab	**75 %**
	Kurdish	**17 %**
	Other	**8%**
Religions	Shi'a Muslim	**62%**
	Sunni Muslim	**35 %**
	Christian and Other	*3%*

LIBYA

DESERTS: SAHARA

LIBYA - GENERAL INFORMATION

Area	1,759,540 sq km
Population Growth	3.08 %
Population Density	2 /sq km
Capital City	TRIPOLI

LIBYA - MAJOR CITIES

	Population	Latitude, Longitude	
TRIPOLI	610,000	32.49N,	13.07E
Misurata	320,000	32.07N,	20.04E
Benghazi	103,000	32.23N,	15.06E
Azzawiya	94,000	32.45N,	12.44E
Al-Beida	77,000	32.49N,	12.45E
Agedabia	51,000	30.48N,	20.15E
Dama	48,000	32.40N,	22.35E
Sebha	41,000	27.09N,	14.29E
Tubruq	35,000	32.05N,	20.30E
Al-Marj	30,000	32.25N,	20.30E
Zeleiten	26,000	32.28N,	14.34E

LIBYA - LANGUAGES, ETHNIC GROUPS AND RELIGIONS

Languages	Arabic	
	Italian	
	English	
Ethnic Groups	Arab Berber	97%
	Other	3%
Religions	Sunni Muslim	97%
	Other	3%

SYRIA

DESERTS: SYRIAN

SYRIA - GENERAL INFORMATION

Area	185,180 sq km
Population Growth	3.74 %
Population Density	62 /sq km
Capital City	DAMASCUS

SYRIA - MAJOR CITIES

	Population	Latitude, Longitude	
DAMASCUS	1,250,000	**33.30N,**	**36.19E**
Aleppo	1,200,000	**36.14N,**	**37.10E**

Helms	430,000	34.44N, 37.17E
Latakaia	240,000	35.31N, 35.47E
Hama	200,000	35.05N, 36.40E
Deir ez-Zor	106,000	35.20N, 40.05E
Rakka	90,000	35.57N, 39.03E
Hasakeh	74,000	36.32N, 40.44E
Tartous	53,000	34.55N, 35.52E
Edleb	52,000	35.56N, 37.21E
Dera'a	50,000	32.37N, 36.06E

SYRIA - LANGUAGES, ETHNIC GROUPS AND RELIGIONS

Languages	Arabic	
	Kurdish	
	Armenian	
	French	
	English	
Ethnic Groups	Arab	90%
	Kurdish and Other	10%
Religions	Sunni Muslim	74 %
	Other Muslim	16 %
	Christian	10%

JORDAN

DESERTS: SYRIAN

JORDAN - GENERAL INFORMATION

Area	91,880 sq km
Population Growth	3.62 %
Population Density	31/sq km
Capital City	AMMAN

JORDAN - MAJOR CITIES

	Population	**Latitude, Longitude**
AMMAN	813,000	**31.57N, 35.56E**
Zarqa	277,000	**32.04N, 36.06E**
Irbid	141,000	**32.33N, 35.51E**
Ajhm	47,000	**32.20N, 35.45E**
Jarash	32,000	32.17N,35.45E
Madaba	26,000	31.44N, 35.48E

JORDAN - LANGUAGES, ETHNIC GROUPS AND RELIGIONS

Languages	Arabic	
	English	
Ethnic Groups	Arab	**98%**
	Circassian	**1%**

	Armenian	1%
Religions	Sunni Muslim	95%
	Christian	5%

DESERTS: ARABIAN

SAUDI ARABIA - GENERAL INFORMATION

Area	2,149,690 sq km
Population Growth	4.16 %
Population Density	7 /sq km
Capital City	RIYADH

SAUDI ARABIA - MAJOR CITIES

	Population	Latitude, Longitude
RIYADH	1,976,000	24.39N, 46.46E
Jeddah	1,084,000	21.30N, 39.10E
Mecca	810,000	21.26N, 39.49E
Taif	300,000	21.15N, 40.21E
Medina	250,000	24.30N, 39.35E
Dammam	175,000	26.25N, 50.06E
Hufuf	120,000	25.20N, 49.34E
Tabouk	80,000	28,22N, 36.32E
Buraidah	75,000	26.20N, 43.59E
Al-Mobarraz	60,000	25.26N, 49.37E
Kharmis-Mushait	54,000	18.19N, 42.45E

SAUDI ARABIA - LANGUAGES, ETHNIC GROUPS AND RELIGIONS

Languages	Arabic	
Ethnic Groups	Arab	90%
	Afro-Asian	10%
Religions	Muslim	99%
	Other	1%

SOUTH AFRICA

BOTSWANA

DESERTS: KALAHARI

BOTSWANA - GENERAL INFORMATION

Area	**231,804 sq mi**
Population 1988	1,190,000
Population Growth	3.49 %
Population Density	5 /sq mi
Capital City	GABORONE

BOTSWANA - MAJOR CITIES

	Population	Latitude, Longitude
GABORONE	95,000	24.45S, 25.55E
Francistown	34,000	21.07S, 27.32E
Selebi-Phikwe	33,000	21.58S, 27.48E
Serowe	25,000	22.25S, 26.44E
Mahalapye	22,000	23.05S, 26.51E
Molepolole	21,000	24.25S, 25.30E
Kanye	20,000	24.59S, 25.19E
Lobatse	20,000	25.11S, 25.40E
Mochudi	19,000	24.28S, 26.05E
Mauri	15,000	20.00S, 23.25E
Ramotswa	14,000	24.56S, 25.50E

BOTSWANA - LANGUAGES, ETHNIC GROUPS AND RELIGIONS

Languages	English	
	Setswana	
Ethnic Groups	Batswana	95 %
	Bushmen	4%
	European	1%
Religions	Indigenous Beliefs	50 %
	Christian	50 %

NAMIBIA

DESERTS: KALAHARA

NAMIBIA - GENERAL INFORMATION

Area	318,261 sq mi
Population Growth	5.30 %
Population Density	4 /sq mi
Capital City	WINDHOEK

NAMIBIA - MAJOR CITIES

	Population	Latitude, Longitude
WINDHOEK	100,000	**22.34S, 17.06E**
Tsumeb	13,000	**19.13S, 17.42E**
Keetmanshoop	11,000	**26.36S, 18.08E**
Otjiwarongo	9,000	**20.29S, 16.36E**
Luderitz	8,000	**26.38S, 15.10E**
Swakopmund	7,000	22.40S, 14.34E

NAMIBIA - LANGUAGES, ETHNIC GROUPS AND RELIGIONS

Languages	Afrikaans
	German
	English

	African Languages	
Ethnic Groups	Black	**86%**
	White	**7%**
	Mixed	**7%**
Religions	Christian	**60%**
	Indigenous Beliefs	**40%**

SOUTH AFRICA

DESERTS: KALAHARI

SOUTH AFRICA - GENERAL INFORMATION

Area	471,443 sq mi
Population Growth	2.20 %
Population Density	74 /sq mi
Capital City	PRETORIA

SOUTH AFRICA - MAJOR CITIES

	Population	**Latitude, Longitude**
PRETORIA	800,000	**25.45S, 28 12E**
Cape Town	860,000	**33 56S, 18 2 8E**
Durban	810,000	**29 53S, 31 00E**
Johannesburg	770,000	**26 10S, 28 02E**
Port Elizabeth	300,000	**33 58S, 25 3 6E**
Roodepoort	210,000	**26 10S, 27 5 3E**
Umhlazi	180,000	**29 53S, 31 00E**
Germiston	140,000	**26 15S, 28 10E**
Pietermaritzburg	130,000	**29 36S, 30 24E**
Boksburg	120,000	**26 13S, 28 15E**
Bloemfontein	110,000	29.07S,26.14E

SOUTH AFRICA LANGUAGES, ETHNIC GROUPS AND RELIGIONS

Language	Afrikaans	
	English	
	Zulu	
	Xhosa	
	Sotho and Other	
Ethnic Groups	African	**70%**
	White	18%
	Mixed	9%
	East Indian	3%
Religions	Christian	**60%**
	Hindu	2%
	Muslim and Other	**38%**

APPENDIX B

EMPLOYMENT OF AIRCRAFT IN DESERT OPERATIONS

The desert is probably the most severe of all environments in which aviation units must operate. Standard operating procedures for desert operations are different from areas having an abundance of contrasting terrain and substantial vegetation. This appendix describes some special considerations when employing aircraft in desert operations.

AIR OPERATIONS

Air combat operations that rely on heavy vegetation and varying contour terrain need to be flexible enough to incorporate different methods of camouflage and terrain flight techniques. The varying types of sand have a tremendous effect on operations-vast flat areas afford unlimited visibility, dunes are hard to distinguish at night, blowing sand impairs visibility and presents flight and maintenance problems, and surface composition affects the choice of landing zones, maintenance sites, FARPs, and operating bases. Additionally, low-hovering and taxiing aircraft generate blowing sand and dust that can cause aircrews to lose outside visual reference, and, if performed near other equipment, present additional maintenance problems for that equipment.

Air operations are not the only area affected by the desert environment. Aviation ground operations may require flexibility and modification to work around the heat and the effects of the terrain. Aircrews must resort to instrument flight during duststorms. Sand also causes excessive wearing, pitting, and eroding of aircraft components.

In certain areas, the desert, with its relatively level terrain and shallow compartments, contains few highly distinguishable terrain features to mask aviation forces. Formations of two or more aircraft can be seen 10 kilometers away because the dark airframes contrast against the desert sand. Aviation units normally deploy their aircraft along routes and may need to consider widely dispersed formations. Aviation forces can make maximum use of deception techniques during periods of limited visibility.

Air cavalry assets can conduct reconnaissance and security operations over great distances in the desert because of the lack of vegetation and relief. Even when they are sand painted, armor vehicles stand out starkly against the sand. When combined with traditional target acquisition principles, such as dust signature and movement, these factors make it easier to acquire and engage armored and mechanized forces well out of range of their main guns.

Aeroscouts flying nap of the earth (NOE) cannot necessarily find the enemy more easily than ground observers. Stationary targets are the most difficult to see as there is little to draw the observer's attention:

Therefore, aeroscouts must use caution to avoid blundering into enemy air defense weapons. The aircraft should land at a distance of 5-10 kilometers from the area of interest, and the observer should dismount and scan the area for suspected enemy. The observer must remain in contact with the pilot by using a portable radio. The process should be repeated at varying intervals until contact is made.

Attack helicopter battalions (ATKHBs) are a potent force in desert warfare. If they are employed quickly and violently, maximum results can be obtained both in offensive and defensive operations. They are best used where a quick concentration of combat power is needed. A desert environment presents excellent target acquisition and engagement possibilities. Attack assets must remain dispersed to provide security. Mission planning that incorporates flexibility is a key ingredient in the successful employment of ATKHBs.

Terrain flying and desert navigation require continuous concentration. Due to lack of terrain and poor reference points, the aviator may rely on dead reckoning, self-contained navigation equipment, and radio navigational aids. As light decreases, the ability to judge distances accurately is degraded and visual illusions become more common. Because of glare, haze, and frequently blowing sand, it may be difficult to detect changes in terrain and the horizon.

Attack helicopters should move from assembly areas to battle positions (or holding areas if necessary) over attack routes that will provide whatever cover and concealment and prominent terrain features necessary to assist in navigation and to decrease the possibility of detection. Attack helicopters may have multiple routes for ingress and egress. Route reconnaissance, premission planning, and prebriefs will maximize the benefits to route planning in desert operations.

The weather in desert regions can be extremely unpredictable. Sandstorms, accompanied by constantly fluctuating wind speeds, may reduce visibility from in excess of 50 kilometers to zero in less than five minutes. Pilots must be carefully briefed on prevailing weather conditions before takeoff. Warning of any expected variations in conditions must be transmitted immediately to all airborne aircraft.

THE PERFORMANCE OF HELICOPTERS IN HEAT

Aviation personnel must refer to the appropriate aircraft technical manuals to determine aircraft limitations and capabilities in the desert environment. Significant effects on the payload capabilities should be anticipated.

Commanders must develop realistic aircraft utilization procedures based on the environmental effects data provided by aviation staff personnel to obtain the fullest benefit from aviation assets.

Helicopters hovering close to the ground can cause the engine to ingest sand; can cause observation by the enemy due to the formation of dust clouds; or cause disorientation of the pilot due to blowing sand, particularly at night. Helicopters should not be moved under their own power while on the ground, but should be pushed or towed by men or vehicles. Maintenance should be restricted to the minimum time, and should take place on rock or on oiled or wet sand, if available. All apertures (Pitot tubes, for example) of aircraft should be covered when not in use (including helicopter windscreens).

Temperature and humidity have a direct impact on personnel and vehicle performance. Temperature and humidity affect air density. Air density decreases as temperature increases. High temperature and humidity reduce the efficiency of aircraft propulsion and aircraft lift capabilities. Although temperature and humidity may not have a direct effect on a particular operation, extremely high temperatures and humidity will reduce aircraft payloads.

APPENDIX C

TECHNIQUES FOR OPERATING EQUIPMENT IN THE DESERT

The effects of the desert environment on equipment were described in Chapter 1. This appendix describes techniques which, if used while operating equipment in the desert, can save both equipment and lives.

DRIVING

Drivers and track commanders should observe the guidelines in the following paragraphs while operating vehicles in desert areas.

Dusty Conditions

Wear goggles while driving open-hatched regardless of visibility. Clear-lens goggles should be worn at night unless night-vision goggles are used. Bandanas or surgical masks should be worn to avoid breathing heavy dust.

Vehicles in an extended convoy should maintain a dust distance of twice the normal interval, or as specified in the unit SOP to allow time for the dust to dissipate. When driving on extremely dusty roads or trails and if traffic conditions permit, a staggered column formation can be used with vehicles alternately driving on the left and right side of the road.

If the vehicles should become engulfed in dust, the convoy commander should consider adjusting the rate of march accordingly. Any commander of a vehicle engulfed in dust should alert the convoy commander by radio, move to the right side of the road, and stop or slow to allow the dust to dissipate. Extreme caution must be observed to ensure oncoming and following vehicles are not jeopardized. The lead vehicle must warn vehicles to return to column formation if encountering traffic.

Sandy Deserts

Sandy deserts may be relatively flat or interspersed with windblown dunes. When driving in sand, the following techniques should be applied:

- The best time to drive on sand is at night or early morning when the sand is damp and traction is better. However, this is not always the case especially with the newer type military tires with closer tread design. Damp sand packs between the tread in the grooves of these tires resulting in virtually no surface traction.

- Drivers of track vehicles must be wary of a lack of steering response, which indicates sand is building up between the rear sprockets and treads. If the buildup of sand is allowed to continue, it will force the tread off. "Shaking" the vehicle with the steering or backing up will remove the sand.
- Wheel vehicles may gain some traction by reducing the air pressure in the tires. However, prolonged driving on partially deflated bias ply tires will overheat the tires and break down the sidewalls. Vehicles equipped with radial tires or central tire inflation system (CTIS) are not affected by the lower tire pressure if the maximum speed listed in the operator's manual is not exceeded.
- Vehicle loads must be evenly distributed. Rear-wheel drive should be used where possible to prevent the front wheels from digging into the sand and becoming mired.
- Drivers must switch to all-wheel drive or change gears before a vehicle bogs down in the sand.
- Before entering the sand drivers should select a gear that will allow the vehicle to maintain as much torque as possible without causing the wheels to spin and to minimize changing gears.
- Large-wheeled vehicles, such as 5,000-gallon tankers, should have a designated "puller". The designated vehicle should be preconfigured to assist these vehicles when they become bogged down in loose sand.

Some sand areas will be covered by a surface crust. This is caused by chemicals in the ground cementing sand particles together, In many cases it will be possible to drive on top of this crust and minimize dust signature and the chance of bogging down. Consider the following techniques when driving on a crust:

- Use staggered columns to facilitate movement. As a general rule vehicles should not follow one behind the other.
- Ensure vehicles maintain a minimum speed (determined from experience) below which they will break through the crust.
- Avoid sharp turns and abrupt starts or stops that could cause a vehicle to break through the crust.
- Reconnoiter patches of the crust that are a different shade to ensure they are not softer than the surrounding crust.

Crossing Dunes

Crossing dunes requires careful reconnaissance. Normally, the upwind side of the dune will be covered with a crust and have a fairly gradual slope. The downwind side will be steeper and have no crust. Prior to crossing a dune, the driver should climb it on foot checking crust thickness, the angle at the crest to ensure the vehicle will not become bellied at the top, and the degree of slope and softness of the downwind side. If he is satisfied his vehicle can climb the dune, he should drive the vehicle straight up it at the best speed, crest it, and maintain a controlled descent on the other side.

Little hills may be formed by the wind blowing sand around small shrubs. Wheel vehicles should not attempt to move through areas where this has occurred without engineer assistance.

Cacti or thorn bushes will cause frequent tire punctures. Increase the number of tires carried in the unit's PLL when operating in areas covered with this type of vegetation.

Rocky Areas

Rock and boulder-strewn areas, including lava beds, may extend for many miles. Desert rocks, eroded and sharp-edged, vary in size and are so numerous that it is almost impossible to avoid any but the largest. The subsequent harsh jolting fatigues individuals and causes extreme wear on tracks, wheels, springs, and shock absorbers. Rocks and stones can become lodged between the tires on vehicles equipped with dusk that can cause severe damage to tires and brake components. Vehicles can follow one another in this type of terrain and it may be feasible to reconnoiter and mark a route. Drivers should achieve a "rolling" effect as they cross large rocks by braking as the vehicle's wheels ride over a rock so the axle settles relatively gently on the far side.

Salt Marshes

Salt marshes (called sebkha) are normally impassable, the worst type being those with a dry crust of silt on top. Marsh mud used on desert sand will, however, produce an excellent temporary road. Many desert areas have salt marshes either in the center of a drainage basin or near the seacoast. Old trails or paths may cross the marsh, which are visible during the dry season but not in the wet season. In the wet season standing water indicates trails due to the crust being too hard or too thick for the water to penetrate. However, such routes should not be tried by load-carrying vehicles without prior reconnaissance and marking.

RECOVERY

Track vehicle recovery methods are the same in the desert as in temperate climates. The techniques described in the following paragraphs will assist wheel vehicle recovery operations in sand crusts or salt marshes. To assist in recovery, wheel vehicles should carry the following items:

- Steel or aluminum channels, at least for the driving wheels. These are pierced to reduce weight and ribbed for strength. Pierced steel planking (PSP) or galvanized iron maybe used as a substitute.
- Sand mats made of canvas, preferably with lateral strips of metal to give strength and increase the traction of the wheels.
- Jacks and jack blocks.
- Tow rope(s).
- Shovels.

Once a vehicle becomes mired, excavate the ground under the vehicle in a gradual slope towards the direction of recovery to a point where no part of the underside is touching the ground. Channels or spurs and mats are laid under or against the wheels facing the direction of recovery. Tire pressure may be reduced to increase traction, but this also lowers the vehicle. It maybe necessary to lift the wheels with a jack if the vehicle is resting on its frame or axles.

When the vehicle begins to move, any faltering will cause it to sink again. Once out, the driver must maintain speed until the vehicle has reached the nearest hard area. At this point the tires are reinflated, the vehicle inspected for damage, and recovery equipment collected.

Vehicles equipped with winches can winch themselves out using ground anchors. The ground anchor may consist of a tarpaulin full of sand placed in a hole and the winch cable attached to it, or it may be one, or preferably two spare wheels well dug in.

A rubberized fabric balloon may be used on light vehicles to lift them free of broken crust. The balloon is placed under the vehicle and blown up with the vehicle exhaust.

If a lone vehicle breaks or bogs down in the desert, the crew must stay with it. A vehicle is much easier to find than a lone man.

MAINTENANCE

Equipment directly affected by heat, such as aircraft and radios, are equally affected by all deserts. However, power trains and suspension systems are affected in proportion to trafficability and soil texture. Most damage to equipment can be avoided by careful driving and by careful observation by vehicle commanders.

Track tension must be correct as constant driving on rocky plateau deserts will reduce the life of the track. Suspension units will require frequent replacement of torsion bars and suspension arms. To prevent damage to internal parts of the idler and suspension arms caused by the terrain, direct-support maintenance units must be provided with equipment capable of tapping and removing the sheared bolts.

To prevent problems that can result when desert vegetation clogs engine oil coolers and cylinder cooling pins, place a small-mesh wire screen over the top grille doors. It may still be necessary to remove packs about every 10 days to clean the engine cooling fins. The wire screening should be periodically checked, removed, and cleaned.

Maintenance personnel must inspect and adjust transmission bands frequently, especially on vehicles operating in hot, barren mountains. This will help reduce transmission oveheating.
Extra stocks of air-cooled generators are necessary because high-ambient temperatures limit their ability to maintain the proper operating temperature and contribute to premature failure.

NOTE: The M113A1 is especially susceptible to overheating problems in desert conditions. This includes the transmission and the solid-state voltage regulator, which is more prone to overheating

WHEELED VEHICLES

Wheeled vehicles are subject to brake system component failures and power steering leaks on rocky deserts. Vehicles equipped with manual transmissions are prone to clutch failure caused by drivers slipping the clutch. Vehicles with automatic transmissions tend to overheat; therefore, stop frequently to allow the transmission time to cool. The M54 5-ton truck is prone to air hydraulic cylinder failure and power steering seal leaks on rocky deserts. All vehicles of the 1/4- to 5-ton range are prone to clutch failure caused by drivers "riding" the clutch pedal. Tire consumption is very high, so all vehicles must carry one, or preferably two spare tires, and the unit's PLL of tires should be considerably increased. Approximately one vehicle in every three to four should carry slave cables to provide for battery failure.

Vehicles should be equipped with the following:

- Extra fan belts.
- Two spare tires.

Extra oil.

Extra radiator hoses.

Heavy duty tape.

Extra air and fuel filters.

- Jack stand support plate.
- Sand ladders (fabricated) and matting.
- Tow rope/cable.
- Extra water cans.
- Siphoning hose and funnel.
- Slave cables.

and early failure than the older mechanical type.

RADIOS

Radios, regardless of type, must be kept cool and clean. They must be in the shade whenever possible and should be located in a ventilated area (or even in an air-conditioned can). If water is available, wrap the radio in a damp towel, ensuring that the air vents are not blocked. Additional radios should be available in vital communications centers, such as tactical operations centers, to allow immediate replacement if the set in use shows signs of overheating.

It is essential that antennas be cut or adjusted to the length of the operating frequency. Directional antennas must be faced exactly in the required direction; approximate azimuth produced by guesswork will not do. A basic whip antenna relies on the capacitor effect between itself and the ground for efficient propagation. The electrical ground may be very poor, and the antenna performance alone may be degraded by as much as one-third if the surface soil lacks moisture, which is normally the case in the desert. If a ground-mounted antenna is not fitted with a counterpoise, the ground around it should be dampened using any fluid available. Vehicle-mounted antennas are more efficient if the mass of the vehicle is forward of the antennas and is oriented towards the distant station.

Desert operations require dispersion, yet the environment is likely to degrade the transmission range of radios, particularly VHF (FM) fitted with secure equipment. This degradation is most likely to recur in the hottest part of the day, approximately 1200 to 1700 hours.

If stations start to lose contact, especially if the hotter part of the day is approaching, alternative communication plans must be ready. Alternatives include the following:

- Using relay stations, including an airborne relay station (the aircraft must remain at least 4,000 meters behind the line of contact). Ground relay stations or RETRANS are also useful and should be planned in conjunction with the scheme of maneuver.
- Deploying any unemployed vehicle with a radio as a relay between stations.
- Using alternative radio links such as VHF multichannel telephones at higher levels, or HF (SSB) voice.
- Using wire. Normally wire will not be used as operations will be fluid, but it maybe of some value in some static defensive situations.
- Using a unit such as all or part of the task force scout platoon for messenger service, Although it is

undesirable to use such a unit in this manner, it may be necessary to maintain communications.

GENERAL TIPS

General tips for operating equipment in the desert include the following:

- Check track tension daily.
- Check drive belt adjustment frequently.
- Lubricate suspension items daily, and clean grease fittings.
- Reduce sand ingestion by stretching nylon stockings over air cleaners.
- Emphasize proper engine cooldown and shutdown procedures, especially diesels.
- Adjust battery specific gravity to the environment (refer to TMs).
- Set voltage regulators at lower end of specifications.
- Start up vehicles regularly to prevent discharge of batteries.
- Increase stocks of oils and lubricants.
- Use high-grade 20W-50 oil; it serves well under desert conditions.
- Compensate for increased pressure due to severe heat in closed pressurized systems.
- Check lubrication orders and TMs for the correct viscosity of lubricants for higher temperatures.
- Keep lubrication to the absolute minimum on exposed or semiexposed moving parts; this includes working parts of weapons.
- Erect screens against blowing sand in maintenance areas.
- Cover the gap between the fuel nozzle and the fuel tank filler neck opening during refueling operations.
- Protect exposed electrical cables and wires with electrical tape.
- Keep optics covered; clean them with a soft paintbrush or a low-pressure air system (this works well for weapons also).
- Clean sand and dirt from hulls of armored vehicles.
- Check tire pressures and fuel levels at the midpoint of the temperature range during the day.
- Ground all refueling equipment-STATIC ELECTRICITY KILLS.
- Replenish radiators with potable water whenever possible.
- Determine battery shortages early and requisition early.
- Drain fuel lines at night and in the morning due to condensation.
- Increase PLLs for the following parts due to high failure rates:

  - Tires.

  - All track components.

  -All suspension components for both wheel and track vehicles.

  - Brake shoes.

  - Bearings and bushings.

-Plastic and rubber parts, including seals.

- All filters.

- Generator components.

- Deploy with plastic bags to cover weapons and protect other equipment during maintenance or when not in use.
- Bring muzzle plugs.
- Prepare all vehicles for desert operations in accordance with the appropriate TMs.
- Issue small paintbrushes to all soldiers/marines for weapons cleaning and other equipment maintenance. The paintbrush is one of the more valuable tools available to the soldier/marine for maintenance.

APPENDIX D

EFFECTS OF THE ENVIRONMENT ON NBC WEAPONS

The primary purpose of using NBC weapons and smoke is to produce casualties, destroy or disable equipment, and generally disrupt operations. NBC weapons and smoke are employed in coordination with conventional weapons. FM 3-3, FM 3-4, FM 3-5, and FM 3-50 describe NBC defense and smoke operations. This appendix addresses the effects of a desert environment on NBC weapons.

WEATHER AND TERRAIN

Both desert weather and terrain affect the behavior of NBC and smoke weapons, and to some extent influence their tactical employment. The following paragraphs describe the effects of the weather and the terrain on the behavior of NBC weapons.

Weather

The effectiveness of NBC and smoke weapons is directly proportional to air stability. Air stability is a result of temperature variations at different levels of the air. The span of desert day and night temperatures causes extremes of air stability. At night and early morning the desert air is very stable. This may be the best time for NBC and smoke employment because of extensive downwind drift and area coverage. Desert air is very unstable during the late morning through afternoon. This may be the worst time for NBC and smoke employment because of quick and irregular dissipation. This may lead to ineffective target area coverage and possible danger to friendly troops. Temperature differences are determined by comparing the air temperature differences every 3.5 meters above the ground. Three types of temperature differences influence NBC and smoke operations:

- Unstable (lapse). Unstable conditions exist when air temperature decreases with altitude. In the desert, this mostly occurs between late morning and early evening.
- Neutral. Neutral conditions exist when air temperature does not change with altitude, In the desert, this mostly occurs during early morning and early evening.
- Stable (inversion). Stable (inversion) weather conditions exist when the air temperature increases with altitude. In the desert, this mostly occurs between late evening and early morning.

High desert temperatures in the middle of the day result in a decrease in air density, so nuclear blast waves move faster. Some chemical agents that come in contact with the skin are more effective during high temperatures due to perspiration. Desert sunlight is lethal to the effectiveness of most biological agents, but most are unaffected at night.

High winds are common in certain desert seasons and affect the dissemination of biological, chemical, smoke, and radioactive clouds. High winds may break up clouds in an odd fashion and force them in the wrong direction and possibly endanger friendly troops.

Terrain

Flat desert terrain lacking vegetation and under stable air conditions moves NBC and smoke effects evenly and steadily in all directions. Nuclear induced radiation is greater in sandy soil due to the high silica content. Desert soil below the surface crust is a fine powder, so the blast and suction effects of a nuclear burst will cause considerable dust clouds.

NUCLEAR WEAPONS IN THE DESERT

Many potential threat nations maintain nuclear weapons and additional nations continue to obtain nuclear capability. Nuclear weapons can be delivered by missiles or bombs and can be exploded in the air, on the ground, or below the ground. Depending upon the radiation dosage received, lethal effects could be felt immediately or delayed for days. Detonated nuclear weapons release energy that affects military troops and equipment in three forms: blast, nuclear radiation, and thermal radiation (heat and light). These effects are discussed in the following paragraphs.

Blast

Nuclear blasts, even from an air burst, raise considerable amounts of desert sand and dirt, which inhibits observation and maneuver for a long time. Lighter desert air density causes a drop in static overpressure, but a more rapidly expanding blast (shock) front. This increases the danger to aircraft and helicopters flying in the area. The radius of damage is normally smaller in desert climates so dug-in personnel and equipment are safer. Desert trafficability is degraded in the immediate area of the strike, especially for wheeled vehicles, due to the destruction of the sand crust.

Nuclear Radiation

Immediate nuclear radiation is a function of weapon yield and changes little with desert temperature. Residual radiation is high in the case of low air burst or ground burst weapons. Residual radiation in the desert is affected by the time of day and the wind. At night, with no wind, residual radiation may be evenly distributed around the point of burst. At night, with a steady wind, residual radiation may drift downwind for many miles. In the late afternoon, residual radiation may drift in a totally irregular pattern and direction due to desert air instability. Induced radiation is prevalent in desert sand due to the amount of silica in the soil. Constant radiation monitoring and reconnaissance are vital to protect troops, especially from contaminated water sources.

Thermal Radiation

The effective range of thermal radiation (heat and light) increases in the desert where there is little terrain masking. This increases the danger of troops receiving severe burns. Blinding light (dazzle), out to 50 kilometers, must be considered in the desert when determining the radius of warning for friendly troops.

BIOLOGICAL WEAPONS IN THE DESERT

The United States has renounced the use of biological agents, but many potential threat nations maintain biological weapons. Biological agents are living microorganisms that multiply inside the human body and cause disease. Biological agents may be disseminated as a liquid or vapor using rockets, bombs, or aerosol generators.

Depending upon the agent, they can produce lethal effects in 1-24 days from the time of exposure. High desert temperatures (120 degrees Fahrenheit and higher) and sunlight may destroy most biological agents. Cooler nighttime temperatures and the lack of sunlight provide biological agents a chance to enter the body. High desert winds will disperse biological agents more rapidly than low winds. Normally, the most effective wind speeds for effective target coverage are from 8-18 knots (14-32 kmph). Stable desert air conditions (night or early morning) provide the greatest agent concentration and area coverage. Unstable desert air conditions (late morning and afternoon) promote atmospheric mixing and lower agent concentration, reducing effective target coverage.

Biological weapons are best suited for strategic rather than tactical use in the desert. Test water and food sources frequently to ensure they are not contaminated. Sanitation, personal hygiene, and immunizations must be instituted to ensure individual protection.

The United States also renounces the use of toxins but some potential threat countries continue to develop toxins for possible military use. Toxins are extracted from natural biological sources and are not chemically produced. They can be disseminated as a liquid, vapor, or powder, and delivered by aerosol generators, artillery, rockets, or bombs. Depending upon the toxin, they can produce lethal effects in one minute to 12 hours from the time of exposure. In the desert, toxins can effectively be used as strategic or tactical weapons.

CHEMICAL WEAPONS IN THE DESERT

The United States has reserved the right to retaliate in kind with chemical agents against enemies who use them frost. There are six major types of chemical agents: nerve, blood, blister, choking, incapacitating, and tear. They are classified as persistent and nonpersistent. Chemical agents may be delivered by mines, rockets, artillery, or bombs. Depending upon the agent type, lethalities can occur between minutes to an hour from the time of attack.

High daytime temperatures of the desert increase incapacitating effects of liquid agents, which rely on skin penetration. Desert air instability (late morning and early afternoon) will usually cause quick, vertical and irregular dissipation of chemical agents. This reduces the lethal concentration and target area coverage. Desert air stability (early morning and evening) will most likely cause adequate agent dissipation and target area coverage. High desert temperatures also cause liquid agents to evaporate more quickly and decrease the time of hazard. If a liquid chemical agent soaks into the desert sand, it will increase the time of hazard. Figures D 1 and D-2 show the evaporation times of chemical agents (HD, GA, GF, VX) on sand at 104 and 68 degrees Fahrenheit for negligible risk levels. Negligible risk levels cause mild incapacitation among no more than 5 percent of unprotected soldiers/marines crossing a contaminated area.

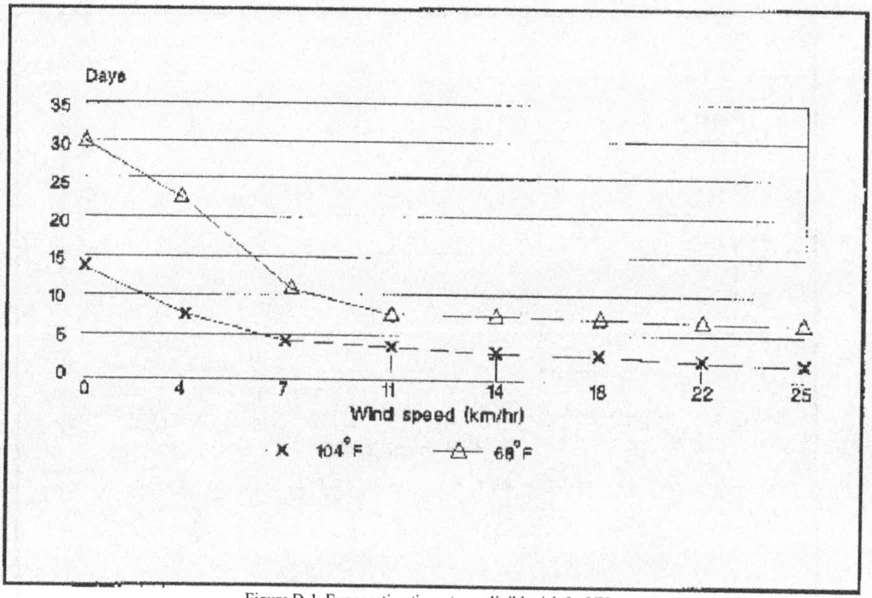

Figure D-1. Evaporation times to negligible risk for VX agents on sand.

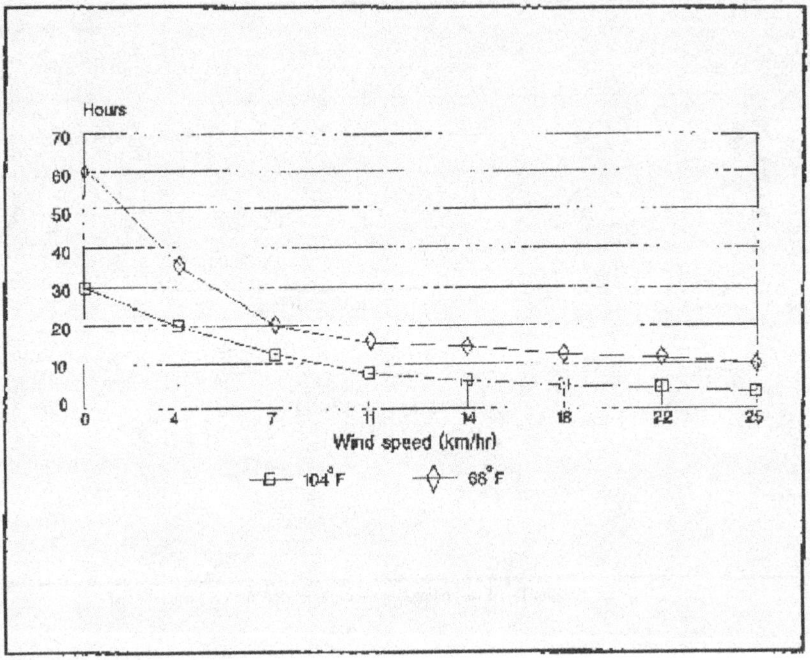

lot GF, HD, and GA agents.

Strong desert winds also increase the evaporation rate of liquid agents. They cause chemical clouds to drift downwind in an irregular direction and concentration. This tends to disperse target area coverage and lethal concentrations, and may endanger friendly troops.

SMOKE SYSTEMS IN THE DESERT

The lack of cover and concealment in flat desert terrain with little vegetation makes the use of smoke more vital to survival.

Desert winds, temperature, humidity, and terrain all affect smoke cloud behavior. The weather condition with the greatest impact on smoke operations is wind. Low desert winds allow smoke to remain on target areas for a longer period of time than high winds, In general, if the wind speed is less than 5 knots (9 kmph) or greater than 10 knots (18 kmph), smoke may not provide good target area coverage.

Temperature

Desert temperature differences have a direct relationship with making effective smoke. Three types of temperature differences influence smoke:

- Unstable (lapse). Lapse desert temperatures tend to break up and dissipate smoke, but are best for producing smoke curtains.
- Neutral. Neutral desert temperatures have limited vertical air currents and are good, but not the best, for producing smoke hazes and blankets.
- Stable. Stable (inversion) desert temperatures have no vertical air currents, and are the best for producing smoke hazes and blankets.

Humidity

Practically all smoke particles absorb moisture from the air. The lack of moisture in the desert air decreases the particle size and density of smoke making it less effective. Desert smoke streamers are shorter and less dense than smoke streamers in humid weather conditions. This increases the time and distance required to build adequate target area coverage.

TACTICS

The effects of NBC weapons and smoke on tactics in desert operations are discussed in the following paragraphs.

Nuclear

Nuclear weapons have a major bearing on tactics because of their ability to contaminate and shape desert terrain by making it impassable. Nuclear weapons can destroy troop and equipment concentrations and command and control centers and are usually considered weapons of mass destruction and combat multipliers; therefore, a smaller force with nuclear weapons may well defeat a much larger force without nuclear weapons.

The high maneuverability of tracked vehicles in the desert produces an endless number of avenues of approach. This creates difficulties for defense operations in the desert. Nuclear explosions could be used to shape desert terrain and canalize an enemy. This would provide profitable targets for other weapon systems. Equally, an attacking force could employ nuclear weapons in the desert to cut off enemy reinforcements, isolating them and making them vulnerable.

Biological Agents/Toxins

High desert temperatures and sunlight prevent the effective use of most biological agents during the daytime. An exception is spore-forming (anthrax) biological agents. During the day, troops crossing or occupying desert terrain face little danger from long-term biological contamination (except from spore-forming agents). But, because of favorable night desert conditions, an aerosol-delivered attack or a liquid biological attack would be effective. During night desert conditions, biological agents could effectively cover hundreds of square kilometers with a small amount of agent. This type of attack could be conducted covertly. These agents could contaminate vast areas of terrain and create mass casualties.

Toxins in the desert have nearly the same military use as chemical agents and are employed in the same manner. However, toxins are more deadly and require lower concentrations for lethalities. Toxins are difficult to detect with today's standard detection equipment.

Chemical Agents

The versatility of chemical agents gives commanders flexibility in desert operations. Commanders must consider how the employment of chemical agents affect offensive and defensive desert operations. Chemical agents can be used to create casualties, degrade performance, slow maneuver, restrict terrain, and disrupt logistical support. An assaulting force could use chemical agents to breach a defense or widen a gap. The best agent to use in this case would be a nonpersistent agent (one that is fast acting and leaves the target area quickly). High explosives could be mixed with the chemical attack to conceal the use of chemicals and complement their effects.

The threat of a chemical attack forces the use of protective masks and clothing. Heat, fatigue, and stress seriously affect the performance of troops. This is especially true with high desert temperatures. Well-trained soldiers/marines tolerate wearing protective gear better than those who are not as well trained. Troops in protective gear fire weapons less accurately, move more slowly, and must rest more often.

The actual or anticipated use of chemical weapons slows down a force and forces troops to take precautions. Desert heat, fatigue, and stress caused by wearing protective equipment slows down unit movement. Chemical agents can be used to create contaminated obstacles to desert maneuver. The best agent to use in this case would be a persistent chemical agent (one that remains on the target area for a time).

Chemical agents could be used to support offensive desert operations. Chemical agents could be used to protect flanks along an axis of advance to slow enemy counterattacks and to slow enemy fire and movement. In defensive desert operations, chemical agents could be used against second echelon forces to separate, slow down, and isolate them.

Chemical agents could be used to supplement conventional obstacles, or they could be used alone to restrict desert terrain. They may slow maneuver and channel attackers into engagement areas. A commander could contaminate a narrow desert mountain pass or bridge with a chemical agent and force the enemy to use an alternate route.

Logistical centers are lucrative targets for desert chemical attacks. Contaminating logistical supplies and equipment reduces the mobility of reinforcements and slows the delivery of supplies and equipment.

Smoke

Smoke is a combat multiplier that enhances the commander's ability to concentrate combat power at the critical time and place. Smoke is a far more significant battlefield factor than ever on flat desert terrain with little cover and concealment. Smoke can defeat enemy binoculars, weapon sites, and laser range finders in the desert. Smoke placed on the enemy at night interferes with enemy operations and observation by defeating enemy night sights and infrared sights.

In the desert offense, smoke can be used to deny the enemy information about the size, composition, and location of friendly maneuver forces. A smoke screen can be placed either to the front or to the flank. When the enemy cannot be screened effectively, obscuring smoke may be required. We must keep the enemy in doubt about the attacking unit's strength, position, activities, and movement. The longer the enemy is in doubt during an operation, the greater the chances are for mission success. Also, smoke can conceal desert breaching operations and river or gap crossings. It can also be used in deception operations.

In the desert defense, smoke is used to deny the enemy information about the size, location, and composition of friendly defensive positions. We must deny the enemy information by concealing the preparation and location of battle positions, artillery units, and reserves. Smoke can be used to support desert defensive positions by slowing enemy maneuver, disrupting command and control, isolating attacking echelons, silhouetting targets, and concealing obstacles.

PROTECTION

The commander of troops in desert operations must choose between the two following options when there is the threat of NBC warfare:

- Troops remain unprotected, with a high-chemical casualty risk, but a lower chance of heat fatigue.
- Troops are fully or partially protected, with a low-chemical casualty risk, but a higher chance of heat fatigue.

A decision on the level of protection is made according to the circumstances. If partial protection is ordered, the pace of physical work will be slower and proficiency reduced. The bulk of strenuous physical activity must be done at night or during the coolest part of the desert day.

In the desert, heat casualties (5 percent minimal) can be expected to occur in 30 minutes while performing heavy work in 90-degree-Fahrenheit temperatures when dressed in MOPP 4. Work/rest periods must be utilized to reduce the chances of heat fatigue. When protective clothing is worn, at least 10 degrees should be added to the WBGT index. Because of higher body temperatures, soldiers/marines in MOPP equipment perspire more than usual. Water must be consumed (2 quarts per hour) during continuous moderate work periods (and in MOPP equipment) when temperatures reach 80 degrees Fahrenheit and above to replace lost fluids or dehydration will follow.

DECONTAMINATION

The main problem of decontamination in the desert is lack of water, Although decontamination takes place as far forward as possible, the lack of water may burden the logistical system. Weathering may be a viable option for chemical contamination. The persistency of nerve agents (GA, GF, VX, GD) and mustard (HD) on chemical agent resistant coating (CARC) painted vehicles is between 4-24 hours at 104 degrees Fahrenheit. Sea water may be used as a substitute for fresh water during normal decontamination operations, but all equipment must eventually be flushed with fresh water to prevent corrosion.

Desert sand can be used for chemical decontamination, but it increases soldier/marine fatigue during its application. Sand removes most, but not all, liquid contamination, and saves valuable water supplies; however, the absorption capacity of desert sand is exhausted in 30-60 seconds after application. Remove the sand by sweeping or brushing the contaminated surface. Chemical agent detection should be conducted to ensure the agent is adequately removed.

APPENDIX E

DESERT CONCEALMENT AND CAMOUFLAGE

In the desert, camouflage problems are encountered that require imagination, ingenuity, and intelligence. The lack of natural overhead cover, the increased range of vision, and the bright tones of the desert terrain place emphasis upon siting, dispersion discipline, and the skillful employment of decoys to achieve deception and surprise. Total concealment is rarely achieved, yet proper camouflage measures can reduce the effectiveness of enemy observation, and consequently enemy operations.

Cover from enemy direct fire may be afforded by dunes, hills, and other irregularities in the desert terrain. Camouflage is an essential part of all operations in the desert and the importance of the concept must be impressed upon fresh units and individual replacements upon their arrival in theater. Poor camouflage may also compromise a high-level plan and lead to an operational failure. One poorly concealed vehicle can compromise an entire task force. Improvisation of available assets is just as important as being able to properly use camouflage systems. As previously described, deserts generally do not offer much natural concealment or means for camouflage; therefore, make maximum use of any artificial means available.

VEHICLES AND AIRCRAFT

Movement of vehicles produces dust, diesel plumes, and distinctive track marks. The slower the speed, the less dust that is produced; however, the need for speed must be balanced against the amount of dust that may be produced. Drivers must avoid harsh use of accelerators, the main cause of diesel plumes.

Shine from optical instruments (which should be kept shaded), and matte paint that has been polished by continual wear, or from tracks, particularly if rubber blocks have been removed, are difficult to camouflage during the desert day. See Figure E-1 for shading optics. Running gear on tracks that has been polished by wear should be covered with burlap when stationary. Windscreens and windows should be removed or lowered to prevent reflection of the sun and heat. Vehicle silhouettes can be reduced in the forward areas by removing cabs and tops.

Figure E-1. shade optics to prevent shine

Disruptive pattern painting for vehicles and aircraft is described in FM 20-3. Local materials can also be used. The color and texture of the local terrain is best represented by placing dirt on vehicles and using a little water to make it stick.

The effects are increased by covering a vehicle with a wide-mesh net and using foliage brackets to attach local vegetation. Twine or wire may be used as an alternative to the mesh net, provided vegetation is available.

Some or all of the equipment listed in the following paragraphs should be available for every vehicle and aircraft, although aircraft will not necessarily be able to carry all of it.

The preferred net is the lightweight camouflage screen system (LCSS), desert version, which provides concealment against visual, near IR, and radar target acquisition/surveillance sensor devices. Additionally, the transparent version of the LCSS allows US units to camouflage radars (less CW type radars) without degrading operations. A desert camouflage net should be a complete cover, as it depends on its limitation of the ground surface, and both color and texture, for its effect. The alternatives to the LCSS in order of priority include the following:

- The specially produced desert-pattern net of the lightweight screen system.
- An open-weave cloth (colored as appropriate to the soil or "patched) stitched to an ordinary wide-mesh net and used with the string uppermost. This provides both color and texture and can be suitably garnished with radar-scattering plastic, such as that used in the lightweight screening system, and with any local vegetation.
- A cover of close-weave cloth, colored as appropriate.
- A standard net garnished solid, threaded in long straight strips that have been colored to harmonize with the terrain. The garnishing must be maintained.

The number of nets issued depends on the size of the equipment to be covered, but should be sufficient to allow a gradual slope of not more than 15 degrees from the top of the equipment to the earth. Each company-size unit should be equipped with a spray gun and various tints of paint to provide for temporary variations in net color to match the terrain.

When using nets for stationary equipment-

- Do not allow nets to touch sensitive items such as helicopter rotor heads and radio antennas which may cause a net to catch fire.
- Do not pull nets so tight that each supporting pole stands out.
- Ensure the net does not prevent the equipment from fulfilling its primary task. In some equipment, such as helicopters, a net must be easily removable to reduce reaction time.
- Avoid straight-edged patterns on the ground, which indicate something is there.
- Use burlap spray-painted in a nondescript desert color to cover all reflecting surfaces (excluding fire control optics) and shadow-producing areas under vehicle bodies, including tank suspensions. Aircraft equipped with windscreen covers will not require it.
- Cut desert scrub in the immediate area.
- Use poles, natural or man-made, to raise the nets from the equipment, thereby hiding its shape. They must be brought into the area of operations by the force and are extremely difficult to replace in the desert if lost or damaged.
- Make a "mushroom" out of thin iron tubing locally, It resembles an open umbrella without its cloth cover and with the end of the spokes joined together. Slotted into a socket that has been welded onto the top of a tank, self-propelled gun, or personnel carrier, it lifts the net above the vehicle, concealing its shape, increasing air circulation, and permitting the crew or team to use the top hatches.
- Hook and hold a camouflage net to the ground away from the vehicle by using wooden pegs or long steel pins, depending on soil consistency.
- Use mallets to drive pegs and pins into the ground.

After dismounting local security, camouflage is the first priority when a vehicle halts. Actions to be taken are-

- Site in vegetation or shadow, if available.
- Cover shiny surfaces and shadow areas with burlap screens.
- Drape the net.
- Add any available vegetation to the net.
- Blot out vehicle tracks for 50 meters behind vehicles.

Stationary aircraft take a relatively long time to conceal as they are fragile in comparison with other equipment, have a considerable heat signature, and must also be readily accessible for maintenance. The more they are concealed, the greater their response time is likely to be. Tactical flying is discussed in Appendix B, but take the following actions in sequence when approaching a landing site where aircraft will stay for some time:

- Ensure aircraft approach the site terrain-masked from enemy surveillance.

- Close down aircraft as soon as possible.
- Cover all reflective surfaces.
- Move aircraft into shadow if it can be towed or pushed.
- Shift the main rotor (depending on the type) until it is at a 45-degree angle with the fuselage and drape a net over the rotor and fuselage. The rotor must be picketed to the ground.
- Conceal the remainder of the aircraft.

POSITION SELECTION

Position selection is critical at all levels. One of the fundamentals of camouflage in any environment, but particularly the desert, is to fit into the existing ground pattern with a minimum of change to the terrain. A wadi bottom with vegetation or a pile of boulders that can be improved with grey burlap and chickenwire are good examples. Sites chosen must not be so obvious that they are virtually automatic targets for enemy suppressive frees, and antennas must be screened against the enemy, if possible.

Shadows, particularly in the morning and evening, identify objects; so equipment must be placed in total shadow (rarely found), or with its maximum vertical area facing the sun so that minimum shadow falls on the ground ("maximum vertical area" is the rear of a 5-ton truck with canopy, but the front of an M88, for example). See Figure E-2 for the effects of shadows. The shadow can be broken up, which is normally achieved by siting equipment next to scrub or broken surfaces, such as rocks. Equipment should not be sited broadside to the sun, and it is usually necessary to move as the sun moves. Digging in reduces the length of any shadow that is cast (on the principle that the lower the object, the shorter the shadow).

Figure E-2.Camoutlags: the effects of shadows.

Vehicles passing over pebbles or heavy ground surfaces press the pebbles or gravel into the soil, causing track marks to be prominent when viewed from the air. Avoid such areas if possible. Use existing trails and blend new trails into old ones whenever possible.

Soil texture suitable for digging must be a consideration when reconnoitering for battle positions. Holes must be covered to avoid shadows being cast. If vehicles will be in position for more than a day, trenches should be dug for them.

In forward areas, tactical operations centers are probably the most difficult positions to hide although their need for concealment is great. They require strict camouflage discipline. Vehicles and aircraft should not be allowed to approach closer than 300-400 meters. They must be dispersed and concealed so nets may have to be readily available for aircraft. Pay special attention to lights and noise at night.

Generators will have to be dug-in and allowed adequate air space for cooling. Radios and antenna systems must be remoted as far out as possible, and in different directions. Whenever possible, dig in the entire command post. Use engineer assets to build a berm around the perimeter and to help break up the silhouette and to enhance security. Other equipment should not be placed too close to minimize the possibility of the enemy's attention being attracted to the site.

Engineer activity often precedes operations, which makes it important that such work be concealed from enemy surveillance. The following guidelines should be used to conceal engineer activity:

- Employ the minimum number of equipment and personnel.
- Keep equipment well away from the site, and dispersed and concealed if not in use.
- Complete all possible preparations well away from the site.
- Follow the ground pattern, if possible.

Combat service and support assets must rely on concealment for most of their protection. The following guidelines will assist unit commanders in concealing trains while stationary or on the move:

- All vehicles of a given type should look alike. This will make it difficult for an enemy to pick out critical vehicles, such as water and fuel trucks, in a column. Canopies over fuel trucks disguise them and prevent radiant heat from striking the fuel containers.
- Vehicles should follow the tracks of the preceding vehicle if it is possible to do so without breaking through the crust, as this reduces the possibility of an enemy intelligence analyst to calculate how many vehicles have passed.
- Screen exhaust systems to reduce heat signature.
- Vehicles must never form a pattern, either when stationary or on the move.

SUPPLY POINTS

A supply point is likely to be in a location where its main threat of detection will be either by the eye or by photograph. Normally, greater emphasis can be placed on selecting supply positions from the point of view of concealment rather than for tactical efficiency, particularly in situations where air defense cover may be limited. The following guidelines should be used when setting up supply points:

- The location should be selected where trails already exist. Vehicles must use existing trails to the extent possible.

- Stocks should be irregularly spaced, both in length and depth, to the maximum extent possible so that there is no definite pattern.
- Stocks should be piled as low as possible and preferably dug-in. For example, a stack of gasoline cans should be only one can high.
- The shape of the area should not be square or rectangular, but should follow the local ground pattern.
- Stocks should be covered with sand, gravel, burlap, netting, or anything else that harmonizes with the local terrain, and the sides should be gradually sloped with soil filled to the top of the dump.
- The contents of each supply point should be mixed so that the destruction of one supply point will not cause an immediate shortage of a particular commodity.

APPENDIX F

OPERATIONS IN MOUNTAINS

Mountain operations are described in detail in FM 90-6. This appendix describes special conditions associated with operating in mountains such as those in the southern Sinai and on the shores of the Red Sea. It does not address tactics and techniques for mountain operations that are equally applicable to all mountains, except for the purpose of clarity.

TERRAIN

Mountains are high and rugged, with very steep slopes. Valleys running into a range become more and more narrow with the sides becoming gradually steeper. Valleys are usually the only routes that allow ground movement of men and equipment at any speed or in any quantity. Water is nonexistent on hilltops and unusual in valleys except during flash floods after rains. Lateral ground communications are limited unless the force is moving across the spines of mountain ranges. Navigation may be difficult, as maps are likely to be inaccurate.

PERSONNEL

Troops operating in mountainous country must be in peak physical condition. Regardless of their normal physical condition, personnel operating in mountainous areas require additional stamina and energy. They must also possess the ability to conduct sustained physical exertion and recover from it quickly.

Acclimatization is described in Chapter 1. Acclimatization to height, which varies much more among individuals than that for heat, must also be considered for operations in mountains. Lack of oxygen at high altitudes can cause unacclimatized troops to lose up to 50 percent of their normal physical efficiency when operating in altitudes over 6,000 feet. Mountain sickness may occur at altitudes over 7,800 feet and is usually characterized by severe headache, loss of appetite, nausea and dizziness, and may last from 5 to 7 days. Troops can acclimatize by appropriate staging techniques. It may take several weeks to become completely acclimatized, depending on altitude and the individual's personal physical reactions.

The risk of sunburn, particularly to the uncovered face, is greater in mountains than on the desert floor due to thinner atmosphere. Use antisunburn ointment and keep the face in shade around midday, using face nets or sweat rags. An individual camouflage net or scarf is particularly useful for this purpose. Recognition of heat illnesses in higher altitudes may not be as apparent as at lower altitudes because sweat evaporates very quickly. Measures to avoid dehydration and salt loss are extremely important. Daily temperature variations may be considerable making it necessary to ensure troops do not become chilled at night. Layering of clothing is essential. Troops who have been sweating heavily before the temperature starts to drop should take their wet shirts off and place them over relatively dry shirts and sweaters. Soldiers/marines should add layers of clothing as it gets colder and remove them as needed. This may have to be leader supervised and disciplined in the same manner as water consumption.

Requirements for hygiene areas important in mountainous areas as in the desert itself. Normal rocky ground will make it extremely difficult to dig any form of latrine so cover excrement with rocks in a specially marked area.

GENERAL CONSIDERATIONS

Infantry is the basic maneuver force in mountains. Mechanized infantry is confined to valleys and foothills (if these exist), but their ability to dismount and move on foot enables them to reach almost anywhere in the area. Airmobile infantry can also be extensively used. Consideration should be given to modifying the TOE of infantry units operating in barren mountains. A strong antitank platoon may not be necessary. However, the infantry requires extra radars and radios for the number of observation posts and separate positions that they may expect to occupy.

Mountains are not a good environment for tank and armored cavalry operations, because tanks and armored cavalry are unable to maximize their mobility, flexibility, and firepower.

Avenues of approach at ground level are few. Roads or trails are limited and require extensive engineer effort to maintain. Off-road trafficability varies from relatively easy to very difficult. Most movement and maneuver in this type of terrain is either by air or on foot. Unnecessary vertical foot or vehicle movement should be avoided. Rock slides and avalanches, although not as common as in high cold mountains, do exist and can restrict movement.

Air cavalry is the major reconnaissance means but they must guard against being ambushed by ground troops located at their own altitude or even higher. Security of units must include observation, especially at night, of all avenues of approach including those within the capabilities of skilled mountaineers.

It is relatively easier to conceal troops in barren mountains than on the desert floor due to rugged ground, deep shadows (especially at dawn and dusk), and the difficulties an observer encounters when establishing perspective. Carefully placed rocks can be used to hide equipment however, rocks can chip and splinter under small arms fire. The normal type camouflage net, which breaks up outline by shadow, maybe used rather than the overall cover normally used in the desert.

Helicopter units of all types can be used, although they maybe slightly inhibited by altitude and rugged terrain. Payloads and endurance are degraded due to density and attitude. Winds are turbulent with considerable fluctuations in air flow strength and direction, particularly on the lee side of mountains. These winds, combined with the terrain, produce extra strain on crews as they have little margin for error. Flight crews should receive training in these conditions before flying in operations under these conditions.

When using men on foot for navigation, use all available maps, the lensatic compass, and a pocket altimeter. The pocket altimeter is essentially a barometer, measuring height by means of varying air pressure. If a navigator can only establish his location in the horizontal plane by resection on one point, the altimeter tells him his height, and thus his exact position. The instrument must be reset at every known altitude as it is affected by fluctuations of air pressure. Air photographs can also be helpful if they are scaled and contoured.

Supply of water and ammunition and the evacuation of wounded, especially if helicopters cannot land, can complicate operations. Water and ammunition may have to be transported by unit or civilian porters using A-frames or other suitable devices, or even by animal transport such as camel or mule.

OFFENSIVE OPERATIONS

The objective in mountainous areas of operations is normally to dominate terrain from which the enemy can be pinned down and destroyed. Avenues of approach are normally few, with very limited lateral movement except by helicopter. Reconnaissance must be continuous using all available means, as enemy defensive positions will be difficult to find. Observation posts are emplaced on high ground, normally by helicopter.

When contact is made, airmobile infantry can be used to outflank and envelop the enemy while suppressive fires and close air support are placed on all suspected positions, especially on dominating ground. Engineers should be well forward to assist in clearing obstacles. If airmobile infantry is unable to outflank the enemy, it will be necessary to launch a deliberate attack.

Frontal attacks in daylight, even with considerable supporting frees, have a limited chance of success against a well-emplaced enemy. Flank attacks on foot take a lot of time. The best opportunity is at night or in very poor visibility, but progress of men on foot will be slow and objectives should be limited.

The force should make every effort to secure ground higher than enemy positions to allow the attack to be downhill. Mobile forces should select objectives to the enemy's rear to kill the enemy as they reposition or counterattack. Foot mobile forces must seek adequate terrain (restrictive) to equalize the enemy's mounted mobility advantage.

Air superiority is required to allow a continuous flow of supplies and combat support by helicopter. Friendly mobile units must concentrate to destroy enemy command and control, artillery, service support, and air defense assets. It may be possible to infiltrate to a position behind the enemy, preferably using the most difficult, and hence unlikely route. Although this is very slow, it normally has the advantage of surprise.

The importance of dominating terrain, together with the enemy's knowledge that troops on the objective will be physically tired and dehydrated, makes an immediate counterattack likely. Supporting weapons must be brought forward at once, preferably by helicopter, and casualties removed by the same method.

Airmobile and attack helicopter units are well suited for pursuit operations. They can be used to outflank retreating enemy, and set up positions overlooking likely withdrawal routes. Small engineer parties can be emplaced to block defiles and interdict trails. Close air support and field artillery are used to reinforce airmobile and attack helicopter units and to counter efforts by enemy engineers to create obstacles.

DEFENSIVE OPERATIONS

A defense from a series of strongpoints is normal in hot mountains due to the need to hold dominating terrain and restrictions on ground mobility. Due to the amount of rock in the soil, it takes more time to prepare positions and normally requires engineer support.

It is necessary to hold terrain dominating avenues of approach. Any terrain that dominates a friendly position must either be held, or denied to the enemy by fire. It may be necessary to stock several days' supplies, especially water, ammunition, and medical equipment in a position in case helicopters or supply vehicles are unable to reach it.

When a covering force is used, it is organized around cavalry reinforced with attack helicopters, supported by field and air defense artillery. Airmobile infantry operates on ridge lines. If the enemy closes on a battle position it is difficult to extract airmobile infantry, so sheltered landing sites nearby should be available. In any event, extractions must be covered by air or ground suppressive fires. Stay-behind observers should be used to call down field artillery fires on targets of opportunity or to report enemy activity. When tanks are a threat and terrain is suitable, the covering force is reinforced with tank-heavy units and antitank weapon systems.

Combat in the main battle area is usually a series of isolated actions fought from strongpoints on ridge lines and in valleys. Patrols are used extensively to harass the enemy and prevent infiltration; all possible routes must be covered. If the enemy attempts to outflank the friendly force, he must be blocked by attack helicopters, if available, or airmobile infantry.

Reserves should be kept centrally located and deployed by air to block or counterattack. If this is not possible, reserves may have to be split up and placed behind key terrain where they are available for immediate counterattack.

If retrograde operations are necessary, mountainous terrain is as good a place to conduct them as anywhere. More time is required to reconnoiter and prepare rearward positions, and they should be prestocked as much as possible. Unlike the desert floor where movement between positions is likely to cover relatively great distances, movement in these conditions is usually from ridge to ridge. Routes must be covered by flank guards, especially at defiles or other critical points, as the enemy will attempt to block them or cut off rear guards.

COMBAT SUPPORT

It may be difficult to find good gun positions at lower altitudes due to crest clearance problems-so high-angle fire is often used. The best weapons are light field artillery and mortars that are airmobile and can be manhandled so they can be positioned as high as possible.

Field artillery observation posts are emplaced on the highest ground available, although in low-cloud conditions it will be necessary to ensure that they are staggered in height. Predicted fire may be inaccurate due to rapidly changing weather conditions making observed fire a more sure method for achieving the desired results.

Like field artillery, there is limited use for self-propelled weapons in this environment, although some may be used in valleys. Airmobile towed weapons allow employment throughout the mountainous area of operations.

Major tasks for engineer, even in an airmobile force, are: construction, improvement, and route repair, and their denial to the enemy. Mining is important due to the limited number of routes. Lines of communication require constant drainage and possibly bridging to overcome the problem of flash flooding.

Because of the frequent interdictions of mountainous roadways, military police will experience multiple defile operations. Use temporary traffic signs to expedite traffic movement to the front. The number of stragglers may be expected to increase in this environment. Because of difficulty in resupply, the supply points for water, POL, food, and ammunition will become especially lucrative targets for enemy attack. Military police rear area security elements must develop plans for relief and for augmenting base defense forces.

COMBAT SERVICE SUPPORT

Air transportation is the best CSS means in mountain operations due to its mobility. It may be limited by the weather, enemy activity, or the scarcity of landing sites, so there should be alternative means available. Terrain permitting, wheel-vehicle transportation should be employed as far forward as possible, using high-mobility vehicles off main mutes. Beyond the limits of wheel transport the only alternatives to CSS transport are animals (which may need to be acclimatized) or porters.

The composition and employment of trains in mountain operations are described in FM 90-6. Brigade trains should locate near an airstrip that can handle USAF tactical airlift. They are an obvious target for enemy air attack or artillery, or raids by enemy deep patrols, so adequate air defense and a coordinated area defense plan are necessary. Guards must be placed on all dominating terrain around the area, equipped with ground surveillance radars and STANO devices, and patrols should be employed outside the perimeter.

Supply points may be set up in the brigade trains area to operate distribution points for Class I, III, and V supplies. However, where routes are limited it may be necessary to resupply totally by air from the DISCOM area.

The variations of supplies in demand in the desert are very much the same as for those in temperate climates and are described in Chapter 4. The differences are described below:

- Class I. Mess trucks are not practical in this terrain. Food is either eaten cold, or heated on can heaters. Each soldier/marine should carry a one-day supply of emergency rations to be used if the daily resupply does not arrive.
- Class II. There is a high demand for footwear. Combat boots may be expected to last approximately two weeks in the harsh rocky terrain.
- Class III. Individual vehicle consumption will be greater than normal. Aircraft fuel requirements are greater, but it should be possible for much of their refueling and servicing to take place well to the rear where resupply is relatively easy.
- Water continues to carry a high priority. Demand for water is approximately 9 quarts per day, per man, as a minimum, and sometimes considerably more. Troops should carry four canteens of water, and every effort should be made to prestock water in positions or along routes.

First aid at squad arid platoon level is very important as medics will not necessarily be able to reach individual isolated positions. It is easy to lose casualties in this terrain so a buddy system to keep watch on each individual should be a matter of SOP. Medical evacuation is most often by air. It is a comparatively long distance to the nearest helicopter landing site, so teams of stretcher bearers will be required.

APPENDIX G

WATER USAGE IN DESERT OPERATIONS

Combat operations in the desert pose a number of unique problems. Because there is so little water and because our troops and much of our equipment cannot survive without it, water is a critical item of supply in the desert. Forces trying to survive in the desert without adequate water supplies have always met with disaster. Finding and keeping water sources may be the most crucial issue in desert conflicts. At the very least, water sources will be critical.

PROPER USES OF WATER

Water must be used to support immediate and future missions. There will be times when there is enough water to fill all requirements, but there will also be times when command direction and considerable thought are necessary to decide how to make the best use of available water. The first priority must go to the survival of the force and accomplishment of the immediate mission, and second priority to the maintenance and sustainment of the force. A general priority for uses of water is--

- Personnel (drinking only).
- Medical treatment.
- Vehicle and equipment cooling systems.
- Personnel (uses other than drinking).
- Decontamination.
- Food preparation.
- Laundry.
- Construction.

How far the available water will stretch depends upon your evaluation of the local situation and mission; how you set priorities for water; and how careful you are in using your limited supply.

INTEGRATED BATTLEFIELD EFFECTS ON WATER USE

Water needs increase dramatically on a nuclear or chemical battlefield. Decontamination of men and equipment requires large quantities of water. Being "buttoned up" in vehicles or in NBC protective gear locks in heat and perspiration and makes men sweat more. This increases the need for drinking water to avoid dehydration and heat casualties. Replacement of fluids lost through sweat is a critical use of water and generally comes ahead of all other uses.

LEADER RESPONSIBILITIES

Your primary responsibility is to accomplish your mission. Water is essential to do that. You must estimate how much you need and when you need it, just as you do for any type of essential supply. When the supply is limited, you must adjust your plans. During the accomplishment of your mission you must monitor the supply and ensure that it is used according to your plan.

One of the biggest water-related problems you may face is that as troops recognize how valuable water is to their survival, they may hoard it and not drink enough to sustain the efforts you expect of them. Before periods of activity, have your troops drink as much as they can. During the activity, take positive steps to make soldiers/marines replace the water lost by sweating. Thirst is a poor indicator of the body's need for water and may be ignored during hard work or in the heat of battle. Squad and section leaders must make their men drink regularly.

EFFECTS OF HEAT AND LACK OF WATER

Objects absorb heat from the sun and the air. In the desert, heat from these two sources is extreme. The clear, low-humidity air lets most of the heat from the sun through. As the sunlight strikes an object, such as a soldier/marine or the ground, much of the heat is absorbed. The ground, heated by the sun, in turn heats the air, often to temperatures well over 100 degrees Fahrenheit. A man, or any other object in the sun, absorbs heat from both sources. A man in the shade only has to contend with the heat from the air. The third major source of heat is the body itself. Like an engine, the body generates excess heat as it functions. The more work performed, the more heat that is generated. There are four ways that heat leaves an object:

- Radiation heat radiates from an object to a cooler object through a medium.
- Conduction heat flows from a hot object to a cooler object through direct contact.
- Convection heat flows from a hot object into a cooler surrounding medium like air.
- Evaporation heat is absorbed in changing a liquid (like water) into a vapor.

The heat losses from radiation and conduction are relatively small for people and equipment. If the air is much cooler than the surface of an object, such as a vehicle radiator or a person's skin, convection can remove significant amounts of heat. As the two temperatures get closer, this loss becomes smaller. If the air is hotter than the object, as is often the case in the desert, heat is gained from the air. In hot, dry desert climates, the primary method of losing excess heat from the body is the evaporation of sweat. The rate of sweating depends on the amount of excess heat the body needs to lose. Hard work in hot climates can result in 1-1/2 to 2-1/2 quarts of sweat lost per hour.

Supporting medical units measure the combined effects of the sun, air temperature, and humidity on dismounted troops in open terrain. This combined effect, measured as the Wet Bulb Globe Temperature (WBGT) or the Wet Globe Temperature (WGT), is translated by the supporting medical unit to a heat condition.

Men cannot be expected to routinely perform more than about 5 hours of heavy work per day in heat condition BLACK. However, using a 20-minute work/40-minute rest cycle, it will take 15 hours to do 5 hours of work. Similarly, in heat condition RED, a maximum of 6 hours of heavy work can be expected; in heat condition YELLOW, 7 hours; and in heat condition GREEN, 8 hours.

The heat condition experienced by individual soldiers/marines may differ from the general heat conditions in the area. The activity level in closed vehicles (such as tanks and armored personnel carriers) will probably be less than that of troops outside, but the air temperature may be 20 to 30 degrees Fahrenheit higher, thus increasing the soldier's heat gain from the air and, as a result, his sweat rate and need to drink water. As a general rule, increase the water intake to that for two or three heat conditions higher for troops in enclosed vehicles. On the other hand, troops performing light duty in the shade are not as severely stressed and will need less water and less rest time.

In practice, keeping track of the current heat condition and applying it to troops working in a wide variety of conditions will be difficult, if not impossible. This data gives you an understanding of how much water is required by troops when they work hard in the desert heat and to point out the limits of an individual's efforts in extreme heat. Work done in the heat of the day takes much longer and is more fatiguing than work done under relatively cooler conditions. Work schedules planned to take advantage of the cooler times of day-early morning, late evening, and night-not only increase productivity and reduce water use, but they are also easier on the men and better for their morale.

EFFECTS OF WATER LOSS

The body has a small reserve of water and can lose some without any effects. After a loss of about 2 quarts (which represents about 2.5 to 3.0 percent of body weight), effectiveness is impaired. Soldiers/marines may begin to stumble, become fatigued and unable to concentrate clearly, and develop headaches. Thirst will be present but not overpowering. So unless well trained, or reminded or goaded to drink, troops may not replace the water loss.

As dehydration continues, the effects will become more pronounced. The soldier/marine will become less and less effective and more likely to become a heat casualty. Some soldiers/marines will experience heat cramps, others will develop heat exhaustion or heatstroke. Heat cramps and heat exhaustion can be treated with good success and the soldier/marine returned to duty in a few days; however, without prompt medical attention, heatstroke can be fatal. Even if the man survives, he will probably not be returned to duty. In any case, a heat casualty is lost for some time. Preventing casualties is much easier than treating and replacing the casualties.

WATER SUPPLY PLANNING

Water planning is complicated because water is heavy (about 8.3 pounds per gallon) and may be considered perishable. Water stored in small containers gets hotter than water stored in large containers. As water gets hotter, it loses its disinfectant and becomes less desirable to drink. These facts make it difficult to carry an adequate supply of water, and frequent resupply is often required. The following questions must be answered when planning the unit's water supply:

- How much water is needed?
- Where is it needed?
- When is it needed?
- How will water get to the unit?
- How does water supply affect the mission?
- How does the mission affect water requirements?
- What measures need to be taken to ensure water is properly used?

There are several requirements for water. Some requirements, such as water for radiators, are reasonably constant. Some, such as water for food preparation or showers, are prescribed by the situation. Others, such as water for drinking and personal hygiene, depend on how the mission is accomplished. Planning water requirements for centralized service support functions (shower, laundry, medical treatment, maintenance, and construction) is the responsibility of the supporting organization. The largest and most critical planning factor is drinking water. The quantity required depends on the environment and the difficulty and intensity of individual activity.

When calculating water requirements for a whole day, you need to consider other requirements, such as shaving, brushing teeth, and helmet baths. On the average, these functions require almost 2 gallons of water per man per day. When B rations are issued, plan for 1 gallon of water per meal for the mess kit laundry, and 0.5 gallon per meal for food preparation and kitchen cleanup if the unit prepares its own B rations. The water used to heat individual combat rations can be reused for washing and shaving.

When calculating water requirements for individual details, plan to use 2 quarts of water per hour of hard work per man (including rest periods) during the heat of the day, or 1 quart of water per hour of hard work per man in the cooler parts of the day. These quantities are intended to satisfy requirements for drinking water as well as for the water that men will pour over their heads when they are hot. Experience with local conditions and the work performed may change these estimates. It is important to remember that water lost by sweating will be replaced sometime during the day, but men work best if the water is replaced as it is lost.

WATER REQUIREMENTS

Some examples of water planning calculations are provided in the following paragraphs.

Example 1. A 10-man squad performing heavy work that will take about 2 hours at a Continental United States (CONUS) base.

If the squad works in the late morning, with a 30-minute work, 30-minute rest cycle (heat condition RED), the work will take almost 4 hours. At 2 quarts of water per man per hour, the squad will require 20 gallons of water or 8 quarts (2 gallons) per man.

If the squad works at night or in the very early morning (heat condition GREEN or cooler), with a 50-minute work, 10-minute rest cycle, the work will take about 2 hours. At 1 quart of water per man per hour, the squad will require 5 gallons of water or 2 quarts (0.5 gallon) per man.

Example 2. A 40-man platoon doing a variety of work over an entire day.

The platoon requires 4 gallons of water per man per day for drinking, and 2.5 gallons per man per day for personal hygiene, or 6.5 gallons per man per day. For the whole platoon, 260 gallons (6.5 x 40) per day are required. (Meals not considered.)

Example 3. A 160-man company doing a variety of work and eating two B. ration meals and one MRE or MCI ration per day delivered from the battalion consolidated kitchen.

The company requires 4 gallons of water per man per day for drinking, 2.5 gallons per man per day for personal hygiene, and 1 gallon per meal for the mess kit laundry.

6.5 gallons (4 + 2.5) per man per day x 160 men	= 1,040 gallons
+ 1 gallon per B-ration meal x 320 rations*	= 320 gallons
	Total 1,360 gallons per day

*Not required if an expedient means can be used and utensils to avoid having to wash plates

NOTE: If company personnel operate their own field kitchen, 0.5 gallon per B-ration meal is required. In this example, an additional 160 gallons (0.5 x 320) would be needed.

Example 4. A 750-man battalion performing a variety of tasks, operating a battalion field kitchen, and feeding two B-ration meals and one MRE or MCI per day.

The battalion requires 6.5 gallons of water per man per day for drinking and personal hygiene, and 1.5 gallons per B-ration meal served.

6.5 gallons per man per day x 750	= 4,875 gallons
+ 1.5 gallons per B-ration meal x 1,500	= 2,250 gallons
	Total 7,125 gallons per day

The situation will usually dictate whether an element will pick up its own water or have it delivered to it. Normally, water is either produced at or delivered to a water supply point in the brigade rear. Forward units pick up their water at the water supply point and move it forward using organic transportation. Water distribution is normally planned and coordinated at company or battalion level.

Long distances and increased consumption compound water transportation problems in the desert. Several specialized pieces of equipment are available to overcome these problems. They have been designed and allocated to permit units to move and store the larger than normal volumes of water required in the desert. They are also capable of delivery by ground and air transportation, airdrop, or low altitude parachute extraction system (LAPES). Air transportation, while possible, is usually limited by aircraft availability and weight of water required. It is normally used only when ground transportation is not feasible. Water shortages can severely limit your unit's mission capabilities. If you cut water use to absolutely essential requirements, you can temporarily overcome limited water shortages, but severe shortages will limit your unit's mobility and capability. Water shortages may make some daylight operations or hard work infeasible or unsupportable. Each course of action must be analyzed with respect to water support requirements and the capacity of troops to sustain their efforts under severe heat conditions. Water supplies are equally important to the enemy. Taking and keeping water sources, or denying or destroying enemy water supplies can critically alter the options available to the enemy.

You and your unit can do several things to use water to your advantage. First, make sure that enough water is available for the most critical use-drinking. Furthermore, water must be available right where it is needed. Troops working in the desert should not have to walk more than a short distance to a water source. During rest periods some soldiers/marines would rather sit than walk to get water, especially if they are not thirsty. It must be easy for them to get water. Similarly, beverages must be readily available in dining facilities. There should be enough beverage dispensers that troops do not have to wait long in lines.

In addition to having water readily available, you must also make sure that soldiers/marines actually drink all the water they need. Since the signs of dehydration are not obvious until a person is close to heat injury, leaders at the squad and platoon level must keep track of water consumed and take the following measures:

- Have troops drink an extra quart of water before hard work. Storing water in your stomach gives you an extra quart of reserve capacity.
- Keep track of how much water each man drinks at the squad level; at the platoon level, monitor the use rate at each squad.
- Have troops take breaks as often as the heat condition requires, and during breaks remind or require the troops to drink.
- Make sure water is kept as cool as possible so it will be as palatable as possible to drink.
- Watch the troops for the first signs of heat stress and reduced effectiveness, such as stumbling and slurred speech.
- Have your soldiers/marines check their urine. A lack of the need to urinate and dark-colored urine are signs of dehydration.
- Use the buddy system within the squad to help ensure soldiers/marines are drinking enough.
- Make sure troops wear their uniform correctly. Shirts should be on with sleeves rolled down, scarf around neck, and hat on. The uniform should be worn as loosely as possible.

In order for water to be useful for its most critical purpose in the desert, it must be protected not only from enemy action but also from heat and contamination. The larger quantities of water required for drinking in the desert increase the importance of the quality of the water. Water can carry minerals, microbiological organisms, and toxic materials. The body can handle only so much of these contaminants before its natural defense mechanisms become swamped and health and effectiveness deteriorate.

When water has been treated and distributed to water points, it has already been checked for contamination by water purification unit operators and medics; however, these checks do not ensure the water will not become contaminated somewhere in the unit distribution system before a soldier/marine drinks it.

When water is purified and distributed, it is usually disinfected by adding chlorine to a level prescribed by the Command Surgeon. The chlorine not only kills the microbiological organisms presently in the water, but some also remains in the water to kill any bacteria that might get into the water later. Such contamination can be minimized by using common sense at the unit level. Store water only in clean containers intended for water, and do not let anything get in the water that you would not want to drink.

However, even if your unit is very careful, water may eventually become contaminated. As the water is handled or gets hot, the chlorine's disinfecting power disappears. Your company's field sanitation team measures chlorine levels in unit water containers. The team also has ampules of a chlorine compound that are used to replace chlorine in small containers. Each soldier/marine is issued a small bottle of iodine tablets to disinfect water in his canteen if he must take water from an expedient, untreated source.

Heat is another contaminant of water in the desert. When water is warmer than 75 or 80 degrees

Fahrenheit, it becomes difficult to drink. Bad tastes in water become more pronounced as water becomes warmer and people will not want to drink it. Small-unit leaders will have a very difficult task trying to get troops to drink all the water they need to replace losses if the water tastes bad. Water tastes best, and it is easier to drink large quantities of it, if it is between 50 and 70 degrees Fahrenheit. There are three ways to avoid the problems of drinking hot water:

- Drink it before it gets hot.
- Keep it in a cool place or in the shade.
- Cool it to a palatable temperature.

Drinking water before a period of heavy work or before leaving the unit area on a mission gets valuable water into the body before the water has a chance to heat up. It also provides an additional reserve of water that is easy to carry.

Water in containers absorbs heat from three sources-the air, the ground, and sunlight. There is no easy way to cut absorption from the air, but you can reduce heat absorbed from the ground and sun by keeping containers in as much shade as possible. There are certainly many ways of getting shade for water containers; with a little American ingenuity and available materials, it can be done.

Small, uninsulated water containers heat up more quickly than larger or insulated containers. Fortunately, the water in small containers can be used more quickly before it heats up. Water in uninsulated 5-gallon cans starting at a cool 60 degrees Fahrenheit can heat up to unpalatable temperatures in three to four hours on a hot (greater than 90 degrees Fahrenheit) day in the full sun, but it will take seven to eight hours if kept in the shade. Unshaded 55-gallon drums will heat to unpalatable temperatures in about one day of full sun, but will stay drinkable for two days if shaded. Insulated 400-gallon water trailers in or out of the sun will keep water cool for several days.

Water must be cool to start with if it is to be kept cool. One of the supplemental items available to company-size units is a small mobile water chiller. The chiller is designed to provide cool (about 60 degrees Fahrenheit), palatable water for company-size units. It can be used to cool and dispense water from any container into canteens, water cans, or any other container at a rate of at least 0.5 gallon per minute. It should be used in all cases where individuals must be encouraged to drink large quantities of water, and to cool water being taken with or to elements working away from the company area.

SURVIVAL TECHNIQUES

The water support logistics structure is designed to provide enough water to retain maximum force effectiveness, especially on a battlefield. It would be foolish to expect that the large quantities of water required for a combat force will always be available. Water is a critical supply in desert combat. Having less than full supply for all needs does not necessarily spell disaster for operations, but it may force changes in plans. There are several options available to you if you are forced to operate on reduced water supplies. The first part of this chapter outlines those options. The second part (described in the following paragraphs) describes actions and techniques for small groups or individuals to take when totally cut off from normal water supply.

The first and most obvious option is to eliminate, reduce, or postpone water uses not immediately required for survival or mission accomplishment. Showers, laundry, personal hygiene, and B-rations can usually be eliminated for several weeks without severe impact on troop health or combat effectiveness, although this tactic will eventually reduce troop health and morale. Also, construction requiring water can often be avoided.

The one water use that cannot be denied without significant risk is individual drinking. However, significant reductions in individual drinking requirements are possible by-

- Limiting all but absolutely essential work to the cooler parts of the day (early morning, late evening, and night).
- Keeping individuals in the shade as much as possible.
- Severely limiting all activity.

All of these actions reduce the effectiveness or capability of a force, but they help it to survive.

What can individuals or small groups do when they are totally cut off from normal water supply? If you are totally cut off from the normal water supply, the first question you must consider is whether you should try to walk to safety or stay put and hope for rescue. Walking requires 1 gallon of water for every 20 miles covered at night, and 2 gallons for every 20 miles covered during the day. Without any water and walking only at night, you may be able to cover 20 to 25 miles before you collapse. If your chance of being rescued is not increased by walking 20 miles, you may be better off staying put and surviving one to three days longer. If you do not know where you are going, do not try to walk with a limited supply of water.

If you decide to walk to safety, follow the following guidelines in addition to the general conservation practices listed in the next section:

- Take as much water as you have and can carry, and carry little or no food.
- Drink as much as you can comfortably hold before you set out.
- Walk only at night.

Whether you decide to walk or not, you should follow the principles listed below to conserve water in emergency situations:

- Avoid the sun. Stay in shade as much as possible. If you are walking, rest in shade during the day. This may require some ingenuity. You may want to use standard or improvised tents, lie under vehicles, or dig holes in the ground.
- Cease activity. Do not perform any work that you do not have to for survival.
- Remain clothed. It will reduce the water lost to evaporation.
- Shield yourself from excessive winds. Winds, though they feel good also increase the evaporation rate.
- Drink any potable water you have as you feel the urge. Saving it will not reduce your body's need for it or the rate at which you use it.
- Do not drink contaminated water from such sources as car radiators or urine. It will actually require more water to remove the waste material. Instead, in emergencies, use such water to soak your clothing as this reduces sweating.
- Do not eat unless you have plenty of water.

Do not count on finding water if you are stranded in the desert. Still, in certain cases, some water can be found. It does rain sometimes in the desert (although it may be 20 years between showers) and some water will remain under the surface. Signs of possible water are green plants or dry lake beds. Sometimes water can be obtained in these places by digging down until the soil becomes moist and then waiting for water to seep into the hole. Desert trails lead from one water point to another, but they may be further apart than you can travel without water. Almost all soils contain some moisture.

APPENDIX H

FRATRICIDE REDUCTION

The problem of fratricide is as old as warfare itself. It is a complex problem that defies simple solutions. Fratricide is defined as "the employment of friendly weapons and munitions, with the intent to kill the enemy or destroy his equipment or facilities that results in unforeseen and unintentional death or injury to friendly personnel." This is obviously a broad definition. This appendix focuses on actions leaders can take with current resources to reduce the risk of fratricide.

MAGNITUDE OF THE PROBLEM

The modern battlefield is more lethal than any in history. The pace of operations is rapid, and the non-linear nature of the battlefield creates command and control challenges for all unit leaders.

Our ability to acquire targets using thermal imagery exceeds our ability to accurately identify targets as friend or foe. The accuracy and lethality of modem weapons make it possible to engage and destroy targets at these extended acquisition ranges.

Added to this is the problem of battlefield obscuration. Rain, dust, fog, smoke, and snow degrade the ability to identify targets by reducing the intensity and clarity of thermal images. The effects of battlefield obscuration must be considered when thermal identification is relied upon.

On the battlefield, positive visual identification cannot be the sole engagement criteria at ranges beyond 1,000 meters. Situational awareness is key and must be maintained throughout an operation.

The following are recommended actions to take at crew and leader level in the event the crew are victims of friendly fires:

- React to contact until you recognize friendly fire.
- Cease fire.
- Report on the next higher unit net—
  - That you are receiving friendly fire.
  - The location and direction of the firing vehicle.
- Provide a visual recognition signal to cease fire.
- Protect troops, request medical assistance as needed.
- Do not return fire when you positively identify the firing unit as friendly.

The following are recommended actions to take at crew and leader level when the crew are engaging friendly forces:

- Cease fire.
- Report on next higher net¬
  - The engaged friendly force (if unknown, report number and type of vehicles).
  - The location.
  - The direction and distance to victim.
  - The type of fire.
  - The target effects.

The following are recommended actions to take at crew and leader level in the event the crew observes a friendly fire incident:

- Seek cover and protect self.
- Report on next higher net¬
  - The friendly force engaged.
  - The location of the incident.
  - The direction and distance to victim/firer.
  - The type of fire.
  - The target effects.
- Provide a visual friendly recognition signal.
- Provide assistance (when safe to do so) as needed.

Leader actions should focus on identifying and stopping the friendly fire incident and establishing controls to prevent its recurrence. Some recommended actions for identifying and stopping friendly fire incidents are-

- Find and stop firing.
- Conduct in-stride risk assessment.
- Implement controls to preclude recurrence.

PREVENTIVE MEASURES

Reduction of fratricide risk begins with the planning phase of an operation and continues through the execution of the operation. The following are considerations for identifying fratricide risks in the planning, preparation, and execution phases of a given operation:

- Planning phase. A good plan that is well understood helps to minimize fratricide risk. The following considerations help indicate the potential for fratricide in a given operation:
  - The clarity of the enemy situation.
  - The clarity of the friendly situation.

- The clarity of the commander's intent.
- The complexity of the operation.
- The planning time available to all levels.

• Preparation phase. The following additional fratricide risks may become evident during rehearsals:

- Number and type of rehearsals.
- Training and proficiency levels of unit/individuals.
- The habitual relationships between units conducting the operation.
- The endurance of the troops conducting the operation.

• Execution phase. During execution, in-stride risk assessment and reaction are necessary to overcome unforeseen Fratricide risk situations. The following are factors to consider when assessing fratricide risks:

- Intervisibility between adjacent units.
- Amount of battlefield obscuration.
- Ability or inability to positively identify targets.
- Equipment similarities and dissimilarities between enemy and friendly vehicles.
- Vehicle density on the battlefield.
- The tempo of the battle.

Graphics are a basic tool that commanders at all levels use to clarify their intent, add precision to their concept, and communicate their plan to subordinates. As such, graphics can be a very useful tool in reducing the risk of fratricide. Commanders at all levels must understand the definitions and purpose of operational graphics and the techniques of their employment. See FM 100-5-1 for the definitions of each type of graphic control measure.

Briefbacks and rehearsals are primary tools in identifying and reducing fratricide risk. The following are some considerations on briefbacks and rehearsals to aid in reducing fratricide:

- Briefbacks ensure subordinates understand their commander's intent. They often highlight areas of confusion, complexity, or planning errors.
- The type of rehearsal conducted impacts on the risks identified.
- Rehearsals should extend to all levels of command and involve all key players.
- Use briefbacks or rehearsals to ensure subordinates know where fratricide risks exist, and what to do to reduce or eliminate the risk.

Maintaining situational awareness at all levels is key to fratricide reduction. Units must develop techniques to gain and maintain situational awareness in SOPs. Techniques could include-

- Eavesdropping on next higher net.
- Cross talk on radio between units.
- Accurate position reporting and navigation.

- Training and use/exchange of LOS.

Risk assessment must be conducted at all levels during the planning, preparation, and execution phases of all operations. Identification of fratricide risk factors is conducted at every level and the results should be clearly communicated up and down the chain of command.

Figure H-1 provides a worksheet for considering fratricide risk in the context of mission requirements. The worksheet lists 26 mission-accomplishment factors that affect the risk of fratricide. Assess the potential risk in each area as low, medium, or high, and assign a point value to each (one point for low risk, two for medium risk, three for high risk). Add the point values for the overall fratricide assessment score. Use the resulting score only as a guide, however. Your final assessment must be based both on observable risk factors like those on the worksheet and on your "feel" for the intangible factors affecting the operation. Note that descriptive terms are listed only in the low- and high-risk columns of the worksheet. Your assessment of each factor will determine whether the risk matches one of these extremes or lies somewhere between them as a medium risk.

The following fratricide reduction measures are provided as reminders for prudent/appropriate actions to reduce fratricide risk. They are not directive in nature, nor intended to restrict initiative. Apply the following measures to METT-T situations as appropriate:

- Identify and assess potential fratricide risk in estimate of the situation. Express this risk in the OPORD or FRAGO.
- Maintain situational awareness-current intelligence; unit locations/dispositions; denial areas (minefields/FASCAM); contaminated areas (e.g., ICM and NBC); SITREPs; and METT-T.
- Ensure positive target identification. Review vehicle/weapons ID cards. Know at what ranges and under what conditions positive ID of friendly vehicles/weapons is possible.
- Establish a command climate that stresses fratricide prevention. Enforce fratricide prevention measures; use doctrinally sound tactics, techniques, and procedures to ensure constant supervision of execution of orders and performance to standards.
- Recognize the signs of battlefield stress. Take quick effective action to deal with it to maintain unit cohesion.
- Conduct individual and collective (unit) fratricide awareness training; target identification/recognition training; fire discipline; and leader training.
- Develop a simple decisive plan.
- Give complete and concise mission orders.
- Use SOPS that are consistent with doctrine to simplify mission orders. Periodically review and change SOPS as needed.
- Strive for maximum planning time for you and your subordinates.
- Use common language/vocabulary and doctrinally correct standard terminology and control measures, such as, fire support coordination line (FSCL), zone of engagement, restrictive fire line (RFL), and others.
- Ensure thorough coordination is performed.
- Plan for and establish good communications.
- Plan for collocation of command posts, as appropriate to the mission (e.g., passage of lines, and so

forth).

- Establish and designate liaison officers (LO) as appropriate.
- Make sure rules of engagement (ROE) are clear.
- Consider the effect of fratricide on key elements of terrain analysis (observation and fields of fires, cover and concealment, obstacles and movement, key terrain, and avenues of approach).
- Conduct rehearsals whenever the situation allows time to do so.
- Be in the right place at the right time. Use position location/navigation (GPS) devices; know your location and the locations of adjacent units (left, right, leading and follow on); and synchronize tactical movement.
- Include fratricide incidents in after-action reviews (AAR).

FACTORS	LOW (1)	MEDIUM (2)	HIGH (3)
1 UNDERSTAND PLAN			
Commander's Intent	Clear		Foggy
- Complexity	Simple		Complex
- Enemy Situation	Known		Unknown
Friendly Situation	Clear		Unclear
- ROE	Clear		Unclear
2 ENVIRONMENT			
- Intervisibility	Favorable		Unfavorable
- Obscuration	Clear		Obscured
- Battle tempo	Slow		Fast
- Positive target ID	100%		0%
3 CONTROL MEASURES			
- Command relationships	Organic		Joint/Combined
- Audio	Loud/Clear		Jammed
- Visual	Well Seen		Obscured
- Graphic	Standard		Not understood
- SOPs	Standard		Not used
- LOs	Proficient		Untrained
Location/Navigation	Sure		Unsure
4 EQUIPMENT (Compared to US)			
- Friendly	Similar		Different
- Enemy	Different		Similar
5. TRAINING			
Individual proficiency	MOS Qual		Untrained
Unit proficiency	Trained		Untrained
- Rehearsal	Multiple		None

[illegible]	[illegible]		[illegible]
– Habitual relationship	Yes		No
– Endurance	Alert		Fatigued
6 PLANNING TIME (1/3 -2/3 Rule)			
– Higher HQ	Adequate		Inadequate
– Own HQ	Adequate		Inadequate
– Lower HQ	Adequate		Inadequate
OVERALL FRATRICIDE ASSESSMENT	LOW 26–46%*	MEDIUM 42–62%*	HIGH 58–78%*

* Commander may use numbers as the situation dictates. Numbers alone may not give accurate fratricide risk.

Figure H-1 Fratricide risk assessment worksheet.

FRATRICIDE RISK CONSIDERATIONS

This format, which parallels the five-paragraph OPORD, contains key factors and considerations in fratricide reduction, This is not a change to the OPORD format; rather, it should be used during OPORD development to ensure fratricide reduction measures are included in the order. It is not a strict guide. The factors and considerations are listed where they would likely appear in the OPORD, but they may warrant evaluation during preparation of other paragraphs.

1. Situation.

a. Enemy forces.

(1) Are there similarities between enemy and friendly equipment and uniforms that could lead to fratricide?

(2) What languages do enemy forces speak? Could these contribute to fratricide risk?

(3) What are the enemy's deception capabilities and its past record of deception activites?

(4) Do you know the locations of enemy forces?

b. Friendly forces.

(1) Among the allied forces, are there differences (or similarities with enemy forces) in language, uniform, and equipment that could increase fratricide risk during combined operations?

(2) Could differences in equipment and uniforms among US armed forces increase fratricide risk during joint operations?

(3) What differences in equipment and uniforms can be stressed to help prevent fratricide?

(4) What is the friendly deception plan?

(5) What are the locations of your unit and adjacent units (left, right, leading, follow-on)?

(6) What are the locations of neutrals and noncombatants?

c. Own forces.

(1) What is the status of training activities? What are the levels of individual, crew, and unit proficiency?

(2) Will fatigue be a factor for friendly forces during the operation? Has an effective sleep plan been developed?

(3) Are friendly forces acclimatized to the area of operations?

(4) What is the age (new, old, or mix) and condition of equipment in friendly units? What is the status of new equipment training (NET)?

(5) What are the expected MOPP requirements for the operation?

d. Attachments and detachments.

(1) Do attached elements know the above information regarding enemy and friendly forces?

(2) Are detached elements supplied the above information by their gaining units?

e. Weather.

(1) What are the expected visibility conditions (light data and precipitation) for the operation?

(2) What effect will heat and cold have on troops, weapons, and equipment?

f. Terrain.

(1) Do you know the topography and vegetation (such as urban, mountains, hilly, rolling, flat, desert, swamp/marsh, prairie/steppe, jungle, dense forest, open woods) of the expected area of operations?

(2) Have you evaluated the terrain using the factors of OCOKA?

2. Mission. Is the mission, as well as all associated tasks and purposes, clearly understood?

3. Execution.

a. Task organization.

(1) Has the unit worked under this task organization before?

(2) Are SOPS compatible with the task organization (especially with attached units)?

(3) Are special markings or signals (for example, cats' eyes, chemlites, or panels) needed for positive identification of uniforms and equipment?

(4) What special weapons and/or equipment are to be used? Do they look or sound like enemy weapons and/or equipment?

b. Concept of the operation.

(1) Maneuver. Are main and supporting efforts identified to ensure awareness of fratricide risks and prevention?

(2) Fires (direct and indirect).

(a) Are priorities of fires identified?

(b) Have target lists been developed?

(c) Has the fire execution matrix/overlay been developed?

(d) Have locations of denial areas (minefields/FASCAM) and contaminated areas (ICM, NBC) been identified?

(e) Are the locations of all supporting fires targets identified in the OPORD/OPLAN overlays?

(f) Are aviation and CAS targets clearly identified?

(g) Has the direct-fire plan been developed?

(h) Have final protective fires (FPF) been designated?

(i) Have you identified and verified sector limits?

(3) Engineer tasks.

(a) Are friendly minefield, including FASCAM and ICM dud-contaminated areas, known?

(b) Are obstacles identified, along with the approximate time needed for reduction/breaching of each?

(4) Tasks to each subordinate unit. Are friendly forces identified, as appropriate, for each subordinate maneuver element?

(5) Tasks to CS/CSS units. Have locations of friendly forces been reported to CS/CSS units?

(6) Coordinating instructions.

(a) Will a rehearsal be conducted? Is it necessary? Are direct and indirect fires included?

(b) Is a briefback necessary?

(c) Are appropriate control measures clearly explained and illustrated in the OPORD and overlays? Have they been disseminated to everyone who has a need to know? What is the plan for using these control measures to synchronize the battle and prevent fratricide?

(d) Have target/vehicle identification drills been practiced?

(e) Do subordinate units know the immediate action, drill, or signal for "cease fire" or "I am friendly" if they come under unknown or friendly fire? Is there a backup action?

(f) Is guidance in handling dud munitions (e.g., ICMs and CBUs) included?

4. Service Support.

a. Are trains locations and identification markings known by everyone?

b. Do medical and maintenance personnel know the routes between train units?

5. Command and Signal.

a. Command.

(1) What is the location of the commander and key staff?

(2) What is the chain of command?

b. Signal.

(1) Do instructions include signals for special and emergency events?

(2) Do instructions include how to identify friendly forces to aircraft?

(3) Do instructions include backup code words and visual signals for all special and emergency events?

(4) Are signal operation instructions (SOI) distributed to all units with a need to know (e.g., higher, lower, adjacent, leading, follow-on).

A2C2	Army airspace command and control
AA	assembly area
AAA	antiaircraft artillery
AAFAD	all army for air defense
AFCE	Allied Air Farces Central Europe
A AG	army artillery group
AAR	after-action review
AC	analysis console
ACA	airspace coordination areas
ACC	air component commander
ACP	Allied communication Publication
ACR	armored cavalry regiment
ACT	air cavalry troop
AD	armored division
ADA	air defense artillery
ADCOORD	air defense coordinator
ADW	air defense warming
AH	Air Force
AFAC	airborne forward air controller
AFCENT	Allied Forres Central Europe
ARO	aerial fire support officer
AFV	armored fighting vehicle
AG	adjutant general
AGOS	air-ground operations system
AGOSOP	Air-Ground Operations Standard Operating Procedures
AHB	attack helicopter battalion
AI	air interdiction
A/L	administrative/logistics
ALO	air liason officer
AM	amplitude modification
ANGLICO	air and naval gunfire liaison company
AO	area of operation
AOE	authorized organization equipment
AP	antipersonnel (mine)
APC	armored personnel carrier
APERS	antipersonnel (animation)

APT	armor-piercing incendiary
ARNG	Army National Guard
ARSOA	Army special operations aviation
ARSOF	Army special operations forces
ARTEP	Army Training and Evaluation Program
ASOC	air support operations center
ASP	ammunition Supply point
AT	antitank
ATGM	antitank guided missile
ATHS	automatic target hand-off system
ATKHB	attack helicopter battalion

ATP ammunition transfer point

AVLJJ	armored vehicle launched bridge
AVUM	aviation unit maintenance
AXP	ambulance exchange point
BAT	battlefield air interdiction
RAS	battalion aid station
B-BAG	battle bags
BDA	battle damage assessment
BDAR	battle damage assessment and repair
BRI	brigade
BDO	battle dress overgarment
BFV	bradley fighting vehicle
BHL	battle handover line
BII	basic issue items
BMNT	beginning morning nautical twilight
BMD	battalion maintenance officer
bn	battalion
Bn/TF	battalion/task force
BP	battle position
BSA	brigade support area
C2	command and control
C2 node	command and control junction

C3 command, control and communication

CA counterair

CAA	combined arms Army
CAB	combat aviation brigade
CAM	chemical agent monitor
CARC	chemical agent resistant coating

CAS	close air support
CBR.	chemical, biological, radiological
CBU	cluster bomb unit
CM	commander
CE	chemical energy; communication-electronics
CEB	clothing exchange and bath
CESO	communications electronics signal officer
CEV	combat engineer vehicle
CFA	covering force area
CFL	coordinated fire line
CFV	cavalry fighting vehicle
CH	cargo helicopter
CLAMMS	cleared lane mechanical making system
cm	centimeter(s)
CMT	company maintenance team
COA	change of assignment course of action
COLT	combat observation lasing team
COMMZ	communications zone
COMSEC	communications security
CONUS	continental United States
COP	command and observation post
COSCOM	corps support command
CP	checkpoint: command post
CPX	command post exercise
CRP	combat reconnaissance patrol
CRT	combat ray tube
CS	combat support
CSB	corps support battalion
CSM	command sergeant major
CSR	controlled supply rate
CSS	combat service support
CSSE	combat service support element
CTA	common table of allowances
CW	continuous wave
D-day	deployment day
DA	Department of the Army
DAG	divisional artillery group
DAO	division ammunition officer
DBDU	desert battle dress uniform
DD	Department of Defense
DEW	director! energy wanton

DISCOM	division support command
DKIE	decontamination kit individual equipment
DLIC	detachment left in conflict
DMD	digital message device
DMMC	division material management center
DNVT	digital nonsecure voice terminal
DP	decision point
DPICM	dual-purpose improved conventional munition
DS	direct support
DSA	division support area
DST	decision support template
DSVT	digital subscriber voice terminal
DTG	date-time group
DTOC	division tactical operations center
DZ	drop zone

EA	engagement area
EAC	echelons above corps
EC	electronic combat
ECM	electronic counter measures
EEFI	essential elements of friendly information
EEI	essential elements of information
EENT	erd evening nautical twilight
EMP	electromagnetic pulse
EN	enemy (in illustration)
EPW	enemy prisoner of war
ETAC	enlisted terminal attack controller
ETL	effective translational lift
EW	electronic warfare
EXTAL	extra time allowance

1SO	first sergeant
FA	field artillery
FAAO	field artillery air observer
FAC	forward air controller
FAAR.	forward area alerting radar
FARP	forward arming refuel point
FASCAM	family of scatterable mines
FAX	facsimile
FDC	fire direction center
FEB A	forward edge of battle area

FED forward entry devices
FFAR folding fin aircraft rocket
FIST fire support team
FIST-V fire support team vehicle
FKSM fort knox supplemental Material
FLIR forward looking infrared
FLOT forward line of own troops
FM field manual; frequency modulation
FO forward observer
FOOGAS petroleum-based expedient Inflammable material
FPF final protective fire
FRAG-HE fragmentary high-explosive (ammunition)
FRAGO fragmentary order
FRG Federal Republic of Germany
FS fire support
FSB forward support battalion
FSCL Fire support coordination line
FSCOORD Fire support coordinator
FSE fire support element
FSO fire support officer
ft feet

G1 assistant chief of staff (personnel)
G3 assistant chief of staff (operations and plans)
G3-Air air operations and planning officer
G4 assistant chief of staff (logistics)
GA gas agent
Gator (a mine delivered by fixed-wing aircraft)
GBU guided bomb unit
GEMSS ground em placed mine scattering system
GLD ground laser designator
GMRD ground motorized rifle division
GPS global positioning systems
GS general support
GSR ground surveillance radar

HE high explosive
HE-APERS high explosive antipersonnel (ammunition)
HEI high explosive incendiary
HEMTT heavy expanded mobile tactical truck

HET	heavy equipment transport
HF	high frequency
HHC	headquarters and headquarters company
HHT	headquarters and headquarters troop
HIMAD	hi-to-medium-altitude air defense
HMMWV	high mobility multipurpose wheeled vehicle headquarters
hq	
HTO	high technology observer
ICM	improved conventional munitions
ID	identification
IDS	infrared discrimination system
IEW	intelligence and electronic warfare
IEWSE	intelligence and electronic warfare support element
IFF	identification, friend or foe
IFFN	identification, friend, foe, or neutral
INTSUM	intelligence summary
IP	initial point
IPB	intelligence preparation of the battlefield
IR	infrared
ITV	improved TOW vehicle
I2	I squared (image intensification)
JAAT	joint air attack team
JTIDS	joint tactical information distribution system
KE	kinetic energy
KIA	killed in action
km	kilometer(s)
kph	kilometer(s) per hour [as a unit of measure only)
kmih	kilometers in the hour
kmph	kilometer(s) per hour (as a unit of measure that indicates motion)
KY	Kentucky
LANTIRN	low-altitude navigation and targeting infrared for night
LAPES	low-altitude parachute extraction system
LBE	load-bearing equipment
LC	line of contact
LCC	land component commander
LCSS	lightweight camouflage screen system
LD	line of departure; low drag
LD/LC	line of departure is line of contact

LGB	laser guided bomb
LO	liaison officer
LQA	light observation aircraft limit of advance
LOC	line(s) of communication
LOG	logistics
LOGPAC	logistics package
LRF	laser range finder
LRP	logistics release point
MAC	military air command
MAGTF	Marine air around task force
medic(s)	medical person: medical personnel
MEF	Marine expeditionary force
MEMO	mission-essential maintenance only
METT-T	mission, enemy, terrain, troops, and time available
MEU	Marine expeditionary unit
MHE	material-handling equipment
MI	military intelligence
MIC	mid-intensity conflict
MICLIC	mine clearing line charge
mib	miles in the boar
MLC	main line of communication
MLRS	multiple-launch rocket system
mm	millimeter(s)
MMS	multi mission ship
MOPP	mission oriented protection posture
MORT	mortar (in illustration)
MOS	military occupational specialty
MOUT	military operations on urbanized Terrain
MP	military police
MPF	maritime pre- positioning force
mph	miles per hour
MPP	mobile pre-positioning
MPSRON-1	maritime pre-positioning squadron one
MPSRON-2	maritime pre- positioning squadron two
MPSRON-3	maritime pre- positioning squadron three
MPT	mobile pay team
MR	motorized rifle
MRB	motorized rifle battalion
MRC	motorized rifle company

MRD motorized rifle division
MRE meals, ready to eat
MRL multiple rocket launcher
MRP motorized rifle platoon
MRR. motorized rifle regiment
MRS motorized rifle squad
MRS muzzle reference system
MSB main support battalion
MSE mobile subscriber equipment

MSR main supply route.
MSRT mobile subscriber radiotelephone terminal
MST maintenance support team
MTF' medical treatment facilities
MTLR moving target locating radar
MTOE modification table of organization and equipment
MTP mission training plan
MULE modular universal laser equipment

NA1 named areas of interest
NATO North Atlantic Treaty Organization
NBC nuclear, biological, chemical
NBCWRS NBC warning and reporting system
NCO noncommissioned office
NCOIC non commissioned officer in charge
NCS net control station
NET new equipment training
NGLO naval gunfire liaison officer
NLT no later than
NOE night observation device
NOR nap of the earth
NVD night-vision device
NVG night vision goggles
NZO standing barrier fire

OCOKA observation and fields of fire, cover and concealment, obstacles, key terrain, avenues of approach
Ol operations and intelligence
OIC officer in charge

OIR	other intelligence requirements
OMG	operational maneuver group
OP	observation post
OPCON	operational control
OPLAN	operational plan
OPORD	operation order
OPSEC	operations security
ORP	objective rally point
P&A	personnel and administration
PAC	Personnel and Administrative Center
PADS	position and azimuth determining system
PDS	personnel daily summary reports
PERINTREP	periodic intelligence report
PEWS	platoon early warning system
PIR	priority intelligence requirement
PL	phase line(s)
PHI (	a hybrid PLRS)
PLL	prescribed land list
PLRS	position location reporting system
plt	platoon
POL	petroleum, oils, and lubricants
PMCS	preventive maintenance checks and services
POMCUS	prepositioning of materiel configured to unit sets
PP	passage point(s)
PRF	pulse repetition frequency
PSG	platoon sergeant
PSNC0	personnel services noncommissioned officer
PSO	personnel services officer
PSP	pierced-steel planking
PSS	personnel service support
PST	pass time
PWRS	prepositioned war reserve stock
PX	post exchange
PZ	pick up zone
PZO	rolling harrier fine
QSTAG	Quadripartite Standardization Agreement

RAG	regimental artillery group
RAOC	rear area operations center
RAP	rocket-assisted projectile
RAU	radio access units
REG	repair and evacuation group
REMS	remotely employed sensor
RETRANS	relay station; retransmit
RFL	restrictive fire line
ROE	rules of engagement
ROWPU	reverse osmosis water purification unit
RP	release point
RPG	rocket-propelled grenade
R&S	reconnaissance and surveillance
RSR	required supply rate
RT	right; route
RX	repairable, exchange
2IC	second in command
S&S	supply and service
S1	adjutant
S2	intelligence officer
S3	operation and training officer
S3-Air	assistant battalion S3 (air operations)
S4	supply officer
S AFAD	small arms far air defense
SAM	surface-to-air missile
SCT	scout (in illustration)
SGM	sergeant major
SITREP	situation report
SLGR	small lightweight GPS receiver
SME	subject matter expert
SOI	signal operation instructions
SOP	standing operating procedure
SP	shut point
SPG	self-propelled grenade
sq km	square kilometer(s)
S&S	supply and service
SSB	single side hand
STANAG	Standardisation Agreement
STB	supertropical bleach

TAACOM	Theater Army Air Defense Command
TAC-A	tactical air coordinator-airborne
TACCP	tactical command post
TACCS	Tactical Army Combat Service Support (CSS) Computer System
TACFIRB	tactical line direction system
TACP	tactical air control party
TAI	target areas of interest
TAMMS	The Army Maintenance Management System
TAR	tactical air reconnaissance
TSC	training support center
TC	tank commander
TCF	tactical combat force
TCP	tactical computer processor
TCT	tactical computer terminal
TD	tank division
TDIS	time-distance
TEWT	tactical exercise without troops
TF	task force
TIRS	terrain index reference system
TTS	thermal imaging system
TOC	tactical operations center
TOE	table(s) of organization and equipment
TOT	Time on target
TOW	tube-launched, optically tracked, wire-guided missile
TPU	tank pump unit
TR	tank regiment
TRP	target, reference points
TSOP	tactical SOP
UAV	unmanned aerial vehicle.
UH	Utility helicopter

UHF	**ultra high frequency**
UMCP	**unit maintenance collection point**
UMT	**unit maintenance team; unit ministry team**
US	**United States (of America)**
USAF	**United States Air Force**
USMC	**United States Marine Corps**
USN	**United States Navy**
VFMED	**variable format message entry device**
VHF	**very high frequency**
VINSON	**encryption device**
vpkm	**vehicle per kilometer**
VTR	**vehicle/tank retriever**
WCS	**weapons control status**
WESS	**weapons effect signature simulator**
WIA	**wounded in action**
WO	**warning order**
WP	**white phosphorus**
WSM	**weapon system manager**
WSRO	**weapons systems replacement operations**
WTR	**wide temperature range**
XO	**executive officer**

REFERENCES

DEPARTMENT OF DEFENSE FORMS (DD Form)

DD Form 565 Statement of Recognition of Deceased. August 1984.

DEPARTMENT OF THE ARMY FORMS (DA Form)

DA Form 581 Request for Issue and Turn-in of Ammunition. August 1989.

DA Form 1155 Witness Statement of Individual. June 1966.

DA Form 1156 Casualty Feeder Report. June 1966.

DA Form 2404 Equipment Inspection and Maintenance Worksheet. April 1979.

DA Form 2765 Request for Issue or Turn-in. April 1966.

DA Form 2765-1 Request for Issue or Turn-in. April 1966.

DA Form 5368-R Quick Fire Plan. December 1984.

FIELD MANUALS (FM)

FM 3-3 NBC Contamination Avoidance. May 1987.

FM 3-4 NBC Protection. October 1985.

FM 3-5 NBC Decontamination. June 1985.

FM 3-50 Smoke Operations. December 1990.

FM 3-100 NBC Defense, Chemical Warfare, Smoke, and Flame Operations. May 1991.

FM 5-34 Engineer Field Data. September 1987.

FM 5-100 Engineer Combat Operations. November 1988.

FM 5-101 Mobility. January 1985.

FM 5-102 Countermobility. March 1985.

FM 5-103 Survivability. June 1985.

FM 6-20 Fire Support in the AirLand Battle. May 1988.

FM 6-20-1 Tactics, Techniques, and Procedures for the Field Artillery Cannon Battalion.

November 1990.

FM 6-20-40 Tactics, Techniques, and Procedures for Fire Support for Brigade Operations (Heavy). January 1990.

FM 7-7J The Mechanized Infantry Platoon and Squad (Bradley). February 1986.

FM 7-10 The Infantry Rifle Company. December 1990.

FM 7-20 The Infantry Battalion. April 1992.

FM 7-30 Infantry, Airborne, and Air Assault Brigade Operations. April 1981.

FM 7-70 The Light Infantry Platoon/Squad. September 1986.

FM 7-90 Tactical Employment of Mortars. June 1985.

FM 17-15 Tank Platoon. October 1987.

FM 17-95 Cavalry Operations. September 1991.

FM 17-98 The Scout Platoon. October 1987.

FM 17-98-1 Scout Leader's Handbook. September 1990.

FM 20-3 Camouflage. November 1990.

FM 20-32 Mine/Countermine Operations. December 1985.

FM 21-18 Foot Marches. June 1990.

FM 21-75 Combat Skills of the Soldier. August 1984.

FM 21-76 Survival. March 1986.

FM 22-100 Military Leadership. July 1990.

FM 31-12 Army Forces in Amphibious Operations (The Army Landing Forces). March 1961.

FM 34-1 Intelligence and Electronic Warfare Operations. July 1987.

FM 34-2 Collection Management. October 1990.

FM 34-3 Intelligence Analysis. March 1990.

FM 34-10 Division Intelligence and Electronic Warfare Operations. November 1986.

FM 34-80 Brigade and Battalion Intelligence and Electronic Warfare Operations. April 1986.

FM 34-130 Intelligence Preparation of the Battlefield. May 1989.

FM 43-5 Unit Maintenance Operations. September 1988.

FM 43-11 Direct Support Maintenance Operations (Nondivisional). September 1991.

FM 43-12 Division Maintenance Operations. November 1989.

FM 44-1 U S Army Air Defense Artillery Employment. May 1983.

FM 44-3 Air Defense Artillery Employment Chaparral/Vulcan/Stinger. June 1984.

FM 44-16 Chaparral/Vulcan/Stinger Platoon Combat Operations. May 1987.

FM 44-100 United States Army Air Defense Operations. November 1988.

FM 63-3J Combat Service Support Operations - Corps. August 1985.

FM 63-20 Forward Support Battalion. February 1990.

FM 71-1 The Tank and Mechanized Infantry Company Team. November 1988.

FM 71-2 The Tank and Mechanized Infantry Battalion Task Force. September 1988.

FM 71-3 The Armored and Mechanized Infantry Brigade. May 1988.

FM 71-100 Division Operations. June 1990.

FM 71-123 Tactics and Techniques for Combined Arms Heavy Forces: Armored Brigade, Battalion/Task Force, and Company/Team. August 1992.

FM 90-2 Battlefield Deception. October 1988.

FM 90-4 Air Assault Operations. March 1987.

FM 90-6 Mountain Operations. June 1980.

FM 90-10 (HTF) Military Operations in Urban Terrain (MOUT) (How to Fight). August 1979.

FM 90-10-1 (HTF) An Infantryman's Guide to Urban Combat (How to Fight). September 1982.

FM 90-13 (HTF) River Crossing Operations (How to Fight). November 1978.

FM 90-13-1 Combined Arms Breaching Operations. February 1991.

FM 90-14 Rear Battle. June 1985.

FM 90-26 Airborne Operations. December 1990.

FM 100-2-1 The Soviet Army: Operations and Tactics. July 1984.

FM 100-2-2 The Soviet Army: Specialized Warfare and Rear Area Support. July 1984.

FM 100-2-3 The Soviet Army: Troops, Organization, and Equipment. June 1991.

FM 100-5 Operations. May 1986.

FM 100-25 Doctrine for Army Special Operations Forces. December 1991.

FM 100-26 The Air-Ground Operations System. March 1973.

FM 100-103 Army Airspace Command and Control in a Combat Zone. October 1987

FM 101-5 Staff Organization and Operations. May 1984.

FM 101-5-1 Operational Terms and Symbols. October 1985.

FM 101-10-1/2 Staff Officers Field Manual-Organizational, Technical, and Logistical Data, Planning Factors (Volume 2). July 1990.

FLEET MARINE FORCE MANUALS (FMFM)

FMFM 1-5 Maritime Pre-positioning Force (MPF) Operations. Coordinating Draft.

FMFM 4-1 Combat Service Support. September 1981.

FMFM 5-1 Organization and Function of Marine Aviation. October 1991.

FMFM 5-35 Assault Support Helicopter Tactical Manual. March 1992.

FMFM 6-1 Marine Division. March 1978.

FLEET MARINE FORCE REFERENCE PUBLICATIONS (FMFRP)

FMFRP 0-52 Remote Sensing Field Guide, Desert. October 1990.

FMFRP 0-53 A foot in the Desert. October 1990.

FMFRP 0-54 The Persian Gulf Region, A Climatological Study. October 1990,

FMFRP 0-55 Desert Water Supply. November 1990.

FMFRP 0-56 Southwest Asia: Environment and its Relationship to Military Activities. November 1990.

FMFRP 0-57 A Study of Windborne Sand and Dust in Desert Areas. October 1990.

FMFRP 0-58 Problems in Desert Warfare. October 1990.

FMFRP 0-59 The Environment and its Effects on Materiel, Personnel, and Operations with Special Emphasis on the Middle East. October 1990.

FMFRP 0-60 General Design and Construction Criteria for Kuwait.

FMFRP 0-63 Troop Construction in the Middle East. December 1990.

FMFRP 2-12 MAGTF: A Global Capability. April 1991.

GRAPHIC TRAINING AIDS

GTA 3-6-5 NBC Warning and Reporting System.

STANDARDIZATION AGREEMENTS (STANAG)

STANAG 2025 Basic Military Road Traffic Regulations.

STANAG 2050 Statistical Classification of Diseases, Injuries, and Causes of Death.

STANAG 2128 Medical and Dental Supply Procedures.

TRAINING CIRCULARS (TC)

TC 90-1 Military Operations on Urbanized Terrain Training. June 1986.

TECHNICAL MANUALS

TM 3-6665-312-12&P Operator's and Organizational Maintenance Manual Including Repair Parts and Special Tools List for M1A1 Automatic Chemical Agent Alarm. March 1985.

COMMAND PUBLICATIONS

Command publications cannot be obtained through Armywide resupply channels. Determine availability by contacting the installation shown.

FKSM 71-2-8 Battalion Task Force Combat Semite Support Doctrinal Sketch. May 1986.

FKSM 71-8 Close Combat Heavy (CCH) Maneuver Organizations. February 1989.

FKSM 71-10 Movement Coordination Exercise (MCX). August 1989.

ST 17-17 Combined Arms Heavy Battalion/Risk Force SOP. May 1990.

ST 100-3 Battle Book.

For information on the preceding references, contact:

Commander, USAARMC
ATTN: ATZK-IMO-RA
Fort Knox, KY 40121-5000

SH 7-176 Infantry Reference Data Book

For information on the preceding reference, contact:

Commandant, USAIS
ATTN: ATSH-AT
Fort Benning, GA 31905-5000

ST 100-9 The Command Estimate

For information on the preceding reference, contact:

Commander, USACGSC
ATTN: ATZL-SWN-L
Fort Leavenworth, KS 66027-6900

Made in the USA
Monee, IL
05 November 2026

67886768R00233